酒水文化与技艺
（双语版）

主　编　徐　倩

副主编　严璐璐　安　洁　刘　颖
　　　　章　荃　朱佳露　秦　琴
　　　　陆　娟　邵昌明

北京理工大学出版社
BEIJING INSTITUTE OF TECHNOLOGY PRESS

内容提要

本书依据职业教育国家教学标准体系，紧密对接职业标准和岗位（群）能力要求，实现"岗课赛证"融通，教材每个专题都增加了实操技能的内容，设置了课后"赛证直通"，丰富了岗位综合实践内容。随着旅游、酒店、餐饮及邮轮等行业国际化水平的提高及酒水文化所具有的国际化特征，教材采用中英双语编写并配套中英双语微课程，有利于提高学生的语言技能也更契合行业发展需求。本书共分为三大模块，模块一酒（共包括七个单元：酒水认知、中国白酒、世界六大基酒、配制酒、鸡尾酒、葡萄酒、其他酿造酒）；模块二茶（共包括四个单元：茶及茶具认知、茶的泡饮及创意制作、中国茶艺表演、茶的品鉴及茶文化）；模块三咖啡及可可（共包括两个单元：咖啡、可可）。

本书适用于旅游管理、酒店管理与数字化运营、餐饮智能管理、民宿管理与运营、葡萄酒文化与营销、茶艺与茶文化、休闲服务与管理、空中乘务等专业，也可供鸡尾酒爱好者、葡萄酒文化爱好者、茶艺与茶文化爱好者、咖啡爱好者及调饮爱好者使用。同时，还可作为酒吧、茶馆、咖啡吧及综合性休闲吧等经营场所的培训教材。

图书在版编目（CIP）数据

酒水文化与技艺：英、汉 / 徐倩主编 . -- 北京：
北京理工大学出版社，2025.1.
ISBN 978-7-5763-5007-4

Ⅰ. TS971.22；TS971.21

中国国家版本馆 CIP 数据核字第 202563AM01 号

责任编辑：李　薇　　　　　　文案编辑：李　薇
责任校对：周瑞红　　　　　　责任印制：王美丽

出版发行 / 北京理工大学出版社有限责任公司

社　　址 / 北京市丰台区四合庄路 6 号

邮　　编 / 100070

电　　话 /（010）68914026（教材售后服务热线）
　　　　　　（010）63726648（课件资源服务热线）

网　　址 / http：//www.bitpress.com.cn

版 印 次 / 2025 年 1 月第 1 版第 1 次印刷

印　　刷 / 河北鑫彩博图印刷有限公司

开　　本 / 787 mm × 1092 mm　1/16

印　　张 / 21

字　　数 / 514 千字

定　　价 / 89.00 元

前 言

随着经济的快速发展，旅游、酒店、餐饮、民宿、邮轮等行业对相关从业人员的专业素养和综合素质的要求也在不断提高。在此背景下，"酒水文化与技艺"作为高等院校旅游管理、酒店管理与数字化运营、餐饮智能管理、民宿管理与运营、葡萄酒文化与营销、茶艺与茶文化、休闲服务与管理、空中乘务等专业酒水类课程的配套教材，能够帮助学生构建完整的酒水知识体系、掌握全面的中外酒水知识、形成广博的文化视野、培育良好的专业素质。本教材具有如下特点。

1. 体系完整，内容全面

教材内容结构突破了传统教材的篇章结构，突出学生能力培养，构建三大模块，十三个单元。全方位涵盖了酒、茶、咖啡等的基础知识、历史文化、生产发展、工艺技术、品鉴方法、调制技艺、文化礼仪、服务实操等内容。既能使学生奠定良好的理论知识基础，又能准确指导学生技能实操。

课程导学

2. 岗课赛证，共通共融

教材编写参考了世界技能大赛餐厅服务赛项、全国职业院校技能大赛酒水服务赛项的考核标准，融入了调酒师、侍酒师、调饮师、茶艺师、咖啡师等国家职业技能标准。不仅在每个专题都增加了实操技能的内容，为学生开展练习提供目标、步骤、方法、标准、职业规范；还在每个专题设置"赛证直通"版块，丰富了学生的岗位综合实践内容，进一步加强学生的实践能力训练，提高学生运用理论知识和基本原理的综合能力和操作技能。

3. 专业为基，文化为核

教材编写始终以提高学生专业理论水平为基础，以涵养学生文化底蕴为内核。因此，既有全面深入且针对性强的专业知识，更有底蕴深厚且延展性广的文化素养知识，两者有机结合，有利于培养和提高学生的综合职业素养。

4. 双语并举，职教出海

教材内容结合行业国际化发展特点，采用中英双语编写并配套相应教学资源，帮助学生提高语言技能和综合素养，契合行业需求，不仅打造中国特色职业教育，更是在创造"出海"契机，为全球培养、输送人才，服务国际产能合作。

5. 技术创新，与时俱进

本教材力求创新，与时俱进，配套微课视频、知识拓展、专业术语等，学生可以通过扫描书中二维码随时随处学习，既可以进行课前预习，也可以实现课后巩固。同时，学生可根据教师教学安排及自身学习需要，随时登录"智慧职教 MOOC"平台学习配套的数字化课程。该课程被评为省级优质课程（精品课程），共含视频资源 1125 分钟，资源总数 1254 个。

配套在线
精品课

6. 思政升华，立德树人

为深入贯彻党的二十大精神，落实习近平总书记在党的二十大报告中的要求，本教材的编写始终以文化自信为基石、以立德树人为宗旨，增加了中国传统文化的知识内容，强化了家国情怀、文化素养、道德修养等思政内容，打造培根铸魂、启智增慧，适应时代要求的精品教材，让学生在学习中树立正确的世界观、人生观、价值观。

　　本书由贵州职业技术学院徐倩担任主编，严璐璐、安洁、邵昌明及贵阳职业技术学院刘颖担任副主编。同时，还得到了葡萄酒领域专家、中国国家一级酿酒师（葡萄酒）章荃老师，世界咖啡与烈酒大赛中国区评委、世界精品咖啡协会 TSCA 认证高级咖啡师朱佳露老师，珍珀酒店餐饮运营总监秦琴女士，趣游吧旅游服务有限公司总经理陆娟女士等的大力支持。他们都作为副主编参与了本教材的编写工作、在此一并致谢。

　　由于编者水平有限，书中难免存在缺漏之处，敬请广大读者和专家批评指正。

编　者

目 录

模块三　咖啡及可可

专业术语

模块一　酒

单元一　酒水认知

绿蚁新醅酒，红泥小火炉。晚来天欲雪，能饮一杯无？

——唐·白居易《问刘十九》

单元导入 Unit Introduction

酒水，作为人类饮食文化中的重要组成部分，不仅具有悠久的历史和深厚的文化底蕴，更是人们日常生活中不可或缺的一部分。从古代的琼浆玉液到现代的各式鸡尾酒、啤酒、葡萄酒等，酒水在人类的生活中扮演着越来越重要的角色。通过学习，使同学们对酒水有全面而深入的认识，不仅能够更好地欣赏和品鉴酒水，更能在实际生活中灵活运用酒水知识，为自己的学习生活增添更多的色彩和乐趣。让我们一起踏上这场关于酒水的探索之旅吧！

Beverages, as an important part of human food culture, not only has a long history and profound cultural heritage, but also an indispensable part of people's daily life. From ancient beverages to modern cocktails, beer, wine, etc., drinks are playing an increasingly important role in human life.

Through learning, I hope to enable students to have a comprehensive and in-depth understanding of beverages, not only to better appreciate and taste beverages, but also to flexibly use beverages knowledge in real life, to add more color and fun to their study life. Let's go on this journey of discovery about beverages!

学习目标 Learning Objectives

➤ 知识目标（Knowledge Objectives）

（1）能描述酒水的基本概念，并能区分不同种类的酒。

To describe the basic concept of beverages and distinguish different kinds of beverages.

（2）掌握发酵、蒸馏、熟化等酒类制作工艺的含义。

Master the meaning of fermentation, distillation, maturation and other liquor production processes.

（3）掌握白酒品鉴的原理。

Master the principle of liquor tasting.

➤ 能力目标（Ability Objectives）

（1）能根据所学知识进行酒与水的分类。

Students are able to classify beverages and water according to what you have learned.

（2）能根据所学知识讲述中国白酒及六大基酒的制作工艺。

Students can tell the production process of Chinese liquor and six basic liquors according to the knowledge.

（3）能根据所学知识掌握中国饮酒礼仪。

Students are able to master the Chinese drinking etiquette according to the knowledge.

➤ 素质目标（Quality Objectives）

（1）从学习中感受大国工匠的精神，培养学生具备良好的职业精神和专业精神。

Let students feel the spirit of artisans in great countries from learning, and cultivate students with good professional spirit and professionalism.

（2）感受中华民族的礼仪之美、和谐之道及智慧之光，增强文化自信。

Let students feel the beauty, harmony and wisdom of the Chinese nation and enhance cultural self-confidence.

【微课】
包罗万象：
酒水概述

专题一
包罗万象：酒水概述

图中的饮品属于酒水吗（图 1-1-1）？

_____

Is the drinks in the picture belong to beverage (Figure 1-1-1)?

_____

牛奶（milk）　　豆奶（soybean milk）　　酸奶（yogurt）

图 1-1-1　饮品（drinks）

酒水是指包括所有含酒精的饮品和不含酒精的饮品。所谓"酒"是指含有酒精的饮品，而"水"则指那些不包含酒精的饮品。在一般情况下，含酒精的饮品被划分到酒（hard drink）这一类别，没有酒精的饮品则被归为软饮料（soft drink），这也就是我们常说的"水"。

The concept of beverage is the general designation of all alcoholic drink and non-alcoholic drink. "Beverages" refers to alcoholic drink and "water" means non-alcoholic drink. In general, we refer to the alcoholic drink as hard drink, and non-alcoholic drink as soft drink, which is also the "water" as we said.

牛奶、豆浆、酸奶、咖啡、茶、可乐等都属于酒水，它们是不含酒精的饮料。不含酒精的饮品也被称作软饮料。在欧美地区，软饮料被定义为人工制作的饮品，其乙醇含量不会超过 0.5%。

Hence, you are very clear that milk, soybean milk, yogurt, coffee, tea, cola, etc. are all beverages, and they are non-alcoholic drinks. We refer to these non-alcoholic drinks as soft drinks. In European and American countries, soft drinks are defined as: artificially prepared drinks with ethanol content not exceeding 0.5%.

一、软饮料的基本概念（Basic Concept of Soft Drinks）

不含酒精或其酒精含量不超过 0.5% 的提神解渴饮品被归类为软饮料，即可日常饮用，也可作为佐餐饮品，更是各种酒的搭档，也是鸡尾酒的重要组成部分。

Non-alcoholic drinks, also known as soft drinks, are refreshing and thirst-quenching beverage products containing no alcohol or edible alcohol content not exceeding 0.5%. It can not only be drunk daily, but also used as a table drink, a partner of various beverages and an important part of cocktails.

二、软饮料的种类（Categories of Soft Drinks）

（一）碳酸饮料（Carbonated Drinks）

碳酸饮料是一种特殊的饮品，它主要由经过纯化的饮用水制成，并通过特定的工艺将二氧化碳（CO_2）气体压入其中。这种饮品因其特有的气泡和口感，通常被称为"汽水"。

Carbonated drinks are special drinks that are mainly made from purified drinking water, and carbon dioxide (CO_2) gas is pressed into it by a specific process. This drink is commonly known as soda because of its unique bubbles and taste.

（二）果蔬汁饮料（Fruit and Vegetable Juice Drinks）

果蔬汁饮料，是一种以新鲜水果和蔬菜为主要原料制成的无酒精饮品。由于其独特的口感和营养价值，深受消费者喜爱，因此，在市场上呈现出了丰富多样的种类。

Fruit and vegetable juice is a non-alcoholic drink made of fresh fruits and vegetables as the main raw materials. Because of its unique taste and nutritional value, it is loved by consumers, so it has shown a rich variety in the market.

（三）乳制品饮料（Dairy Beverage）

乳制品饮料一般是牛乳或以牛乳为原料制成的各种饮料，如纯牛奶、脱脂牛奶等。

Dairy beverages are generally milk or various beverages made from milk, such as pure milk, skimmed milk, etc.

（四）水（Water）

水是维持人体正常运作不可或缺的关键成分。每天摄入大约 2 500 毫升的水量，对于促进人体的新陈代谢至关重要。适量饮水对于保持身体健康有显著益处。

Water is essential for the proper functioning of our bodies. A daily intake of about 2 500 milliliters of water is essential for boosting the body's metabolism. Drinking water in moderation has significant benefits for maintaining good health.

三、酒的基础知识（Basic Knowledge of Liquor）

酒即含有酒精的饮料。确切地说，凡酒精含量在 0.5% 以上的饮料都是酒精饮料，通常称作酒。酒是多种化学成分的混合物，乙醇和水是主要成分。另外，还有酸、酯、醛、醇等众多化学物质，它们的含量较小但决定着酒的质量、特色与风味。

First of all, let's talk about the concept of alcoholic beverages. Alcoholic beverages refers to a beverage containing alcohol. To be exact, any beverage containing more than 0.5% alcohol is an alcoholic beverage, usually called alcohol. Alcoholic beverages is a mixture of many chemical components, with ethyl alcohol and water as the main components. In addition, there are also acids, esters, aldehydes, alcohols and many other chemicals, their content is small but determines the quality, characteristics and flavor of alcoholic beverages.

酒精也称"乙醇"。乙醇的物理特性表现为：在常温状态下，呈现为无色透明的液体，具有易燃和易挥发的性质。其沸点与汽化点均为 78.3 ℃，而冰点达到 -114 ℃。此外，乙醇能很好地溶于水，并具备强大的杀菌能力。在酿酒工业中，乙醇主要来源于葡萄糖的转化过程。

Alcohol also called "ethyl alcohol". The physical characteristics of ethanol are as follows: at room temperature, it appears as a colorless transparent liquid, with flammable and volatile properties. The boiling point and vaporization point are both 78.3 ℃, and the freezing point reaches −114 ℃. In addition, ethanol is well soluble in water and has a strong bactericidal ability. In the beverages industry, ethanol is mainly derived from the conversion process of glucose.

（一）酒精度（Alcoholic Content）

酒精度，即酒中乙醇的浓度，是衡量酒品中酒精含量的一个重要指标。具体地说，它是指在 20℃的温度条件下，每 100 毫升酒液中所含有的乙醇的毫升数或百分比。以五粮液为例，当我们说其酒精度为 52 度时，这意味着在 20℃的条件下，每 100 毫升的五粮液中所含的乙醇量为 52 毫升。这种表述方式直观地反映了酒品中乙醇的含量，为消费者提供了明确的参考信息。

Alcohol content, that is, the concentration of ethanol in beverages, is an important index to measure the alcohol content in alcoholic beverages. Specifically, it refers to the milliliter number or percentage of ethanol contained in every 100 mL of liquor at a temperature of 20℃. Taking Wuliangye as an example, when we say that its alcohol content is 52 degrees, it means that under the condition of 20℃, the amount of ethanol contained in every 100 mL of Wuliangye is 52 mL. This expression directly reflects the content of ethanol in alcoholic beverages, and provides clear reference information for consumers.

此外，各个国家在标识酒精浓度时可能采用不同的标准体系。例如，美国采用了一种特定的酒精度表示方法，即美制酒度标准，以 Proof 为单位进行表示。根据美国的计量标准，1Proof 相当于 0.5 度的酒精含量。

In addition, countries may use different standard systems for labeling alcohol concentrations. For example, the United States has adopted a specific method of representing alcohol, the American Alcohol Standard, which is expressed in units of Proof. According to American standards of measurement, we know that 1Proof is equivalent to 0.5 degrees of alcohol content.

（二）酒的分类（Classification of Alcoholic Beverages）

按不同的分类方式酒有很多类。

Alcoholic beverages can be divided into many categories in accordance with different classification methods.

1. 按生产工艺分类（Divided According to Production Technique）

（1）蒸馏酒是一种特殊的酒品，它经过发酵过程后，再通过蒸馏技术提高其酒精浓度。通过这种工艺制成的酒有人们熟知的中国白酒，以及国际知名的威士忌、白兰地、金酒、朗姆酒和伏特加等。

Distilled beverages is a special kind of beverages, it goes through the fermentation process, and then through distillation technology to increase its alcohol concentration. The beverages made through this process include the well-known Chinese liquor, as well as the internationally renowned Whisky, Brandy, Gin, Rum and Vodka.

（2）发酵酒，也被称为酿造酒，是通过原料的发酵过程直接提取或通过压榨法获得的酒品。这类酒一般酒精度较低，常见的发酵酒包括葡萄酒、啤酒、清酒和黄酒等，它们各自以其独特的酿造工艺和风味特点深受人们喜爱。

Fermented beverages, also known as brewing beverages, is a beverages obtained directly by the fermentation process of raw materials or by pressing method. This kind of beverages is generally low in alcohol, and the common fermented beverages include wine, beer, sake and rice beverages, etc., which are deeply loved by people for their unique brewing technology and flavor characteristics.

（3）配制酒是一种经过精心调配的酒品，其制作过程通常是将蒸馏酒与发酵酒按特定配方混合，或者将蒸馏酒或发酵酒与香料、果汁等原料进行勾兑。这种独特的酿造方法赋予了配制酒丰富多样的风味和口感，使其在市场上占有一席之地。常见的配制酒包括味美思（Vermouth）、利口酒、中国药酒以及露酒等。

The preparation of beverages is a carefully prepared beverages, the production process usually involves the distillation of beverages and fermented beverages according to a specific recipe, or the distillation of beverages or fermented beverages with spices, fruit juices and other raw materials. This unique brewing method gives the mixed beverages a rich variety of flavor and taste, making it a place in the market. Common blended spirits include Vermouth, liqueur, Chinese medicinal Beverages and dew Beverages.

（4）鸡尾酒是一种独具魅力的混合酒品，它遵循特定的酒谱，巧妙地将烈性酒、葡萄酒、果汁、汽水等液体原料与调色和调香的辅助成分混合或勾兑而成。鸡尾酒的主要构成包括基酒、辅酒以及辅料等，这些成分在调酒师的巧手下相互融合，创造出丰富多样的口感和色彩。

这些鸡尾酒不仅在味觉上令人陶醉，同时也能带来视觉上极大的享受。口感上，鸡尾酒可以展现从清新爽口到浓郁香醇的多样变化，每一口都能让饮用者感受到不同风味层次的交织与融合。色彩上，鸡尾酒则通过不同酒液、果汁、糖浆等原料的混合，展现绚丽多姿、引人入胜的色彩搭配，使人们在品尝美酒的同时，也能享受视觉的盛宴。这种对口感和色彩的极致追求，正是鸡尾酒调制艺术中不可或缺的魅力所在。

A cocktail is a unique blend of beverages that follows a specific beverages spectrum and skillfully blends or blends liquid ingredients such as spirits, beverages, juices, and sodas with auxiliary ingredients for color and aroma. The main composition of the cocktail includes the base beverages, the auxiliary beverages and the auxiliary ingredients, etc. These ingredients are combined under the skillful hands of the bartender to create a rich variety of taste and color, these cocktail works not only make people intoxicated on the taste, but also bring great enjoyment on the visual. In taste, the cocktail can show a variety of changes from fresh and refreshing to rich and mellow, each bite can let the drinker feel the interweaving and fusion of different flavor levels. In terms of color, cocktails show colorful and attractive color matching through the mixture of different beverages, juices, syrups and other raw materials, so that people can enjoy a visual feast while tasting beverages. This ultimate pursuit of taste and color is the indispensable charm of the art of cocktail making.

2. 按酒精度分类（Divided According to Alcohol Content）

（1）高度酒：酒精度高于 40 度。

High-alcohol beverages: Alcohol content is higher than 40 degrees.

（2）中度酒：酒精度在 20 度～ 40 度的酒品。

Medium-alcohol beverages: beverages with an alcohol content of 20-40 degrees.

（3）低度酒：酒精度在 20 度以下。

Low-alcohol beverages: Alcohol content is below 20 degrees.

3. 按制酒原料分类（Divided According to Raw Material for Beverages Making）

（1）植物酒：这类酒品采用植物作为主要原料，经过发酵和蒸馏工艺精心酿制而成。如源自墨西哥的著名酒品 —— 特基拉酒（Tequila）。

Plant Beverages: This Beverages uses plants as the main raw materials, and is carefully brewed through fine fermentation and distillation processes. A typical example is Tequila, a famous beverages from Mexico.

（2）粮食酒：这类酒品采用多样化的谷物作为原料，经过发酵和蒸馏工艺酿造而成。其中，中国的白酒、金酒以及威士忌酒等是杰出代表，它们各自以其独特的口感和风味赢得了广泛赞誉。

Food Beverages: this type of beverages uses a variety of grains as raw materials, and is brewed through a careful fermentation and distillation process. Among them, Chinese liquor, Gin and Whiskey are outstanding representatives in this category, each of which has won widespread praise for its unique taste and flavor.

（3）水果酒：这类酒品以丰富多样的水果为基础原料，通过发酵、蒸馏工艺或者独特的调配手法酿造而成。葡萄酒、白兰地以及味美思等都属于这一类别，它们各自展现出独特的果香和口感，深受消费者的喜爱。

Fruit Beverages: This type of beverages is based on a rich variety of fruit as raw materials, through fermentation, distillation or unique preparation methods to brew. Wine, Brandy and Vermouth all fall into this category, each displaying a unique fruity aroma and taste that is loved by consumers.

4. 按西餐用餐习惯分类（Divided According to Western Food Dining Habits）

（1）餐前酒：也被称为开胃酒（Aperitifs），是与开胃菜搭配的酒类，它们能够刺激食欲，为接下来的正餐做准备。常见的开胃酒包括味美思（Vermouth）和鸡尾酒（cocktail）等。

Aperitif: Also known as aperitifs, are drinks that accompany appetizers to stimulate the appetite and prepare for the rest of the meal. Common aperitif drinks include Vermouth and cocktail.

（2）佐餐酒：与主菜搭配的酒类，称之为佐餐酒（Table Beverages）。这类酒品主要包括红葡萄酒和白葡萄酒，它们能够完美衬托主菜的口感，为用餐体验增添一抹优雅。

Table Beverages (main course Beverages)：When talking about the Beverages with the main course, we call it table Beverages. Mainly Red and White Wine, these Beverages complement the main course perfectly and add a touch of elegance to the dining experience.

（3）餐后酒：也被称为甜食酒（Dessert Beverages），通常是在享用甜点时饮用的酒类。虽然这些酒品有时也使用水果作为原料，但它们的酿造方式与一般的果味酒有所不同，包括发酵、蒸馏或配制等多种工艺。常见的餐后酒有利口酒、波特酒等。

After-dinner beverages: also known as dessert beverages, is usually consumed during dessert. Although these beverages sometimes use fruit as an ingredient, they are made in a different way than regular fruity beverages, involving a variety of processes such as fermentation, distillation or preparation. Common after-dinner beverages include wine, brandy and Vermouth.

赛证直通 Competitions and Certificates

一、基础知识部分（Basic Knowledge Part）

二维码 1-1-1：
知识拓展

（一）专业名词解释（Explanation of Professional Terms）

1. 酒水　Beverages：

2. 酒精度　Alcohol by volume：

（二）思考题（Thinking Question）

谈谈对不同酒类生产工艺的理解。

Talk about your understanding of different Beverages production processes.

二、技能操作部分（Skill Operation Part）

口述酒水的分类方法。

Dictate the classification of beverages.

专题二
千态万状：各种酒杯认知

你知道图 1-1-2 中酒杯的名称及用途吗？

Do you know the name and purpose of the goblet in the Figure 1-1-2?

想要进入调酒的世界，只了解酒的种类是不够的！还要知道酒杯要搭配什么酒使用，才算是真正了解一杯酒！好的酒搭配对的酒杯，可以衬托出一杯酒的气质，也可以表达调酒师的情感。

图 1-1-2　酒杯（goblet）

In order to enter the world of bartending, it is not enough to know the type of wine! But also know the glass to use with what wine, is really understand a glass of wine! A good wine with the right glass can set off the temperament of a glass of wine, and it can also allow the bartender to express emotions.

一、常见的酒杯类型（The Types of Common Wine Glasses）

（一）威士忌酒杯（Whisky Glass）

威士忌酒杯有时被称为一口杯（Shot Glass），其中"shot"一词意为"一小杯"，主要用于销售单杯威士忌时。它也被称为净饮杯（Straight Glass），表明这是直接饮用威士忌的理想选择。根据容量大小，它可以分为单酒杯（Single，30 毫升）和双饮杯（Double，60 毫升）两种规格。

This wine glass is often called Shot Glass, where the word "shot" means "a small shot", and is mainly used when selling single shots of whiskey. It is also known as Straight Glass, indicating that this is ideal when drinking whiskey straight. According to the capacity size, it can be divided into Single (30 mL) and Double (60 mL) two specifications.

（二）古典杯（Old Fashioned Glass）

古典杯又被称为摇滚酒杯（Rock Glass），实际上，它由现代平底杯（Tumbler）演变发展而来，一直以来都作为酒器使用，也被誉为经典传承的酒杯。其宽大的杯口设计使得它能够容纳大块的冰块，因此，在享用冰镇威士忌或鸡尾酒时，这款酒杯常常成为首选。不仅如此，它还能够容纳切割成圆柱状的冰块，让饮用者在品酒的同时，欣赏冰块在杯中的美丽姿态。

This glass is also known as the Rock Glass, in fact, it is the predecessor of the modern Tumbler, has long been used as a wine vessel, also known as the classical tradition of wine glasses. Its wide mouth design allows it to hold large chunks of ice, so it is often the glass of choice when enjoying a cold whiskey or cocktail. Not only that, it can also hold ice cubes cut into a cylindrical shape, so that you can enjoy the beautiful posture of the ice cubes in the glass while enjoying the wine.

（三）平底杯（Tumbler）

平底杯即人们常说的杯子（cup），是酒吧里不可或缺的一种器具，常被用于盛放高球鸡尾酒、金汤尼和软饮等饮品。鉴于其高频率的使用，酒吧应多备几个以备不时之需。这款杯子的容量多样，从180毫升到300毫升不等，能满足不同的需求。其中，按照国际标准，其容量为240毫升（相当于8盎司）。

The cup, as it is often called, is an indispensable tool in a bar. It is often used to serve drinks such as highball cocktails, Gin Tonic and soft drinks. Given its high frequency of use, it is recommended that bars keep a few on hand for emergencies. The cups come in a variety of capacities, ranging from 180 mL to 300 mL, to meet different needs. Among them, according to international standards, its capacity is 240 mL (equivalent to 8 ounces).

（四）柯林杯（Collin's Glass）

柯林杯（图1-1-3）为圆筒形状的高身杯，通常被称作烟筒杯（Chimney Glass）或高筒杯（Tall Glass），可装入含有碳酸气泡的鸡尾酒。

Collin's glass (Figure 1-1-3) is cylinder-shaped tall glass, often called Chimney Glass or Tall Glass, it can be filled with a cocktail containing carbonated bubbles.

图1-1-3 柯林杯
（Collin's glass）

（五）啤酒杯（Beer Glass）

一般啤酒杯有三种。一是皮尔森啤酒杯，特点是杯口大、杯底小，呈喇叭形；二是常见啤酒杯，特点是杯口阔，呈郁金香形状；三是有柄啤酒杯，特点是带柄壁厚，容量大。

There are many kinds of wine glasses, including large beer glasses and small wine glasses. It is worth mentioning that small wine glasses with handles are sometimes used as containers for iced coffee or large drinks.

（六）利口酒杯（Liqueur Glass）

这类酒杯特别适合直接品饮利口酒，当然也可以用来直接享用威士忌或其他蒸馏酒。它的标准容量通常为30毫升，以无色透明的款式为佳，这样能更好地展示利口酒的美丽色泽。

This glass is ideal for drinking liqueur directly, but can also be used to enjoy whiskey or distilled spirits directly. Its standard capacity is usually 30 mL, and it is preferred to be colorless and transparent, which can better show the attractive color of the liqueur.

（七）白兰地杯（Brandy Glass）

这类酒杯的杯身圆润饱满，而杯口相对较窄。其容量范围广泛，多为180～300毫升。它不仅适用于直接品饮白兰地，也是品饮葡萄酒或芳香型药草利口酒的理想选择。

The body of the glass is round and full, and the mouth is relatively narrow. Its capacity ranges widely, mostly between 180 and 300 mL. It is not only suitable for direct tasting of brandy, but also ideal for tasting wine or aromatic herbal liqueur.

（八）鸡尾酒杯（Cocktail Glass）

这是一款专为浅饮鸡尾酒设计的酒杯（图 1-1-4），其形状独特，呈现为倒三角形，底部配有杯脚。

This is a glass (Figure 1-1-4) specially designed for shallow drinking cocktails, its shape is unique, presented as an inverted triangle, with a foot at the bottom.

图 1-1-4　鸡尾酒杯（cocktail glass）

（九）雪莉酒杯（Sherry Glass）

在品尝西班牙产的雪莉酒时，这款酒杯尤为适用。相较于利口酒杯，它的尺寸稍大一些，容量为 75 毫升。其设计简洁大方，也同样适用于威士忌，具有广泛的实用性。

This glass is especially useful when tasting sherry from Spain. Compared to the liqueur glass, it is slightly larger in size, with a capacity of 75 mL. Its design is simple and generous, even when enjoying whiskey, it is also applicable and has a wide range of practicality.

（十）香槟杯（Champagne Glass）

香槟杯即香槟酒、发泡性葡萄酒以及香槟类鸡尾酒所配用的酒杯，主要有两种风格：一种是宽口碟形，另一种是细长笛形。

Champagne glass is the glass used for champagne, sparkling wine and champagne cocktails. They are used in two main styles: one is a wide mouth saucer type, and the other is a tall thin flute type.

（十一）葡萄酒杯（Wine Glass）

葡萄酒杯的设计和大小，往往因各国或地区的风俗以及葡萄酒的类型（颜色、气味）差异而有所不同。对于酒吧中的葡萄酒杯，通常只要满足以下几个条件即可：

首先，为了能够更好地欣赏葡萄酒的色泽，应选择无色透明的酒杯；

其次，酒杯的上部应呈现为向内侧弯曲的郁金香形状；

再次，杯口的直径应在 6 厘米以上；

最后，容量应超过 240 毫升。

The design and size of wine glasses often vary according to the customs of various countries or regions, as well as the type of wine (color, smell). For wine glasses in the bar, usually as long as the following conditions can be met:

First, in order to be able to better appreciate the color of the Wine, should choose a colorless transparent glass;

Second, the top of the glass should appear as a tulip shape curved inward;

Third, the diameter of the mouth of the cup should be above 6 cm;

Finally, the capacity should be more than 240 mL.

（十二）玛格丽特杯（Margarita Cup）

玛格丽特杯以其独特的设计引人注目，它是一款高脚、宽口的玻璃杯，杯身呈梯形状，自顶而下逐渐缩小，直至底部的收窄。这款酒杯专为玛格丽特鸡尾酒所设计。

The Margarita cup is notable for its unique design, which is a tall, wide mouth glass with a trapezoidal body that gradually shrinks from the top down until it narrows at the bottom. Designed for Margaritas and other long drinks, this glass has a capacity of between 210 and 270 mL.

（十三）飓风杯（Hurricane Glass）

飓风杯是一款造型优雅、线条流畅的玻璃杯，其名称来源于其形状酷似飓风灯。这款杯子常用于盛装热带风味的饮品，容量范围为 240 ～ 360 毫升。

The Hurricane glass, is an elegant, streamlined glass whose name comes from its resemblance to a hurricane lamp. This cup is often used to hold tropical flavored drinks, and its capacity ranges from 240 to 360 mL.

二、酒杯的选择方法（The Selection of drinkware）

为了应对酒杯选择的多样性和灵活性，在选择酒杯时，应考虑以下几个重点：

首先，从品质与格调出发，可以选择那些享有盛名的名牌酒杯或是具有独特魅力的古董酒杯。

其次，不要拘泥于传统或品牌的限制，勇于尝试那些设计新颖、富有创意的酒杯。

再次，考虑到顾客的喜好，可以选择那些能够引发人们情感共鸣的酒杯。

最后，根据实际需求，决定是选择手工精心打造的酒杯还是高效生产的机器制品。

通过这些考虑和选择，能够确保酒杯的多样性，从而满足不同场合和客人的需求。

In order to cope with this variety and flexibility, it is recommended to consider the following priorities when choosing wine glasses:

First of all, from the quality and style, you can choose those famous brand wine glasses or antique wine glasses with unique charm.

Secondly, do not adhere to the restrictions of tradition or brand, and have the courage to try those innovative design and creative wine glasses.

Moreover, taking into account the preferences of customers, you can choose those with naturalistic colors, can trigger people's emotional resonance of the glass.

Finally, according to the actual needs, decide whether to choose hand-crafted wine glasses or efficient production of machine products. Through such consideration and selection, we are able to ensure the variety of wine glasses to meet the needs of different occasions and guests.

三、酒杯 T.P.O 原则（The T.P.O Principle of Wine Glasses）

酒杯 T.P.O 原则是指酒杯的选择要考虑到时间（Time）、地点（Place）、场合（Occasion）。酒杯的选择体现了正式与休闲之分：带有杯脚的酒杯往往与正式场合相匹配，平底酒杯则更适合休闲时光。

The T.P.O principle of wine glasses means that the choice of wine glasses should take into account the time, place, and occasion. The choice of wine glasses reflects the difference between formal and casual: glasses with glass legs tend to match formal occasions, while flat glasses are more suitable for leisure time.

公元前人们饮酒的器具与现今所见截然不同。考古学家发现，那时人们甚至使用家畜的头角作为酒器。随着时间的推移，青铜和黏土制成的酒器逐渐出现，但它们的形状依然受到原始酒器——家畜头角的影响，呈现出杯口窄、杯身高的特点。为了模仿动物角的形状，人们发明了带有杯脚的酒杯，这样的设计使得酒杯更不易倾倒，也因其复杂的制作工艺而成为高级酒器，常在社交场合中使用。

Back in antiquity, people drank with very different devices than we see today. Archaeologists tell us that people even used the horns of livestock as wine vessels. As time went by, bronze and clay wine vessels gradually appeared, but their shape was still influenced by the original wine vessel——the horn of the animal, showing the characteristics of narrow mouth and height. In order to imitate the shape of animal horns, people invented a glass with a cup foot, such a design makes the glass more difficult to pour, but also because of its complex production process and become a high-level wine vessel, more seen in sacrificial occasions or high social.

另一方面，将酒杯底部磨平并附上底座的设计，因其制作简便且易于量产，迅速成为常用酒器。这种酒器文化的传承，对现代酒杯设计产生了深远影响。在高雅庄重的场合，人们倾向于选择高脚杯；在轻松愉快的氛围中，平底酒杯则更为合适。

On the other hand, the design of smoothing the bottom of the wine glass and attaching the base, because of its simple production and easy mass production, quickly became the daily wine ware of the common people. The inheritance of this wine vessel culture has had a profound impact on modern wine glass design. In elegant and solemn occasions, we tend to choose a goblet; In a relaxed atmosphere, a flat glass is more appropriate.

酒杯的选择，除了考虑场合的正式程度外，还应考虑酒精度数的高低和酒风味的浓淡。如今对于酒杯的分类使用标准已变得相对宽松，更多时候，调酒师的品位和创意成为选择酒杯的关键因素。

The choice of wine glass, in addition to considering the formal degree of the occasion, should also consider the level of alcohol and the flavor of the wine. Nowadays, however, the classification of wine glasses has become relatively loose, and more often, the taste and creativity of the bartender become the key factor in choosing a wine glass.

四、酒杯的清洗和擦拭（The Wine Glass Cleaning and Wiping）

在清洁酒杯时，首先使用蘸有中性清洁剂的海绵细心地洗净酒杯。接着，使用清洁的热水再次冲洗，确保酒杯内外洁净无污。最后，将酒杯倒放，使其自然沥干多余的水分。在酒杯表面温度未降低时，利用麻布或棉麻混纺的专用擦拭巾轻轻拂去剩余的水珠，确保酒杯彻底干燥。

When cleaning the wine glass, we first carefully wash the glass with a sponge dampened with a neutral detergent. Then, rinse again with clean hot water to ensure that the inside and outside of the glass are clean. After that, turn the wine glass upside down and let it drain off the excess water naturally. When the surface of the glass is still warm, use a linen or cotton linen blended special wipe to gently wipe off the remaining water beads to ensure that the glass is thoroughly dry.

　　值得注意的是，即使采用酒杯清洗机，酒杯在自然晾干后也可能留下水珠的痕迹。为了避免这种情况，强烈建议使用柔软的布进行最后的擦干步骤。在收纳酒杯时，专业的调酒师会选择将杯口朝上放置。这是因为酒杯的设计初衷并非为了倒放。同时，酒杯的收纳方式也是酒吧形象的一部分，因此，选择正确的收纳方式同样重要。

　　It is worth noting that even if the glass washing machine is used, the glass may leave traces of water beads after natural drying. To avoid this, it is highly recommended to use a soft cloth for the final drying step.When collecting glasses, professional bartenders will choose to place the glass with the mouth facing up. That's because wine glasses aren't designed to be upside down. At the same time, the way the wine glasses are stored is also part of the image of the bar, so the correct way of storage is equally important.

 ## 赛证直通 Competitions and Certificates

一、基础知识部分（Basic Knowledge Part）

（一）专业名词解释（Explanation of Professional Terms）

1. 古典杯（Classic Cup）：
2. 玛格丽特杯（Margarita Cup）：

（二）思考题（Thinking Question）

谈谈对选择合适酒杯的理解。

Talk about your understanding of choosing the right wine glass.

二维码 1-1-2：
知识拓展

二、技能操作部分（Skill Operation Part）

1. 能分别出实训室中不同的酒杯。

Can distinguish different wine glasses in the training room.

2. 能根据酒种类的不同选择不同的酒杯。

Can choose different glasses according to different types of wine.

专题三

酒中寻味：酒的品鉴

你知道茅台酒主要的香气特征有哪些吗？（图 1-1-5）

【微课】
酒中寻味：
酒的品鉴

_____

Do you know what are the main aroma characteristics of Moutai (Figure 1-1-5)?

图 1-1-5　茅台酒（Moutai）

酒的品鉴，不仅可以品尝到酒的独特魅力，还可以感受中国传统文化的深厚底蕴。根据不同白酒的特点，可以尝试不同的品鉴方法，如搭配不同的美食，感受不同的口感体验；或者邀请朋友一起品尝并分享品味体验。这些不同的品鉴方法可以进一步开发感官体验，让餐酒搭配更加出众。

Wine tasting, you can not only taste the unique charm of wine, but also feel the profound heritage of traditional Chinese culture. According to the characteristics of different baijiu, you can try different tasting methods, such as matching different food, feeling different taste experience, or inviting friends to taste baijiu and share taste experience. These methods can further develop the sensory experience and make the food and wine pairing even more outstanding.

一、酒的品评内容（The Content of Wine Evaluation）

（一）酒色泽的鉴定（Identification of Wine Color）

评估酒的色泽，需要通过肉眼仔细检视酒的外观、色泽深浅、清澈程度以及是否有异物。观察酒的正确方法是：将酒倒入杯中后，将杯子举起，以白纸为背景，迎着光线观察；或者将杯口与眼眉平视进行观察。如果是啤酒，首先应观察其泡沫和气泡的升腾状态。

Various wines have specific color standards, and to evaluate the color of wine, you need to carefully inspect the appearance of the wine by the naked eye, the depth of color, the degree of clarity, and the presence of foreign objects. The correct way to observe the wine is: pour the wine into the cup, lift the cup, with the white paper as the background, facing the light observation; or look at the top of the cup at eye level. If it is beer, the first thing you should observe is the rising state of the foam and bubbles.

（二）酒香气的鉴定（Identification of Wine Aroma）

酒类香气主要源于酿造过程中微生物发酵产生的代谢产物，如酶类等。酒的香气不仅可以通过咽喉进入鼻腔，且会有"回味"。回味的长短可以反映出酒的纯净度（即是否有不良或杂质的气味），以及是否具有刺激性。酒的香气和味道紧密相连，人们对滋味的感觉很大程度上依赖于嗅觉的参与。

The aroma of wine are mainly derived from metabolites produced by microbial fermentation during the brewing process, such as enzymes. The aroma of wine can not only enter the nasal cavity through the throat, and there will be "aftertaste". The length of the aftertaste can reflect the purity of the wine (that is, whether there is a bad or foreign odor), and whether it is pungent. The aroma and taste of wine are closely linked, and our perception of taste is largely dependent on the involvement of smell.

（三）酒口味的鉴别（The Identification of Wine Taste）

酒中蕴含了丰富的味道成分，其中主要包括高级醇、有机酸和羰基化合物等，这些成分的形成与酒的酿造原料、工艺以及贮存方式紧密相关。酒的口味风格是消费者普遍关注的重点，其好坏直接反映了酒品的质量。在评价酒的口感时，通常会用酸、甜、苦、辣、咸等描述。

Wine contains a wealth of flavor components, including higher alcohols, organic acids and carbonyl compounds, the formation of these components is closely related to the brewing raw materials, technology and storage methods of wine. The taste style of wine is the focus of consumers' general concern, and its quality directly reflects the quality of wine. When evaluating the taste of wine, people usually use words such as sour, sweet, bitter, spicy and saltiness to describe it.

在酒的品鉴中，酸味与甜味为相对属性。当酒液中的酸度成分显著超越甜度时，品饮者将明显感知到酸味，这一口感特征在英语中常以"dry"来表述，因此，这类酒被归类为"干型"，如常见的干白葡萄酒和干红葡萄酒等。酸型酒往往具有一种特有的醇厚与干爽的风味，同时，酸度还有助于刺激食欲，促进胃口。

In wine tasting, sour and sweet are relative attributes. When the acidity component of the wine significantly exceeds the sweetness, the drinker will obviously perceive the sour taste, this taste feature is often expressed in English as "dry", therefore, this type of wine is classified as "dry", such as common Dry White Wine and Dry Red Wine. Sour wine often has a characteristic mellow and dry flavor, at the same time, the acidity also helps to stimulate the appetite, promote appetite.

甜味是酒品中很受喜爱的口味之一，许多酒品都以甜为主要特点。这种甜味主要来源于酿酒原料中的麦芽糖和葡萄糖，特别是果酒中的含糖量更高。甜味能给人一种滋润、纯美、浓郁的感觉。

Sweetness is one of the most popular tastes in wine, and many wines feature sweetness as their main feature. This sweetness is mainly derived from the maltose and glucose in the brewing raw materials, especially the sugar content in fruit wine is higher. Sweetness can give people a kind of moist, pure beauty, rich feeling.

苦味是一种独特的酒品风格，虽然在酒类中并不常见，但比特酒等就以苦味为主。此外，啤酒也保留了其特有的苦香味道。适量的苦味能带来净口、止渴、生津和开胃的效果，但苦味也具有较强的味觉破坏力，因此，在使用时需要谨慎。

Bitter taste is a unique style of wine, although it is not common in wine, but Bitters is mainly bitter. In addition, the beer also retains its characteristic bitter flavor. The right amount of bitter taste can bring the effect of purifying the mouth, quenching thirst, promoting fluid and appetizing, but the bitter taste also has a strong destructive taste, so it needs to be careful when used.

辣味也称为辛味，它给人一种冲头、刺鼻的感觉，特别是在高浓度的酒精饮料中最为明显。辣味主要来源于酒液中的醛类物质。

Spicy, also known as pungent flavor, gives a punch, pungent feeling, especially in highly concentrated alcoholic beverages. The spicy taste mainly comes from the aldehydes in the liquor.

咸味在酒中并不常见，但少量的盐类可以增强味觉的灵敏度，使酒味更加浓郁。例如，墨西哥的特基拉酒在饮用时需要加入少量盐，以增加其独特的口感。

Saltiness is not common in wine, but a small amount of salt can enhance the sensitivity of the taste and make the wine taste more intense. For example, Tequila from Mexico needs to be drunk with a small amount of salt to add its unique taste.

（四）酒体（Wine Body）

酒体是酒品整体风格的一种集中体现，评价一款酒的酒体，实质上是对这款酒独特风格的整体感受。优质的酒体追求的是和谐与完美，它应该集色泽、香气、口感于一体，相互映衬，互为补充。每款酒都有其独特的风格，这些风格的形成离不开酒中各种物质的相互作用，如水的纯净度、酸类物质的酸度、酯类物质的香气、醛类物质的口感以及醇类物质的醇厚等。这些物质的种类和含量，都会对酒品的风格和质量产生深远影响。

Wine body is a concentrated reflection of the overall style of wine, evaluation of a wine's body, in essence, is the overall feeling of the unique style of this wine. Quality wine body is the pursuit of harmony and perfection, it should set color, aroma, taste in one, complement each other. Each wine has its own unique style, and the formation of these styles is inseparable from the interaction of various substances in the wine, such as the purity of water, the acidity of acids, the aroma of esters, the taste of aldehydes, and the richness of alcohols. The variety and content of these substances will have a profound impact on the style and quality of the wine.

二、酒的品评规范与要求（Standards and Requirement for Wine Evaluation）

在品味白酒时，通常遵循一定的香型顺序，即先品清香，接着是米香、浓香，然后是其他香型，最后才是酱香。

When tasting liquor, we usually follow a certain flavor order, that is, the first fragrance, followed by rice, Luzhou, then other flavors, and finally the Moutai flavor.

1. 评酒规则（Liquor Rating Rules）

（1）评酒前，评酒员应确保得到充分休息，避免使用有气味的化妆品，并禁止携带任何气味浓烈的食物，以防干扰品评结果。

Before wine evaluation, wine reviewers should ensure adequate rest, avoid the use of smelly cosmetics, and do not bring any strong smelling food, so as to avoid interfering with the results of the evaluation.

（2）在评酒期间，应避免食用具有刺激性的食品，如生姜、生蒜、生葱、辣椒等，以及过甜、过咸、油腻的食品。

During the wine evaluation period, should avoid eating irritating food, such as ginger, raw garlic, green onion, chili, etc., and too sweet, too salty, greasy food.

（3）在评酒前30分钟及品评过程中，评酒员严禁吸烟，并在评酒前进行刷牙漱口，以确保嗅觉和味觉的敏锐度。

30 minutes before and during the wine rating, the wine reviewer should refrain from smoking, and brush his teeth and rinse his mouth before the wine rating to ensure the acuity of smell and taste.

（4）评酒过程中应保持安静，避免大声喧哗和私下交流，确保每位评酒员都能独立思考并准确填写评酒单。

Keep quiet during the wine evaluation process, avoid loud noise and private communication, and ensure that each wine reviewer can think independently and accurately fill out the wine evaluation list.

（5）在正式的品评过程中，为确保评酒过程的公正性和规范性，所有评酒员必须严格遵守相关规定。未经明确许可和授权，任何评酒员均不得擅自进入准备室。

In the formal wine tasting activities, in order to ensure the fairness and standardization of the wine evaluation process, all wine reviewers must strictly abide by the relevant regulations. It is hereby clear that no wine reviewer shall enter the preparation room without express permission and authorization.

（6）在品评酒样时，应将酒液布满舌面，尽量减少吞咽，以品味为主。切勿将评酒当作饮酒，过度饮用是绝对不允许的。

When evaluating wine samples, the liquor should be covered with the tongue, minimize swallowing, and focus on taste. Do not evaluate wine as drinking, and excessive drinking is absolutely not allowed.

2. 评酒室环境（Environment of Wine Evaluation Room）

（1）评酒室内的噪声水平应被严格控制，确保其在40分贝以下，以提供一个宁静的品酒环境。

The noise level in the wine evaluation room should be strictly controlled to ensure that it is below 40 decibels to provide a peaceful wine tasting environment.

（2）室内光线应保持充足而柔和，避免阳光直接照射，同时辅以适当的照明设施，确保品酒过程不受光线影响。

Indoor light should be kept sufficient and soft, avoid direct sunlight, while supplemented by appropriate lighting facilities to ensure that the tasting process is not affected by light.

（3）评酒室应远离嘈杂的环境。室内空气需保持新鲜，禁止有烟雾和异味。每次品酒结束后，都应启动室内排风机进行空气交换。在品酒过程中，室内应保持无风状态，确保温度和湿度稳定，建议室温维持在 15 ～ 22 ℃，相对湿度在 50% ～ 60%。

The wine room should be away from noisy environments. The indoor air must be kept fresh and free of smoke and odor. After each wine tasting, the indoor exhaust fan should be started for air exchange. During the tasting process, the room should be kept without wind to ensure stable temperature and humidity. It is recommended to maintain the room temperature at 15 ℃ to 22 ℃ and the relative humidity between 50% and 60%.

（4）室内的照明应采用白色散射光，品酒桌应铺上白色台布，以减少干扰。评酒室的内部色调应统一，选择中等反射率的色调，反射率为 40% ～ 50%。地板应光滑、清洁、耐水，并保持足够的空间，避免过于拥挤。

Indoor lighting should use white scattered light, and the tasting table should be covered with a white tablecloth to reduce interference. The internal color of the wine room should be unified, and the color of medium reflectivity should be selected, and the reflectivity is about 40% to 50%. The floor should be smooth, clean, water resistant, and keep enough space to avoid overcrowding.

（5）设有专门的洗漱池和洗杯池，并配备热水管。每位评酒员都应拥有专用的桌、椅、酒杯、漱口杯等个人用品。

There is a special bath and cup washing pool, and equipped with hot water pipes. Each wine reviewer should have a dedicated table, chair, wine glasses, mouthwash cups and other personal items.

（6）评酒员在进入评酒室前，应在休息室内等待和休息。为了提高评酒结果的公正性，可以在评酒员之间设置合适高度的挡板。

Evaluators should wait and rest in the lounge room before entering the room. In order to improve the fairness of the results of wine evaluation, it is possible to set the appropriate height of the baffle between the wine reviewers.

3. 评酒时间（Rating Time）

评酒的最佳时间推荐在上午的9点至11点，下午大约是14点。为了保持品评的准确性和效果，建议每次评酒的时间不超过两小时，因为过长的评酒时间可能会导致疲劳，从而对品评结果产生不良影响。

The best time to evaluate wine is recommended between 9 and 11 AM, and around 14 PM. In order to maintain the accuracy and effectiveness of the reviews, it is recommended to review the wine for no more than two hours each time, as a long review time may lead to fatigue, which can adversely affect the results of the reviews.

4. 评酒前的准备工作（The Preparation Before the Wine Evaluation）

（1）酒样分类：为确保品评的公正性和准确性，应对参评的酒样进行细致分类。这包括根据酒的香型、生产工艺以及原料类型进行归类。每组酒样的数量不超过六个，每日最多进行四组酒样的品评。每组品评结束后，评酒员需休息约30分钟以恢复状态。

Wine sample classification: in order to ensure the fairness and accuracy of the evaluation, we will conduct detailed classification of the wine samples. This includes classification according to flavor type, production process and type of raw material. The number of wine samples in each group is not more than six, and a maximum of four groups of wine samples are reviewed daily. After each group of wine reviews, the wine reviewer needs to rest for about 30 minutes to recover.

（2）酒样编号与量度：工作人员应对每一份酒样进行精确编号，并确保酒样编号与酒杯编号完全一致，以防止混淆。在注入酒杯时，每杯酒样的量均为酒杯的3/5，以确保每位评酒员品评到的酒样量是一致的。

Wine sample numbering and measurement: Staffs should accurately number each wine sample and ensure that the wine sample number is exactly the same as the glass number to prevent confusion. When filling the glass, the amount of each wine sample is 3/5 of the glass to ensure that each wine reviewer reviews the same amount of wine sample.

（3）酒样温度控制：酒样的温度对于评酒员的嗅觉和味觉有着直接影响。为了确保品评的公正性和准确性，酒液的品评温度应在15～20℃，这是最适宜的温度范围。

Wine sample temperature control: the temperature of the wine sample has a direct impact on the smell and taste of the wine reviewer. In order to ensure the fairness and accuracy of the evaluation, we stipulate that the evaluation temperature of liquor should be between 15 ℃ and 20 ℃, which is the most suitable temperature range.

三、白酒的品质鉴别（Quality Identification of Liquor）

（一）色泽透明度鉴别（Color Transparency Identification）

白酒的正常色泽应为无色透明，无悬浮物和沉淀物。对于酱香型白酒，由于其特殊的酿造工艺和原料，酒体可能微微发黄。鉴别时，需在光线明亮处，用手举杯对光，以白布或白纸为底，用肉眼观察酒的颜色、透明度及有无悬浮物、沉淀物。

The normal color of liquor should be colorless and transparent, without suspended matter and sediment. For Maotai-flavor liquor, due to its special brewing technology and raw materials, the wine body may be slightly yellow. When identifying, it is necessary to raise a glass to the light in a bright place, with a white cloth or white paper as the base, and observe the color, transparency and whether there is suspended matter and sediment of the wine with the naked eye.

（二）香气鉴别（Aroma Identification）

在品鉴白酒的香气时，推荐使用大肚小口的玻璃杯。将白酒倒入杯中并轻轻摇晃后，将鼻子靠近杯口，仔细感受其散发的香气；或者将几滴酒滴在手掌上，轻轻搓揉后再嗅闻，以此判断香气的浓度与类型。

When tasting the aroma of white wine, it is recommended to use a glass with a large stomach and small mouth. Pour the liquor into the cup and gently shake it, put your nose close to the mouth of the cup, and carefully feel the aroma. Or a few drops of wine on the palm of your hand, gently rub and then smell, in order to judge the concentration and type of aroma.

（三）滋味鉴别（Taste Identification）

在品鉴白酒时，其滋味应展现出丰富多样性，包括浓厚、淡薄、绵软、辛辣、纯净等。白酒咽下后，其回甜与苦辣的特点亦值得品味。为了准确鉴别白酒的滋味，将酒饮入口中，并在舌头和咽部细细感受，以辨别酒味的醇厚程度和整体滋味的优劣。一般而言，上乘的白酒香气四溢，咽下后余香缭绕，口感醇厚、无异味且刺激性适中。在品鉴时，务必让舌尖和喉部充分体验，以准确评估白酒滋味的品质。

When tasting liquor, its taste should show rich diversity, including thick, light, soft, spicy, pure and so on. After swallowing liquor, its sweet and bitter characteristics are also worth tasting. In order to accurately identify the taste of liquor, we should drink the wine into the mouth, and feel it carefully in the tongue and pharynx to distinguish the mellow degree of wine taste and the overall taste. Generally speaking, the superior liquor aroma is overflowing, lingering incense after swallowing, mellow taste, no odor and irritating moderate. When tasting, be sure to let the tongue tip and throat fully experience, in order to accurately evaluate the quality of the liquor taste.

赛证直通 Competitions and Certificates

一、基础知识部分（Basic Knowledge Part）

（一）专业名词解释（Explanation of Professional Terms）

酒体（Wine Body）：

（二）思考题（Thinking Question）

如何做好白酒的品鉴？

How to do a good tasting of liquor?

二、技能操作部分（Skill Operation Part）

品鉴一款白酒并写出其品质特征。

Tasting Guizhou Xijiu and writing down its quality characteristics.

专题四
礼仪之邦：中国酒文化

你知道图 1-1-6 中的物品是什么吗？

_____

Do you know what's the object in the Figure 1-1-6?

_____

中国是酒文化的发源地之一，具有悠久的酿酒历史和文化传统。在中华民族五千多年的辉煌历程中，酒与其独特的文化一直占据着举足轻重的地位。中国的酒文化历经千年，已然形成了独树一帜的风貌，其背后深藏的酒的起源、饮酒礼仪及诸多酒俗，都是值得我们深入探索与学习的宝贵财富。

图 1-1-6　青铜器（bronze ware）

As one of the splendid ancient civilizations in the history of world civilization, Chinese civilization has always been known as the cradle and pioneer of wine culture, and has mastered exquisite wine making skills since ancient times. In the glorious history of the Chinese nation for five thousand years, liquor and its unique culture have always occupied an important position. This drink is not only an indispensable part of people's dietary life, but also plays an important role in the spiritual level. After thousands of years, China's liquor culture has formed a unique style. The origin of liquor, drinking etiquette and many liquor customs behind it are valuable wealth for us to explore and learn.

一、酒的起源（The Origin of Liquor）

（一）仪狄造酒的故事（The Story of Yidi Making Liquor）

仪狄造酒的说法源于古代传说和史籍记载。仪狄被传在夏禹时期发明了酿酒技术，有多处史籍提到他"作酒而美"或"始作酒醪"。具体来说，一种说法是仪狄制作的是酒醪，这是一种由糯米经过发酵而成的醪糟，其性温软、味甜，多产于江浙一带。在现在的一些家庭中，仍有人自制醪糟；另一种说法是"酒之所兴，肇自上皇，成于仪狄"，意味着在上古三皇五帝时期，就已有各种酿酒方法在民间流行，仪狄则是对这些方法进行了归纳总结，使之流传后世。总的来说，"仪狄造酒"的说法在史籍中有所记载，但具体细节和仪狄的身份职责仍有待进一步考证。这些说法不仅反映了古代酿酒技术的发展历程，也体现了古人对酒的看法和态度。

The idea that Yidi made liquor originated from ancient legends and historical records. Yidi is said to have invented liquor-making technology during the Xia Yu period, and many historical records mention that he " made liquor and was beautiful " or " began to make liquor mash ". Specifically, one saying is that Yidi is made of liquor mash, which is a glutinous rice fermented into a glutinous liquor, its nature is soft, sweet, mostly produced in Jiangsu and Zhejiang. In some families today, some people still make their own glutinous rice liquor. Another way of saying is " the rise of liquor, caused by the emperor, into the Yidi ", which means that in the ancient period of Three Emperors and Five Sovereings, there have been a variety of liquor making methods popular in the folk, Yidi is the summary of these methods, so that it passed on to later generations. In general, the claim that " Yidi made liquor " has been recorded in historical records, but the specific details and Yidi's identity and responsibilities remain to be further verified. This statement not only reflects the development of ancient liquor making technology, but also reflects the ancient people's views and attitudes towards liquor.

（二）杜康造酒的故事（The Story of Dukang Making Liquor）

杜康造酒的故事源自中国古代的传说。传说中，杜康是黄帝手下的一位大臣，负责管理生产和保存粮食。由于粮食丰收，没有合适的储存方法，导致粮食霉坏。杜康因此被降职，专职负责粮食保管。在一次偶然的机会中，杜康发现了粮食在潮湿的山洞中发酵后产生了一种液体，这种液体带有独特的香气和味道，这就是最初的酒。"杜康造酒"的故事是中国古代文化中的一个重要传说，它不仅反映了古代酿酒技术的发展历程，也体现了古人对酒的热爱和追求。

According to legend, Du Kang was a minister under the Yellow Emperor who was responsible for the production and preservation of grain. Due to the abundant grain, there is no suitable storage method, resulting in the spoilage of grain. Du Kang was demoted, full-time responsible for food storage. In a chance chance, Du Kang found that the grain in the wet cave mold produced a liquid, this liquid with a unique aroma and taste, this is the original wine. In general, the story of " Dukang's liquor making " is an important legend in ancient Chinese culture, which not only reflects the development of ancient liquor making technology, but also reflects the ancient people's love and pursuit of liquor.

（三）猿猴造酒的传说（The Legends of Apes Making Liquor）

"猿猴造酒"的说法源自古代传说。在远古时代，猿猴们可能发现了果实自然发酵后产生的酒香，并对此产生了浓厚的兴趣。猿猴们可能会将采集到的果实存放在石洼中或树洞中，以备不时之需。然而，当这些果实存放过久，其中的糖分在野生酵母菌的作用下会自然发酵，产生酒精，从而形成了具有酒味的液体。虽然"猿猴造酒"的说法在一定程度上反映了自然界中果实发酵的现象，但严格来说，猿猴并没有有意识地制造酒。这种"猿酒"或"自然酒"更像是自然界的一种巧合产物，与人类有意识地利用果品酿造酒的过程存在本质的区别。不过，"猿猴造酒"的说法既体现了人类对于自然现象的观察和思考，也丰富了酒文化的内涵。

The idea that "apes make liquor" comes from ancient legends. In ancient times, apes may have discovered the aroma of liquor produced by natural fermentation of fruits and developed a strong interest in it. The apes may store the fruit they collect in a hollow rock or in a hole in a tree for a rainy day. However, when these fruits are stored for too long, the sugars in them will naturally ferment under the action of wild yeasts, producing alcohol, which forms a liquor-like liquid. Although the idea that "apes make liquor" partly reflects the phenomenon of fruit fermentation in nature, strictly speaking, apes do not consciously make liquor. This "ape liquor" or "natural liquor" is more like a coincidental product of nature, and there is an essential difference from the process of human consciously using fruit to make liquor. However, the saying that "apes make wine" not only reflects human's observation and thinking of natural phenomena, but also enriches the connotation of wine culture.

二、酒器（Liquor Vessels）

（一）酒器的定义（Definition of Liquor Vessels）

酒器，顾名思义，即用于饮酒的各类器皿。随着中国古代酿酒业的蓬勃发展，各式各样的酒具相继问世，丰富了酒文化的内涵。

Liquor vessels, as the name suggests, are all kinds of vessels used for drinking. With the vigorous development of ancient Chinese liquor industry, a variety of liquor sets came out one after another, which enriched the connotation of liquor culture.

（二）酒器的分类（Classification of Liquor Vessels）

历经不同历史时期，酒器的制作手法与风格各异，形成了多姿多彩的酒器家族。从材料上分类，酒器有使用天然材料如木材、竹制、兽角、海螺、葫芦等制作的酒器；还有陶制、青铜制、漆制、瓷制、玉制、水晶制、金银制、锡制、景泰蓝制、玻璃制、铝制、不锈钢制等多种材质制成的酒器。

Through different historical periods, the production methods and styles of liquor vessels are different, forming a colorful family of liquor vessels. From the material classification, liquor vessels include the use of natural materials such as wood, bamboo products, animal horns, conch, gourd and other liquor vessels. There are also pottery, bronze, lacquer, porcelain, jade, crystal, gold and silver, tin, cloisonne, glass, aluminum, stainless steel and other materials made of liquor..

1. 锡制温酒器（Tin Liquor Warmer）

在明清至现代的历史时期，锡制温酒器广为使用。这种酒器并非直接作为酒杯使用，主要用于温酒，通过加热的方式使酒达到适宜的温度，以便有更好的口感和风味。锡制温酒器以其优良的导热性能和独特的材质美感，在酒器领域中占据重要地位。

In the historical period from Ming and Qing dynasties to modern times, tin liquor warmer was widely used. This liquor vessel is not used directly as a liquor glass, but is mainly used for warming liquor, so that the liquor can reach the appropriate temperature by heating, in order to better taste its taste and flavor. Tin liquor warmer occupies an important position in the field of liquor warmer because of its excellent thermal conductivity and unique aesthetic feeling of material.

2. 夜光杯（Luminous Cup）

"夜光杯"（图 1-1-7）这一名称源于唐代诗人王翰的诗句"葡萄美酒夜光杯"，指的是一种以玉石为原料精心制作的酒杯。该酒杯在古代以其独特的光泽和质感，在饮酒文化中占有重要地位。

Luminous cup (Figure 1-1-7), this name comes from the Tang Dynasty poet Wang Han's poem "grape liquor luminous cup", refers to a kind of jade as raw materials carefully made liquor glasses. In ancient times, the liquor glass played an important role in drinking culture with its unique luster and texture.

图 1-1-7　夜光杯（luminous cup）

3. 倒装壶（Inverted Pot）

倒装壶（图 1-1-8），作为陕西历史博物馆所珍藏的一件北宋耀州窑的杰出作品，其设计之巧妙令人叹为观止。此壶独特之处在于，酒液可通过壶底的小孔注入，而在将壶倒置时，酒液却不会溢出，展现了古代工匠高超的制瓷技艺。

As an outstanding work of Yaozhou Kiln in the Northern Song Dynasty collected by Shaanxi History Museum, the Reflux Pot (Figure 1-1-8) has a stunning design. The unique feature of this pot is that the liquor can be injected through a small

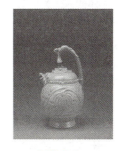

图 1-1-8　倒装壶（Inverted Pot）

hole in the bottom of the pot, but when the pot is turned upside down, the liquor will not overflow, demonstrating the ancient craftsmen's superb porcelain making skills.

4. 鸳鸯转香壶（Yuanyang Turning Incense Pot）

鸳鸯转香壶的独特之处在于其内部结构的巧妙设计，使得一壶之中可以分隔存放两种不同的酒液，并通过特定的机关控制，实现分别倒出两种不同酒液的功能。这一设计不仅增强了饮酒的趣味性，也反映了当时人们对生活美学的追求。

The unique feature of the pot is the ingenious design of its internal structure, which makes it possible to separate two different kinds of liquor in a pot, and realize the function of pouring different liquor separately through a specific organ control. This design not only enhances the fun of drinking, but also reflects people's pursuit of life aesthetics at that time.

5. 九龙公道杯 (Kowloon Fairness Cup)

九龙公道杯 (图 1-1-9)，其名称来源于其独特的设计和装饰。杯内精心雕刻有龙形图案，而杯外则绘制有八条龙纹，因此得名"九龙杯"。此杯的真正独特之处在于其功能性设计。当杯中的酒液超过某一特定量时，酒会通过"龙身"部分产生的虹吸效应，自动且全部流入杯底的底座之中。这一设计巧妙地体现了古人对"公道"这一理念的追求和象征，即在饮酒之时也应遵循公平与公正的原则。

图 1-1-9　九龙公道杯
(Kowloon Fairness Cup)

Kowloon Fairness Cup (Figure 1-1-9) derives its name from its unique design and decoration. The inside of the cup is elaborately carved with a dragon pattern, while the outside of the cup is painted with eight dragon patterns, hence the name "Kowloon Cup". However, that makes this cup truly unique is its functional design. When the liquor in the glass exceeds a certain amount, the liquor automatically flows into the base of the glass through the siphon effect generated by the "dragon body" part. This design cleverly reflects the ancient people's pursuit and symbol of the concept of "fairness", that is, the principle of fairness and justice should be followed when drinking.

6. 渎山大玉海 (Du Shan Da Yu Hai)

渎山大玉海，是一件专为储存酒液而制作的玉瓮，由一整块杂色墨玉精心雕琢而成。其造型雄伟壮观，四周刻有栩栩如生的海龙、海兽图案，展现出极高的艺术价值和工艺水平。该玉瓮容量巨大，可储存大量酒液，具有极高的使用价值。如今，这件珍贵的玉瓮已成为中国古代玉器艺术和酒文化的珍贵遗产，吸引着无数游客前来观赏和研究。

Du Shan Da Yu Hai is a jade urn specially made for the storage of liquor, which is carefully carved by a whole block of variegated ink jade. Its shape is magnificent and magnificent, engraved with lifelike sea dragon and sea animal patterns around, showing a high artistic value and craft level. The jade urn has a huge capacity, can store a large amount of liquor, has a high practical value. Today, this precious jade urn has become a precious heritage of ancient Chinese jade art and liquor culture, attracting countless tourists to watch and study.

三、古代饮酒礼仪 (Ancient Drinking Etiquette)

饮酒礼仪旨在规范人们的饮酒行为，防止因过量饮酒而导致的失序。在古代，文人雅士尤其注重饮酒的情境与雅趣，将其归纳为饮人、饮地、饮候、饮趣、饮禁和饮阑六大要素。

Drinking, as a traditional food activity in Chinese culture, has been accompanied by a set of strict etiquette since antiquity. The formation of these rituals aims to regulate people's drinking behavior and prevent disorder caused by excessive drinking. In ancient times, scholars paid special attention to the situation and taste of drinking, and summarized it into six elements: drinking people, drinking places, drinking time, drinking interest, drinking taboos, and drinking end.

（一）饮人（Drinking People）

在饮酒文化中，"饮人"这一要素强调的是选择合适的陪伴者，即选择志同道合的知己故交作为饮酒的伙伴。这种选择不仅基于彼此的友情与信任，更在于其风度高雅、性情直率的特质。与这样的伙伴一同品酒，能够营造出一种和谐、愉悦的饮酒氛围，使得参与者能更专注于酒的品质与风味，从而更深入地品味到酒中真谛。这种选择不仅体现了饮酒者对饮酒活动的重视，也展现了对友情与人际关系的珍视。

In drinking culture, the "drinker" element emphasizes the choice of the right companion, that is, the choice of like-minded friends as a drinking partner. This choice is based not only on mutual friendship and trust, but also on the quality of elegance and forthright temperament. When tasting liquor with such a partner, it can create a harmonious and pleasant drinking atmosphere, so that participants can focus more on the quality and flavor of the liquor, so as to taste the true meaning of the liquor more deeply. This choice not only reflects the importance of drinking activities, but also shows the value of friendship and interpersonal relationships.

（二）饮地（Drinking Places）

饮地（图1-1-10）指的是专门用于品酒或饮酒的场所，以自然环境优美、景色宜人之地为首选。这样的环境能够极大地提升饮酒的雅趣与品质。具体来说，诸如翠竹环绕的幽篁之地、繁花盛开的花园胜景，亦或是壮丽的名山大川等，均可作为理想的饮地。这些场所不仅能为饮酒者提供视觉上的享受，还能带来心灵上的宁静与愉悦，从而增强饮酒体验的文化内涵与审美价值。

Drinking place (Figure 1-1-10) refers to a place specially used for liquor tasting or drinking, with a beautiful natural environment and pleasant scenery as the first choice. Such an environment can greatly enhance the taste and quality of drinking. To be specific, the land surrounded by green bamboo, the space under the bamboo grove in full bloom, or the magnificent famous mountains and rivers can be the ideal drinking place. These places can not only provide visual enjoyment for drinkers, but also bring peace and pleasure to the soul, thereby enhancing the cultural connotation and aesthetic value of the drinking experience.

图 1-1-10　饮地（drinking places）

（三）饮候（Drinking Time）

饮候，即饮酒的时机选择，强调对时间节点的审慎考量。选择如清秋、雨后、新月等充满诗意的时刻饮酒，能够更好地增添饮酒的韵味与情感深度。这样的时刻不仅具有独特的自然美感，而且能够引发饮酒者的情感共鸣，使其在品酒的过程中领略到更为丰富的文化意蕴与审美享受。因此，合理的饮候选择是提升饮酒体验与品味的重要环节。

Drinking time, that is, the timing of drinking, emphasizes the careful consideration of time nodes. Choosing poetic moments such as clear autumn, after rain, new moon and so on to drink, can better increase the charm and emotional depth of drinking. Such a moment not only has a unique natural beauty, but also can trigger the emotional resonance of the drinkers, so that they can enjoy a richer cultural implication and aesthetic enjoyment in the process of liquor tasting. Therefore, reasonable choice of drinking time is an important link to enhance the drinking experience and taste.

（四）饮趣（Drinking Interest）

"饮趣"这一概念，在文化和艺术领域中具有深远的意义。它指的是通过一系列文化活动，如联吟（即诗词唱和）、清谈（即高雅的谈话或辩论）等，营造和烘托特定的氛围，从而提升饮酒的兴致和体验。这些文化活动不仅能够增强饮酒场合的文化内涵和艺术气息，还能使参与者在享受美酒的同时，感受到心灵的愉悦和满足。通过饮趣的展现，人们能够在饮酒的过程中达到身心愉悦、情感共鸣的境界。

The concept of "drink interest" has profound significance in the field of culture and art. It refers to the creation and enhancement of a specific atmosphere through a series of cultural activities, such as lianyin (that is, poetry chorus) and talk (that is, elegant conversation or debate), so as to further enhance the enjoyment and experience of drinking. These cultural activities can not only enhance the cultural connotation and artistic atmosphere of the drinking occasion, but also make the participants feel the pleasure and satisfaction of the soul while enjoying the liquor. Through the display of drinking interest, people can achieve physical and mental pleasure and emotional resonance in the process of drinking.

（五）饮禁（Drinking Taboos）

饮禁，指在饮酒过程中需要遵守的一系列规范与限制，旨在避免饮酒时出现不愉快或不适宜的情况。这些规范与限制包括但不限于避免恶意劝酒、取笑他人等行为，以确保饮酒环境的和谐与舒适。遵守饮禁不仅有助于提升饮酒体验，更能展现个人的修养与素质，是饮酒文化中不可或缺的一部分。

Drinking Taboos refers to a series of norms and restrictions that need to be observed during the drinking process to avoid unpleasant or inappropriate drinking situations. These norms and restrictions include, but are not limited to, avoiding malicious exhortation and making fun of others to ensure a harmonious and comfortable drinking environment. Complying with the taboos of drinking not only helps to enhance the drinking experience, but also shows personal accomplishment and quality, which is an indispensable part of drinking culture.

（六）饮阑（Drinking End）

饮阑，指的是酒宴接近尾声，宾客可以继续进行的文化娱乐活动，如赋诗吟咏、悠然散步、登高望远或垂钓闲趣，以此延续和深化酒宴带来的欢乐与氛围。这些活动不仅丰富了酒宴的内涵，也展现了宾客的文化素养与高雅情趣，是酒文化中不可或缺的一部分。

Drinking End, refers to the cultural entertainment activities that guests can continue to carry out when the banquet is coming to an end, such as writing poems, taking a leisurely walk, climbing a mountain or fishing, in order to continue and deepen the joy and atmosphere brought by the banquet. These activities not only enrich the connotation of the banquet, but also show the cultural quality and elegant taste of the guests, which is an indispensable part of the liquor culture.

在古代，饮酒礼仪的讲究体现了深厚的文化传统与社交规范。当主人与宾客共同饮酒时，需遵循特定的礼节，相互跪拜以表达深深的尊重。在晚辈与长辈共饮的场合中，晚辈需先行跪拜礼，以示对长辈的恭敬，随后方可入座。在长辈未饮尽之前，晚辈需保持谦逊，不得先行饮酒。

In ancient times, the attention to drinking etiquette reflects profound cultural traditions and social norms. When the host and guests drink together, they are required to follow certain etiquette and bow down to each other to show deep respect. In the occasion of drinking with the elders, the younger generation needs to bow down first to show respect to the elders, and then can take a seat. Before the elders drink up, the younger generation should remain humble and not drink first.

整个饮酒过程被细分为四个步骤：拜、祭、啐、卒爵，每一步都蕴含着特定的寓意。其中，"拜"表示对天地的敬畏，"祭"表达了对主人的尊敬，"啐"为初尝酒意，而"卒爵"则寓意着对美酒的赞美与珍惜。

The whole drinking process is subdivided into four steps: worship, sacrifice, taste, and drain, each step contains a specific meaning. Among them, "worship" expresses the awe of heaven and earth, "sacrifice" expresses the respect of the host, "taste" is the first taste of liquor, and "drain" means the praise and cherish of liquor.

在酒宴之上，主人需向客人敬酒，这一举动被称为"酬"，体现了主人的好客与热情。客人则需回敬主人，此举称为"酢"，展示了客人的礼貌与谦逊。此外，宾客之间还会相互敬酒，这一行为被称为"旅酬"，体现了社交场合中的和谐与融洽。

On top of the banquet, the host needs to toast the guests, which is called "reward", reflecting the host's hospitality and enthusiasm. Guests are required to return the favor to the host, which is called "reciprocation", showing the guests' courtesy and modesty. In addition, guests will also toast each other, which is called "Toast each other", reflecting the harmony and harmony in the social occasion.

在敬酒过程中，双方都需要起立"避席"，这是一种表达敬意的方式，显示了彼此对对方的尊重与重视。一般而言，在古代的酒宴上，普通敬酒以三杯为度，这一传统既体现了对饮酒节制的重视，也展示了酒文化的深厚底蕴。

In the process of toasting, both parties need to stand up and "avoid the table", which is a way to show respect and show the respect and importance of each other. Generally speaking, in the ancient banquet, the ordinary toast with three cups as the degree, this tradition not only reflects the importance of drinking control, but also shows the profound heritage of liquor culture.

四、现代饮酒礼仪（Modern Drinking Etiquette）

（一）斟酒的礼仪（Etiquette of Pouring Liquor）

在酒宴礼仪中，酒水的斟倒是一项重要的仪式，通常在饮用之前进行，以表达对客人的尊重和敬意。当主人亲自为客人斟酒时，客人应当端起酒杯，以表示对主人的感激之情。若情况需要，客人还应起身站立或点头示意，以显示对主人的尊重和感谢。此外，客人还可采用"叩指礼"来回应主人的斟酒。具体操作是，客人右手三指并拢，指尖向下，轻叩桌面数次，这一举动既体现了对主人的敬意，也展示了客人的修养和风度。在斟酒的过程中，主人应选择品质最佳的酒水，并当场启封，以示对客人的诚意和尊重。斟酒时，主人应确保公平及适量，不应偏袒某位客人或过量斟酒，以保证每位客人都能感受到舒适的饮酒体验。

Pouring liquor etiquette, liquor pouring is an important ceremony, usually before drinking, to show respect and respect for the guests. When the host personally pours liquor for the guests, the guests should hold the glass to show their gratitude to the host. If the situation requires, the guest should also stand up or nod to show respect and gratitude to the host. In addition, guests can also use the "tapping ceremony" to respond to the host's liquor. The specific operation is that the guest's right hand three fingers together, fingertips down, tap on the table several times, which not only reflects the respect for the host, but also shows the guest's cultivation and demeanor. In the process of pouring liquor, the host should choose the best quality liquor and open it on the spot to show sincerity and respect for the guests. When pouring liquor, the host should ensure fairness and amount, and should not favor one guest or overpour liquor to ensure that each guest can enjoy a suitable drinking experience.

（二）饮酒的礼仪（Drinking Etiquette）

中国的酒文化源远流长、博大精深，它承载着深厚的文化意义和重要的社交价值。在现代社交场合中，酒已然成了一种独特的文化交流媒介，它代表着礼仪的庄重、氛围的和谐、情感的交融以及心境的抒发。在欢迎宾客的仪式中，酒是表达敬意和热情的媒介；在朋友聚会的场合，酒是增进感情、拉近彼此距离的桥梁；在沟通交流的过程中，酒是打破沉默、促进深入交流的催化剂；在传递友谊的时刻，酒更是情感交流的纽带，让友情在酒香中得以升华。

Chinese liquor culture has a long history, extensive and profound, it carries profound cultural significance and important social value. In modern social occasions, liquor has become a unique cultural communication medium, which represents the solemn etiquette, the harmony of the atmosphere, the integration of emotions and the expression of the mood. In the ceremony of welcoming guests, liquor is the medium of respect and enthusiasm. On the occasion of friends gathering, liquor is a bridge to enhance feelings and narrow the distance between each other. In the process of communication, liquor is a catalyst to break the silence and promote in-depth communication. In the moment of conveying friendship, liquor is the bond of emotional exchange, so that friendship can be sublimated in the aroma of liquor.

在酒宴的筹备中，座次的安排尤为重要，它严格遵循着传统的规矩和礼仪。主要座位的设置中，位于上席中央的座位是东道主的专属，即酒宴的发起者或支付款项的人。在其右手边则是贵宾座，这一位置专门留给最为尊贵的客人，以表达对其的尊重和礼遇。其左手边的座位相对稍次，但同样体现了对客人的尊重和欢迎。两侧则依次坐着陪客和其他受邀的客人。值得强调的是，在入席时，所有客人应遵循一致的原则，即与其他客人同步进入，以避免出现单独行动的情况，这样不仅能维护酒宴的秩序，更能体现出对主人的尊重和对其他客人的礼貌。

In the preparation of the banquet, the seating arrangement is particularly important, which strictly follows the traditional rules and etiquette. In the setting of the main seat, the seat in the center of the upper table is the exclusive seat of the host, that is, the initiator of the banquet or the person who pays the money. On its right hand side is the VIP seat, which is reserved for the most distinguished guests to express respect and courtesy. The seat on the left hand side is relatively inferior, but it also reflects the respect and welcome of the guests. On either side are the chaperones and other invited guests. It is worth emphasizing that when entering the table, all guests should follow the same principle, that is, enter with other guests at the same time, to avoid the situation of acting alone, which can not only maintain the order of the banquet, but also reflect the respect for the host and the politeness of other guests.

在工作与日常生活中，关于饮酒的时机和程度需进行恰当把握。工作前应避免饮酒，以防止酒精可能导致的交流障碍，确保与他人的有效沟通。同时，在休息期间饮酒也应保持适度，以免因过量饮酒而干扰到工作状态，进而影响工作表现和效率。

In professional and daily life, the timing and degree of alcohol consumption need to be properly grasped. Avoid drinking alcohol before work to prevent communication barriers that may be caused by alcohol and ensure effective communication with others. At the same time, drinking during the rest should also be maintained in moderation, so as not to interfere with the work state due to excessive drinking, which will affect work performance and efficiency.

（三）敬酒的礼仪（Etiquette of Toasting）

敬酒，也称祝酒，在正式宴会中扮演着举足轻重的角色。这一仪式不仅体现了主人对来宾的深切敬意与美好祝愿，更是宴会中不可或缺的一环。在敬酒时，主人通常会发表祝酒词，其内容多涉及对来宾身体健康、事业有成的祝愿，以及对双方友谊长存的期许。

Toasting, also known as propose a toast, plays an important role in formal banquets. This ceremony not only reflects the host's deep respect and good wishes for the guests, but also an indispensable part of the banquet. During the toast, the host usually delivers a toast, which mostly involves the wishes for the guests' health and career success, as well as the hope for the lasting friendship between the two sides.

敬酒可以穿插在饮酒的各个环节，但正式的祝酒词应在不影响来宾用餐体验的合适时间发表。这样不仅有助于营造温馨和谐的宴会氛围，还能使来宾在享受美食的同时，感受到主人的热情与诚意。因此，在敬酒时，主人应充分考虑时机，以确保整个宴会流程顺畅、愉快。

The toasting process can be interspersed throughout the drinking process, but the formal toast should be delivered at an appropriate time that does not interfere with the guests' dining experience. This not only helps to create a warm and harmonious banquet atmosphere, but also enables guests to enjoy the food at the same time, feel the host's enthusiasm and sincerity. Therefore, when toasting, the host should fully consider the timing to ensure that the whole banquet process is smooth and pleasant.

中国酒文化，不仅仅是一杯琼浆玉液，更是承载了数千年的历史、智慧与情感的载体。在这份独特的文化传承中，我们感受到了中华民族的礼仪之美、和谐之道和智慧之光。

Chinese liquor culture is not only a cup of liquor, but also a carrier of thousands of years of history, wisdom and emotion. In this unique cultural inheritance, we feel the beauty of Chinese etiquette, the way of harmony and the light of wisdom.

赛证直通 Competitions and Certificates

一、基础知识部分（Basic Knowledge Part）

（一）专业名词解释（Explanation of Professional Terms）

1. 卒爵（drain）：
2. 旅酬（exchange toast）：

（二）思考题（Thinking Question）

谈谈对中国古代饮酒礼仪的理解。

Talk about your understanding of drinking etiquette in ancient China.

二维码 1-1-4：
知识拓展

二、技能操作部分（Skill Operation Part）

在实训室练习斟酒礼仪。

Practice liquor etiquette in the training room.

专题五

彬彬有礼：外国酒文化

【微课】
彬彬有礼：
外国酒文化

你知道图 1-1-11 中描述的是什么场景吗？

_____

Do you know what the scene depicted in the Figure 1-1-11 is?

_____

在世界的每一个角落，都有其独特的酒文化与礼仪。我们将一同深入探索外国酒文化的魅力，感受不同国家酒文化背后所蕴含的深厚内涵。

In every corner of the world, there is its own unique wine culture and etiquette. Together, we will deeply explore the charm of foreign wine culture and feel the profound

图 1-1-11 壁画（mural painting）

connotation behind the wine culture of different countries.

一、酒的起源与发展（Origin and Development of Alcohol）

（一）古代酿造技术的起源与发展（The Origin and Development of Ancient Brewing Technology）

据历史文献和考古发现，公元前 10 世纪，古埃及和古希腊等文明古国已经掌握了一定的酒酿造技术。这里的人们开始使用各种粮食和水果作为原料，通过发酵和简单的加工方法，酿造出具有独特风味的酒品。古埃及人利用面包和椰枣等原料，结合独特的发酵技术，酿造出具有丰富口感的啤酒。古希腊人则通过压榨葡萄并让其自然发酵，制造出葡萄酒。这些早期的酿造技术虽然简单，但为后来的酿酒工业奠定了坚实的基础。

According to historical documents and archaeological findings, by the 10th century BC, ancient Egypt and ancient Greece and other civilizations had mastered certain brewing technology. The people of these civilized regions began to use various grains and fruits as raw materials, through fermentation and simple processing methods, to brew wine with unique flavors. The ancient Egyptians used raw materials such as bread and date palm, combined with unique fermentation technology, to brew beer with rich taste. The ancient Greeks made wine by pressing grapes and letting them ferment naturally. These early brewing techniques, though simple, laid a solid foundation for the later brewing industry.

（二）蒸馏技术的引入与高度酒的诞生（The Introduction of Distillation Technology and the Birth of High Alcohol）

随着酿造技术的不断进步和人们对酒类品质的追求，17 世纪，蒸馏技术被引入酿造业。这一技术的引入极大地推动了酿酒业的发展，使得高度酒开始出现在人们的日常生活中。蒸馏技术通过加热和冷凝的方式，提取出原料中的酒精成分，从而制造出酒精度更高的酒。白兰地、威士忌等高度酒因其独特的口感和较高的酒精度，受到人们的喜爱。这些高度酒的诞生，标志着酿酒技术进入了一个新的发展阶段。

With the continuous progress of brewing technology and people's pursuit of alcohol quality, distillation technology was introduced into brewing in the 17th century. The introduction of this technology greatly promoted the development of the brewing industry, making high alcohol began to appear in people's daily life. Distillation technology, by heating and condensing, extracts the alcohol from the raw material to create a higher alcohol content. Brandy, Whiskey and other high spirits are favored by people because of their unique taste and high alcohol content. The birth of these high alcohol marked that liquor-making technology had entered a new stage of development.

（三）现代酿酒技术的完善与多样化（The Perfection and Diversification of Modern Liquor-making Technology）

经过数百年的发展和技术积累，现代酿酒技术有了显著的进步和创新。现代酿酒技术不仅继承了古代酿造技术的精髓，还融入了现代科技元素，使酒类产品的品质得到了极大的提升。现代的酿酒师们通过精细的原料选择、科学的发酵工艺和精湛的调配技术，能够精确地控制酒品的酒精度、口感和特色。同时，现代酿酒技术还具备配制各种酒品的能力，以满足不同消费者的需求。

After hundreds of years of development and technical accumulation, modern wine-making technology has made remarkable progress and innovation. Modern liquor-making technology not only inherits the essence of ancient liquor-making technology, but also integrates modern scientific and technological elements, which makes the quality of alcohol products greatly improved. Modern winemakers can precisely control the alcohol content, taste and characteristics of alcohol through the fine selection of raw materials, scientific fermentation technology and fine blending technology. At the same time, modern Liquor-making technology also has the ability to formulate a variety of alcohol to meet the needs of different consumers.

在原料方面，现代酿酒师们开始尝试使用非传统的原料酿造特色酒。例如，使用果汁、蜂蜜、香料等添加剂增添酒的风味和口感。此外，一些酿酒师还开始采用有机原料生产酒类产品，以满足消费者对健康和环保的需求。

In terms of ingredients, modern winemakers are beginning to experiment with non-traditional ingredients to make specialty alcohol. For example, juice, honey, spices and other additives are used to add flavor and texture to the wine. In addition, some winemakers have also begun to use organic raw material to produce alcohol products to meet consumer needs for health and environmental protection.

在工艺方面，现代酿酒技术注重精细调控和智能化控制。通过精细调控酿酒过程中的温度、湿度、酵母数量和时间等因素，酿酒师们能够精确地控制发酵过程中的化学反应，从而调整酒的风味和香气。同时，智能化的控制系统能够实时监测和记录酿酒过程中的各种数据，为酿酒师们提供科学的指导和支持。

In the process, modern liquor-making technology pays attention to fine regulation and intelligent control. By fine-tuning factors such as temperature, humidity, number and timing of yeast during the liquor-making process, winemakers can precisely control the chemical reactions during fermentation, thus adjusting the flavor and aroma of the alcohol. At the same time, the intelligent control system can monitor and record all kinds of data in the winemaking process in real time, providing scientific guidance and support for winemakers.

在品牌和文化方面，现代酿酒企业开始注重品牌文化的塑造和传播。通过独特的品牌故事、包装设计和营销策略等方式，提升企业产品的差异化和附加值，吸引更多的消费者。同时，一些企业还通过举办品鉴会、文化沙龙等活动，加强与消费者的互动和沟通，增强品牌的影响力和忠诚度。

In terms of brand and culture, modern liquor-making enterprises began to pay attention to the shaping and dissemination of brand culture. Through unique brand stories, packaging design and marketing strategies, companies can enhance product differentiation and added value to attract more consumers. At the same time, some enterprises also strengthen the interaction and communication with consumers by holding tastings, cultural salons and other activities to enhance brand influence and loyalty.

二、外国饮酒礼仪（Foreign Drinking Etiquette）

（一）英国（UK）

在英国社会中，礼仪教育根深蒂固，是其文化不可或缺的组成部分（图 1-1-12）。一般而言，宴会上的饮酒顺序包括餐前酒、餐中酒和餐后酒。餐前酒通常用于开胃和营造轻松愉快的氛围；餐中酒则与菜肴相搭配，增添饮食的情趣和风味；餐后酒则是在用餐结束后享用，用以回味和放松。这种有序的饮酒顺序，不仅彰显了英国贵族的严谨和讲究，也体现了他们对宴会文化的深入理解和精心安排。

图 1-1-12　服务礼仪
（Service etiquette）

In British society, etiquette education is deeply rooted and an indispensable part of its culture. (Figure 1-1-12). In general, the order of drinking at a banquet includes aperitif, table wine, and dessert wine. Aperitif is usually used to amuse and create a relaxed atmosphere. The table wine is matched with the dishes, adding the interest and flavor of the diet. The dessert wine is enjoyed at the end of the meal for afterthought and relaxation. This orderly drinking order not only highlights the rigor and attention of the British aristocrats, but also reflects their deep understanding of the banquet culture and careful arrangements.

在正式的宴会筹备中，餐前酒具有独特的意义。一般而言，餐前酒通常在客厅被宾客享用，这一环节的主要目的是刺激食欲，同时也为即将到来的用餐体验提供一个缓冲的时间段。通过这样的安排，宾客可以在进入餐厅前，利用这段时间进行轻松的交流，从而避免等待用餐时可能产生的尴尬。餐前酒的饮用时间通常设定在正式用餐前的 30 分钟左右，以确保宾客在享用完餐前酒后，能够适时进入餐厅享用正式餐点。

In the formal banquet preparation, the aperitif is of special significance. In general, aperitif is usually enjoyed by guests in the living room, the main purpose of this part is to prepare and stimulate the appetite of guests, but also to provide a buffer time for the upcoming dining experience. With this arrangement, guests can use the time to communicate easily before entering the restaurant, thus avoiding the embarrassment that may arise while waiting for their meal. The aperitif is usually set around 30 minutes before the formal meal to ensure that guests can enter the restaurant in time to enjoy the formal meal after enjoying the aperitif.

在正式的宴会中，餐中酒的品鉴与主菜的搭配可以展现优雅品位及对细节的关注。葡萄酒作为常见的选择，要重视其搭配原则。一般而言，红酒与"红肉"相互辉映，能够增强食物的口感和风味，同时应避免在红酒中加入冰块以保持其原有的风味。白葡萄酒则与"白肉"相得益彰，在饮用前应先冰镇，使其口感更加清爽宜人。在宾客享用餐中酒之前，侍酒师会将酒瓶置于托盘上，向宾客介绍酒的品牌、年份等信息。随后，侍酒师会为宾客逐一倒酒，确保每位宾客都能品尝到优质的酒品。在倒酒完毕后，主人会首先举杯向宾客示意，引领宾客进行餐中酒的品鉴。宾客应首先欣赏酒液的色泽，透过酒杯观察其清澈度和色泽深浅；接着，轻轻晃动酒杯，使酒香充分释放，嗅闻其独特的香气；最后，轻抿一小口酒液，品味其口感和风味，并慢慢咽下，享受这一过程带来的愉悦感受。

In a formal banquet, the tasting of table wine and the pairing of the main course is an important part of showing elegance and attention to detail. Wine as a common choice, we should pay attention to its matching principles. In general, red wine and red meat complement each other, can enhance the taste and flavor of food, and should avoid adding ice to red wine to maintain its original flavor. White wine and " white meat " complement each other, should be chilled before drinking, make it more refreshing and pleasant taste. Before the guests enjoy the wine, the waiter will place the bottle on the exquisite tray and introduce the wine brand, year and other information to the guests. Afterwards, the waiter will pour the wine for the guests one by one, ensuring that each guest can taste the quality of the wine. After pouring the wine, the host will first toast to the guests and lead the guests to taste the wine. Guests should first appreciate the color of the wine, through the glass to observe its clarity and color. Then, gently shake the glass, so that the wine fully released, smell its unique aroma. Finally, sip the wine, savor its taste and flavor, and swallow slowly, enjoy the pleasure of the process.

在宴会礼仪中，餐后酒的选择同样讲究。其中，白兰地因其独特的口感和香气，成为常见的选择。在享用餐后酒时，必须使用透明的酒杯，以充分展现白兰地酒色的纯净。透明的酒杯能够清晰地反映出酒液的颜色和光泽，使宾客能够更好地欣赏和品味这一美妙的饮品。

In the banquet etiquette, the choice of dessert wine is also exquisite. Among them, Brandy has become a common choice because of its unique taste and aroma. When enjoying the dessert wine, you must use a transparent glass to fully show the pure and true color of brandy. The transparent glass can clearly reflect the color and luster of the wine, so that guests can better appreciate and taste this wonderful drink.

（二）美国（USA）

在美国的饮酒文化中，健康饮酒风尚愈发成为酒桌上的主旋律。随着时间的推移，美国人的口味逐渐从浓郁的深色烈酒转向更为温和的浅色啤酒、葡萄酒和果酒，这些酒品更符合现代健康生活的理念。在正式的宴会场合，美国人倾向于选择葡萄酒作为社交饮品。他们不会像英国人那样严格遵循烦琐的饮酒礼仪，也不会像法国人那样过分讲究葡萄酒的品鉴。美国人更注重与他人的交流和享受饮酒带来的愉悦感，而不是过于追求酒品的品质或礼仪的规范。

In the fast-paced life of the United States, alcohol as a social drink is widely used to relieve the stress of life and work. American culture, however, is unique in its health-oriented approach to drinking. Over time, American tastes have shifted from strong, dark spirits to milder, lighter beers, wines and cider that better fit modern ideas of healthy living. At formal banquets, Americans tend to choose wine as a social drink. They don't follow the rigor of drinking etiquette as the British do, nor do they obsess over wine tasting as the French do. Americans pay more attention to communicating with others and enjoying the pleasure brought by wine, rather than pursuing the quality of wine or etiquette norms.

美国人认为，在宴会上饮用葡萄酒有多重益处。首先，葡萄酒不仅有益健康，而且酒精浓度适中，不易使人醉酒失态。其次，葡萄酒特别是香槟，能有效活跃气氛，使庆祝活动氛围更加欢快融洽。再次，葡萄酒是绝佳的佐餐之选，无论是红葡萄酒、白葡萄酒还是冰葡萄酒，都能为肉类和鱼类增添鲜美口感，使菜肴更加美味可口。此外，葡萄酒还能营造出浪漫而优雅的氛围，既适合商务宴请，也适合朋友间的聚会。在休闲时刻，如烧烤聚会时，美国人则更偏爱啤酒，且注重酒类与食物的搭配。

Americans believe that wine at a dinner party has multiple benefits. First of all, wine is not only good for health, but also moderate alcohol concentration, which is not easy to make people drunk. Secondly, wine, especially champagne, can effectively enhance the atmosphere and make the celebration more cheerful and harmonious. In addition, wine is a great choice for meals, whether red, white or iced wine, can add flavor to meat and fish, making dishes more delicious. In addition, wine can create a romantic and elegant atmosphere, which is suitable for business dinners or gatherings among friends. In casual moments, such as barbecue parties, Americans prefer beer, and they pay great attention to the combination of alcohol and food.

（三）法国（France）

在法国的饮酒文化中，礼仪贯穿于多个环节，尤为显著地体现在倒酒、敬酒和饮酒的方式上。正确的倒酒方式，不仅是一项需要反复练习的技能，更是法国饮酒文化的重要体现。在倒红酒时，法国人通常会将酒液倒至酒杯的三分之一处，这不仅确保了酒液不会溢出，更为重要的是，有助于酒的香气在酒杯中充分保留和散发。这一细节处理，彰显了法国人对饮酒文化的尊重和对品质的追求。

In French drinking culture, etiquette runs through many links, especially in the way of pouring, toasting and drinking. The correct way to pour wine is not only a skill that requires repeated practice, but also an important reflection of French drinking culture. When pouring red wine, the French usually pour the wine to a third of the glass, such an operation not only ensures that the wine will not overflow, but more importantly, it helps the aroma of the wine to be fully retained and distributed in the glass. The handling of this detail shows the French people's respect for drinking culture and their pursuit of quality.

在探讨法国人的饮酒礼仪时，握杯的姿势是一个不可忽视的细节。在品鉴红酒时，法国人通常采用特定的握杯方式，他们倾向于用手指捏住杯身下方的杯杆，或是用拇指和食指捏住杯底。尽管这种握法可能在初看时显得不那么自然，但它背后蕴含了丰富的文化内涵和实用意义。这种握杯方式的首要目的在于防止人体温度对酒温的影响。手掌的温度通常高于室温，直接握住杯身可能会使酒液温度升高，从而影响其口感和风味。通过捏住杯杆或杯底的方式，能够确保酒液保持适宜的温度，从而更好地体验红酒的复杂口感。此外，这种握杯方式还有助于保持杯身的清洁。由于红酒的色泽是品鉴过程中重要的考量因素之一，一个干净的杯身能够更好地展现红酒的颜色和光泽。通过捏住杯杆或杯底的方式，法国人能够避免在品酒过程中手指直接接触杯身，从而保持杯身的清洁和美观。在喝红酒时采用的握杯方式既体现了他们对饮酒文化的尊重，也展现了他们对品酒细节的精益求精。这种握杯方式（图1-1-13）不仅具有实用性，更是一种文化传统的传承和体现。

将杯柄置于拇指和食指间

夹住杯柄

夹住杯柄底部

用大拇指托住杯底

图 1-1-13　正确的持杯方式
（the correct way to hold the glass）

When discussing French drinking etiquette, the

position of holding the cup is a detail that cannot be ignored. When tasting red wine, the French usually use a specific way to hold the glass. They tend to pinch the underside of the cup with their fingers, or the bottom between their thumb and forefinger. Although this grip may seem unnatural at first glance, there is a wealth of cultural and practical significance behind it. The primary purpose of this grip is to prevent the influence of human body temperature on the temperature of the wine. Since the temperature of the palm is usually higher than room temperature, holding the glass directly may heat up the wine, which can affect its taste and flavor. By pinching the stem or bottom of the glass, the French are able to ensure that the wine remains at the right temperature, thus better experiencing the complex taste of red wine. In addition, this way of holding the cup also helps to keep the body clean. Since the color of the wine is one of the important factors in the tasting process, a clean glass body can better show the color and luster of the wine. By pinching the stem or bottom of the glass, the French are able to avoid direct finger contact with the body of the glass during the tasting process, thus keeping the body clean and beautiful. The way the French hold the glass when drinking red wine reflects their respect for the drinking culture, but also shows their meticulous attention to the details of wine tasting. This way of holding the cup (Figure 1-1-13) is not only practical, but also the inheritance and embodiment of a cultural tradition.

在探索外国酒水文化的旅程中，我们感受到不同国家酒文化的独特韵味，领略其背后的丰富内涵。在这个过程中，不仅要学会品鉴美酒，更要学会以"彬彬有礼"的态度去欣赏和尊重不同的文化。继续保持对酒文化的热爱和探索，让"彬彬有礼"成为饮酒时永恒的座右铭。

In the journey of exploring foreign wine culture, we feel the unique charm of different countries' wine culture and appreciate the rich connotation behind it. In this process, we must not only learn how to taste wine, but also learn how to appreciate and respect each beverage culture with a "polite" attitude. Let us continue to maintain the love and exploration of wine culture, and let "polite" become our eternal motto when tasting wine.

赛证直通 Competitions and Certificates

一、基础知识部分 (Basic Knowledge Part)

二维码 1-1-5:
知识拓展

(一) 专业名词解释 (Explanation of Professional Terms)

1. 酿造酒 (Fermented alcoholic drink):
2. 蒸馏酒 (Distilled spirits):

(二) 思考题 (Thinking Question)

谈谈对不同国家饮酒礼仪的理解。

Talk about your understanding of drinking etiquette in different countries.

二、技能操作部分（Skill Operation Part）

在实训室练习正确的持杯方式。

Practice the Proper way of holding cup in the training room.

【单元一　反思与评价】Unit 1　Reflection and Evaluation

学会了：_____

Learned to: _____

成功实践了：_____

Successful practice: _____

最大的收获：_____

The biggest gain: _____

遇到的困难：_____

Difficulties encountered: _____

对教师的建议：_____

Suggestions for teachers: _____

单元二　中国白酒

兰陵美酒郁金香，玉碗盛来琥珀光。

但使主人能醉客，不知何处是他乡。

——唐·李白《留客中行》

单元导入 Unit Introduction

白酒，是中国的传统蒸馏酒，其种类繁多，风格各异。那么中国白酒分为哪些类别？不同类别有什么特点？又有哪些白酒名品呢？这个单元将带你走进中国白酒的世界。

Baijiu, a traditional distilled liquor in China, has many kinds and different styles. So what are the categories of Chinese Baijiu? What are the characteristics of the different categories? What are the famous liquor products? This unit will take you into the world of Chinese Baijiu.

学习目标 Learning Objectives

➤ 知识目标（Knowledge Objectives）

（1）了解白酒的特点。

To understand the characteristics of baijiu.

（2）了解白酒的制作工序和香型分类。

To understand the production process and flavor classification of baijiu.

（3）熟悉中国著名白酒。

To know famous Chinese baijiu.

（4）熟悉不同香型白酒的工艺特点及风味特征。

The process characteristics and flavor characteristics of different fragrance types.

（5）了解中国名人与酒的故事。

Understanding the stories of Chinese celebrities and their relationship with liquor.

➤ 能力目标（Ability Objectives）

（1）能介绍中国白酒的制作工序。

Students can introduce the production process of Chinese baijiu.

（2）能介绍中国名酒及其主要特点。

Students can introduce famous Chinese baijiu and their main features.

（3）能根据所学知识区分白酒香型。

Be able to distinguish the fragrance types of baijiu based on the knowledge learned.

（4）能说出各类香型白酒的代表名品及产地。

Be able to tell the representative famous products and production places of various fragrance type Baijiu.

➤ 素质目标（Quality Objectives）

（1）了解酒文化在社会政治生活、文艺、生活态度、审美情趣等诸多方面的体现。

Let students understand that liquor culture is reflected in social and political life, literature and art, life attitude, aesthetic taste and other aspects.

（2）理解白酒是中国的文化符号之一。

Let the students understand that baijiu is one of the cultural symbols of China.

（3）在饮食文化发展中感受中华民族生生不息的灵魂力量，提升民族自信心和自豪感。

Let students feel the eternal soul power of the Chinese nation in the development of food culture, and enhance national self-confidence and pride.

（4）弘扬中华酿造技艺的传统文化和技术，树立文化自信。

Promote the Chinese traditional culture and technology of brewing skills, and establish cultural confidence.

专题一
东方佳酿：中国白酒概述

你认识图 1-2-1 中的中国知名白酒吗？能说出它们的产地吗？

【微课】
东方佳酿：
中国白酒概述

Do you know the famous Chinese baijiu in the picture (Figure1-2-1)? Can you tell me where they came from?

图 1-2-1　中国知名白酒

一、白酒的概念与特点 (Concept and Characteristics of Baijiu)

白酒是世界八大蒸馏酒之一，以曲类、酒母为糖化发酵剂，利用淀粉质原料酿制成的烈性酒，又称烧酒、老白干、烧刀子等。中国白酒酒质无色（或微黄）透明，气味芳香纯正，入口绵甜爽净，酒精含量较高，经过贮存老熟后，具有以酯类为主体的复合香味。中国各地均有白酒生产，以贵州、四川及山西等地的产品最为有名，各地的名酒都具有其独特的风格。

Chinese Baijiu, as one of the eight major distilled liquors worldwide, is a strong liquor brewed with starchy raw materials, with koji and liquor yeast as saccharifying ferment. It is also called Arrack, Laobaigan, and Shaodaozi. Chinese Baijiu is colorless (or pale yellow) and transparent, with pure aroma, sweet and fresh. Also, it contains higher alcohol content. After storage and maturation, it enjoys a compound aroma with esters as the main body. Chinese baijiu is produced in various regions of China, which is characterized by the products from Guizhou, Sichuan and Shanxi, and the famous liquors nationwide feature their own unique styles.

二、白酒的分类 (Classification of Baijiu)

中华人民共和国成立后，用"白酒"这一名称代替了以前所使用的"烧酒""高粱酒"等名称。由于酿酒原料多种多样，酿造方法也各有特色，酒的香气各有千秋，所以白酒的分类方法有很多。1979 年，在全国第三次评酒会上首次提出按酒的香型可将白酒划分为五种。

After the founding of the People's Republic of China, the name "Chinese baijiu" substituted the previously used names involving "Arrack" and "Kaoliang Spirit." Because of the variety of brewing materials, brewing methods are also different, wine aroma characteristics are different. So, there are many ways to classify Chinese baijiu. Today, we mainly discuss the categories of Chinese baijiu as per the flavor type. In 1979, it was first proposed at the Third National Wine Appraisal Conference that baijiu falls into five flavor types by the flavor of liquor.

（1）酱香型白酒，也称为茅香型白酒，以贵州茅台酒为代表。酱香型白酒的发酵工艺最为复杂，其主要特点为酱香突出、优雅细致、酒体醇厚、回味悠长。

Moutai-flavor liquor, also known as Maoxiang type Liquor, is represented by Kweichow Moutai Liquor. Moutai-flavored Chinese Baijiu has the most complicated fermentation process. It mainly features its prominent Moutai-flavor, elegant and delicate, full-bodied with lingering aftertaste.

（2）浓香型白酒，也称为泸香型、窖香型白酒，以五粮液、泸州老窖等酒为代表。这类白酒的特点为：香、醇、浓、绵、甜、净。在名优酒中，浓香型白酒的产量最大，四川、江苏等地的酒厂所产的酒多是这种类型。

Luzhou-flavor liquor, also known as Luxiang type and Jiaoxiang type liquor, is represented by Wuliangye, Luzhou Old Cellar and other liquors. This type of Chinese baijiu can be described as follows: fragrant, mellow, strong, soft, sweet and clean. Among these famous liquors, Luzhou-flavor liquor has the largest output, which are represented by the liquors produced by breweries in Sichuan, Jiangsu and other places.

（3）清香型白酒，也称汾香型白酒，以山西汾酒为代表。其特点是清香纯正，它的主体香乙酸乙酯与乳酸乙酯协调搭配在一起，更好地发挥出它的独特之处。

Mild aromatic Chinese baijiu, also known as fen-flavor liquor, is represented by Shanxi Fen Wine. It is characterized by a pure fragrance and its main note is blended with ethyl acetate and ethyl lactate to better display its uniqueness.

（4）米香型白酒，也称为蜜香型白酒，以桂林三花酒为代表。其特点是纯大米酿制，米香纯正，口感温柔，入口有着大米的柔和感。

Rice-flavor liquor, also known as Mi-flavored liquor, is represented by Guilin Sanhua Liquor. It is brewed from pure rice, which has pure rice flavor with gentle taste, and one can feel the softness of the rice when put it in the mouth.

（5）其他香型白酒，这类酒的主要代表有西凤酒、董酒、白沙液等，香型各有特征。这些酒的酿造工艺采用浓香型、酱香型或汾香型白酒的一些工艺，有的采用了串香法。

Other flavored liquors are mainly represented by Xifeng Liquor, Dong Liquor, Baishaye, etc., each with its own characteristics. These liquors are brewed through the process of brewing Luzhou-flavor, Moutai-flavor or Fen-flavor liquor, and some use the combined brewing method.

三、中国著名白酒（Chinese Famous Baijiu）

中国白酒品牌众多，其中茅台、五粮液、剑南春等颇具代表性。"茅五剑"是中国高端白酒的代名词，主要体现在历史文化、酿造工艺、风味品质、品牌影响等方面。

There are numerous Chinese baijiu brands, among which Moutai, Wuliangye, and Jiannanchun are particularly representative. "Mao Wujian (the combined name of liquors)" is a synonym of China's high-end liquor, which is mainly reflected in China's historical culture, brewing process, flavor quality, brand influence.

1. 茅台（Moutai）

茅台酒之所以能成佳酿，除了神奇的酿造技术之外，得益于茅台酒出产地茅台镇极其特殊的自然环境和气候条件。酿制茅台酒需要使用赤水河的水，每年雨季来临，两岸丹霞地貌中的泥沙汇入河中，既赋予了河水以赤红色，又带进特有的微生物群落和其他矿物质。赤水河沿岸还孕育了很多著名的白酒，因此，这条河也被称为"美酒河"。

Besides themagical brewing technology that makes Moutai a vintage wine largely lies in its special natural environment and climatic conditions in Moutai Town. To brew Moutai requires using water from the Chishui River. Every year when the rainy season comes, the sand from the Danxia landform on both sides of the river flows into the river, which makes river water a reddish red color, and brings in unique microbial communities and other minerals. Many famous liquors have originated from the Chishui River, so this river is also known as the "Vintage Bank."

2. 五粮液（Wuliangye）

五粮液产自万里长江第一城——酒都宜宾，这座城市是中国酒文化的缩影。有道是"川酒甲天下，精华在宜宾"，宜宾的酒文化距今已有2 000多年的历史，五粮液更是中国酒文化的提炼和结晶。宜宾紫色土上种植的高粱，是五粮液独有的酿酒原料。

Wuliangye is produced in Yibin, the first city of the Yangtze River, which is the epitome of Chinese wine culture. There is a saying that "Sichuan wine is without equal in the world, and the essence is in Yibin." Yibin's wine culture has a history of over 2 000 years, and Wuliangye is the refinement and crystallization of Chinese wine culture. Sorghum grown on Yibin's purple soil can be regarded as Wuliangye's unique liquor brewing material.

3. 剑南春（Jiannanchun）

剑南春是汉族的传统名酒，产于四川省绵竹市。剑南春酒的前身剑南烧春，是正史记载的大唐御酒，是被载入史册的当代中国名酒。相传著名诗人李白为喝此酒竟把皮袄卖掉，留下了"解貂赎酒"的故事。2002年，剑南春酒被中国历史博物馆正式收藏，这是中国历史博物馆继茅台后收藏的唯一历史名酒，并宣布收藏剑南春后将不再收藏任何白酒。

Jiannanchun, a traditional famous wine of the Han nationality, is produced in Mianzhu City, Sichuan Province. Jiannan Shaochun, the predecessor of Jiannanchun, is the imperial liquor of the Tang Dynasty recorded in the official dynastic histories and the famous liquor of contemporary China recorded in the annals of history. It's said that the famous poet Li Bai should sell his fur-lined jacket for drinking this wine, leaving behind the story of "dealing the mink to purchase the liquor." In 2002, Jiannanchun was officially collected by the Museum of Chinese History. This is the unique historical famous wine collected in the Museum of Chinese History Museum followed by Moutai, and the Museum of Chinese History Museum announced that after the collection of Jiannanchun, it will no longer collect any liquor.

四、中国白酒文化（Chinese Baijiu Culture）

酒文化是人类文明中不可或缺的重要部分，每个国家都有自己代表性的酒品。作为世界八大蒸馏酒之一的白酒，承载着数千年来的中华文化，是具有中国特色的酒类饮料。中国的饮食文化、礼仪文化、社交文化、祭祀文化、礼品文化等都与白酒形成完美交融，体现在中国人生活的各种场景之中。

Wine culture is a fundamental part of human civilization, and each country has its own representative wines. As one of the eight major distilled liquors worldwide, Chinese baijiu has carried Chinese culture for thousands of years, which is a hard drink with Chinese characteristics. Chinese food culture, etiquette culture, social culture, and sacrificial culture are perfectly blended with liquor, which can be seen in various scenes of Chinese people's daily life.

1. 饮食文化（Food Culture）

白酒常常与中式菜肴搭配，尤其在川菜、湘菜、黔菜等辣味菜肴中，白酒可以中和辣味，提升食物风味。在中国的宴席上，常常以白酒招待客人，表示对客人的尊重。

Chinese baijiu is often paired with Chinese dishes, especially in spicy dishes like Sichuan, Hunan, and Guizhou cuisines. Baijiu can neutralize the spiciness and enhance the flavor of the food. At Chinese banquets, it is common to serve baijiu to guests as a sign of respect.

2. 礼仪文化（Etiquette Culture）

白酒的饮用文化和礼仪相当丰富。例如，在宴请客人时，敬酒是一种常见的礼仪，通常由主人先敬客人，然后客人回敬，敬酒时通常会有一番敬酒词，表达祝愿或感谢。干杯时，通常要将杯中酒一饮而尽；碰杯时，通常晚辈或职位较低的人的杯口应略低于对方的杯口。

The drinking culture and etiquette of Chinese baijiu are quite rich. For example, when entertaining guests, proposing a toast is a common etiquette. Usually, the host proposes a toast to the guests first, and then the guests reciprocate. When proposing a toast, there is usually a toast speech to express wishes or gratitude. When clinking glasses, it is customary to drink the wine in the cup in one gulp; when clinking glasses, the cup mouth of the younger or lower-ranking person should be slightly lower than the other person's cup mouth.

3. 社交文化（Social Culture）

白酒在商务社交中扮演着重要角色，经常用于商务谈判、合作交流等场合，以促进关系的建立和维护。在朋友聚会中，白酒也是一个重要的社交媒介，通过饮酒进行交流和娱乐。

Chinese Baijiu plays an important role in business socializing, often used in business negotiations, cooperative exchanges, and other occasions to promote the establishment and maintenance of relationships. At friends' gatherings, Chinese baijiu is also an important social medium, used for communication and entertainment through drinking.

4. 祭祀文化（Sacrificial Culture）

在中国的传统节日和祭祀活动中，白酒常用于祭拜祖先、神灵，以表达敬意和祈求保佑。一些重要的节日，如清明节、中元节等，人们会用白酒作为祭品。

In traditional Chinese festivals and sacrificial activities, Chinese baijiu is often used to worship ancestors and deities to express respect and pray for protection. On some important festivals, such as the Qingming Festival and the Ghost Festival, people use Chinese baijiu as a sacrificial offering.

五、白酒饮用要点（Gist of Drinking Chinese Baijiu）

白酒的主要成分为酒精和水，乙醇含量愈高，酒度愈烈，对人体的危害也愈大。适量饮用白酒有利于睡眠，并能刺激唾液和胃液分泌，起到健胃的作用。白酒还有驱寒、舒筋、活血的作用。饮用白酒时如果同时摄入牛奶、脂肪、甜饮料，乙醇的吸收速度会降低。如果饮用碳酸饮料，则会加速乙醇的吸收。因此，为了身体健康，要适度、正确地饮用白酒。

The main ingredients of Chinese baijiu are alcohol and water. The higher the alcohol content, the stronger the alcohol content, and the greater the harm to the human body. Drinking a moderate amount of Chinese baijiu is conducive to one's sleep quality, stimulates the secretion of saliva and gastric juice, and strengthens the stomach function. Chinese baijiu also can dispel cold, relax sinew and enhance blood circulation. If one intakes milk, fat, and sweet drinks while drinking liquor, the absorption rate of ethanol will be reduced. Drinking carbonated beverages nevertheless can help accelerate the absorption of ethanol. So please properly and correctly drink Chinese baijiu for your physical health.

赛证直通 Competitions and Certificates

一、基础知识部分（Basic Knowledge Part）

二维码 1-2-1：
知识拓展

（一）专业名词解释（Explanation of Professional Terms）

1. 酒曲（koji）：
2. 固态发酵（solid-state fermentation）：

（二）思考题（Thinking Question ）

1. 中国有哪些著名的白酒？
What are some famous Chinese baijiu?
2. 白酒按香型可划分为哪几类？
What kinds of Chinese baijiu can be divided into according to flavor type?

二、技能操作部分（Skill Operation Part）

能鉴别酱香型白酒和浓香型白酒。

【微课】
馨香四溢：白
酒香型及名品

专题二
馨香四溢：白酒香型及名品

你知道白酒的香味有什么不同吗？你都听说过哪些白酒香型（图 1-2-2）？

_____

Do you know the difference in the flavor of Chinese Baijiu? Which types of Chinese Baijiu flavors have you heard of（Figure1-2-2）?

_____

图 1-2-2　中国白酒（Chinese baijiu）

一、白酒香型发展史 （The Development History of Chinese Baijiu Fragrance Types）

（一）香型分类 （Classification of Fragrance Types）

1979 年 8 月，中央人民政府轻工业部（现中华人民共和国轻工业部）在大连组织召开第三届全国评酒会。这次评酒会首次按香型、生产工艺和糖化剂划分，对大曲酱香、浓香和清香、麸曲酱香、米香、其他香型及低度等组分别进行评比。这次评酒会是中国评酒史上一个重要的里程碑，以香型、工艺和糖化剂作为评定标准，更加专业，也更深得人心。此次评酒会提出了酱香型、浓香型、清香型、米香型及其他香型五种类型（图 1-2-3）。

In August 1979, the National Light Industry Department (Ministry of Light Industry of People's Republic of China) organized the third National Baijiu Evaluation Conference in Dalian. This conference marked the first time that Chinese baijiu was evaluated and categorized based on its fragrance type, production process, and saccharifying agent. The Chinese baijiu was divided into groups such as Daqu (large fermentation starter) with soy sauce aroma, strong aroma, light aroma, Fuqu (wheat bran fermentation starter) with soy sauce aroma, rice aroma, other aroma types, and low-alcohol content for evaluation. This Chinese baijiu evaluation conference was a significant milestone in the history of Chinese baijiu evaluation. The use of fragrance type, production process, and saccharifying agent as the evaluation criteria made the process more professional and well-received. The conference proposed five types of flavors: Moutai-flavor, Luzhou-flavor, Fen-flavor, rice-flavor, and other flavor (Figure 1-2-3).

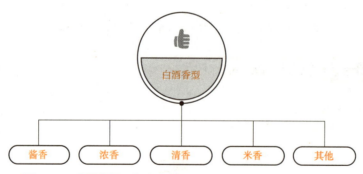

图 1-2-3　中国白酒五大香型（five flavors types of Chinese Baijiu）

（二）发展演变 （Development and Evolution）

随着白酒的发展，现在演变为 12 个香型（图 1-2-4）。12 个香型中，浓香型、酱香型、清香型、米香型为四种基本类型，其他八种则由 1979 年的其他香型演变发展而来。其他八种具体包括：浓酱兼香型、药香（董香）型、特香型、凤香型、芝麻香型、豉香型、老白干香型、馥郁香型。这八种类型基本上都是四种基本类型的一种或多种在工艺融合下衍生的独特香型。

With the development of Chinese baijiu, there are now 12 types of flavors (Figure 1-2-4). Among these 12 flavors, Moutai-flavor, Luzhou-flavor, Fen-flavor, rice-flavor, are the four basic types, while the other 8 types have evolved and developed from other types of flavors in 1979. The other 8 specific types include: Jian-flavor, Dong-flavor, Te-flavor, Feng-flavor, Zhima-flavor,

Chixiang-flavor, Laobaigan-flavor, Fuyu-flavor. These eight types are essentially unique aromatic profiles derived from one or more of the four basic types through the gentle process of craftsmanship.

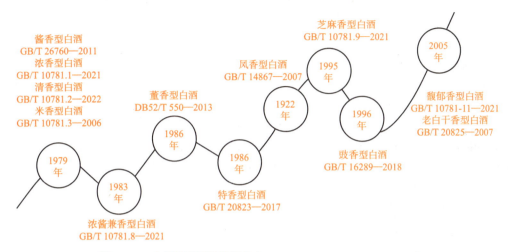

图 1-2-4　中国白酒香型发展史〔History of Chinese Baijiu Flavor〕

二、不同香型的工艺特点〔Process Characteristics of Different Flavor〕

（一）基本香型工艺〔Basic Flavor Process〕

1. 浓香型〔Luzhou-flavor〕

浓香型白酒市场占有率最大，其执行标准为 GB/T 10781.1—2021。按照制作原料划分，浓香型白酒可以分为单粮型和多粮型。单粮酒多以高粱为原料酿造，由于没有其他粮食成分，其香味相对单一，但酿造工艺更加考究，口感更加柔顺。多粮酒的原料非常丰富，除高粱外，还有大米、小麦、糯米、玉米、豌豆等。由于由多种粮食混合酿造，各种粮食的香味相互调和补充，使其粮香味更加丰富、和谐，口感更加柔和、细腻。

The Luzhou-flavor liquor has the largest market share, and its implementation standard is GB/T 10781.1-2021. Depending on the raw materials used, Luzhou-flavor liquor can be divided into single grain type and multi-grain type. Single grain liquors are mainly brewed from sorghum, and due to the absence of other cereal components, their aroma is relatively single, but their brewing process is more refined, and the taste is smoother and cleaner. The raw materials for multi-grain liquors are very rich, including not only sorghum but also rice, wheat, glutinous rice, corn, peas, etc. Because of the mixed brewing of various cereals, the aromas of various cereals complement each other, making the aroma of grains richer and more harmonious, and the taste is more soft and delicate.

浓香型白酒生产工艺分为原窖法工艺、跑窖法工艺和混烧老五甑法工艺。

The production process of Luzhou-flavor is divided into three types: ferment in the same pit oder way, ferment in the different pit order way, distilling raw and fermented material together of old five-pot oder way.

（1）原窖法工艺，也称为原窖分层堆糟法。原窖是指将本窖的发酵糟醅经过加原料、辅料后，再经蒸煮糊化、打量水、摊晾、下曲后仍然放回原来的窖池内密封发酵。分层堆糟是指窖内发酵完毕的糟醅在出窖时按照面糟、母糟分开出窖，面糟蒸酒后做丢糟处理，母糟出窖时由上而下逐层出糟堆放，经取糟配料、拌料、蒸酒、蒸粮、撒曲后仍投回原窖池发酵。原窖法强调保持原窖母糟风格和窖池的等级质量，窖的质量决定大曲酒的质量，也就是俗称的"千年老窖万年糟"。

The process of ferment in the same pit oder way, also known as the original cellar stratified pile fermentation method. The original cellar refers to the fermentation of the fermented grains after adding raw materials and auxiliary materials, and then after steaming gelatinization, sprinking amount of hot water, rapid cooling, and scattering fermentation agent, it is still placed back in the original fermentation pit for sealed fermentation. Stratified piling refers to the separation of the fermented grains in the cellar into refermentation grains and maternal fermented grains when discharged from the cellar. The refermentation grains is processed as spent grains after steaming, and the maternal fermented grains is piled up layer by layer from top to bottom when discharged from the fermentation pit, and after taking the residue for mixing, blending, steaming, steaming grains, and scattering fermentation agent, it is still returned to the original fermentation pit for fermentation. The ferment in the same pit oder way process emphasizes maintaining the style of the original mother residue and the quality level of the fermentation pit, the quality of the fermentation pit determines the quality of the Daqu spirits, also known as the "thousand-year-old cellar and ten-thousand-year-old residue".

（2）跑窖法工艺，也称为跑窖分层蒸馏法工艺。跑窖是指将发酵完成的糟醅从窖中取出，经加原辅料、蒸馏取酒、糊化、打量水、摊晾冷却、下曲粉后装入事先预备好的空窖池中，而不将发酵糟醅放回原窖。窖内发酵完成的糟醅逐甑进行蒸馏，因而称为分层蒸馏。分层蒸馏有利于量质摘酒和按质并坛等措施的实施。

The ferment in the different pit order way process, also known as the ferment in the different pit order way stratified distillation process. Ferment in the different pit order way refers to the fermented grains removing from the fermentation pit, adding raw materials and auxiliary materials, distilling and taking spirits, gelatinization, sprinkling amount of hot water, rapid cooling, and scattering fermentation agent, and then placing it into a pre-prepared empty fermentation pit, instead of placing the fermented grains back in the original fermentation pit. The fermented residue in the fermentation pit is distilled in turns, hence the name stratified distillation. Stratified distillation is conducive to the implementation of gathering distillate according to the quality and quality blending.

（3）混烧老五甑法工艺。混烧是指原料与出窖的酒醅在同一个甑桶同时蒸馏和蒸煮糊化。老五甑法工艺是指在每次出窖蒸酒时，将每个窖的酒醅拌入新投的原料，分成五甑蒸馏，蒸后其中四甑料重回窖内发酵，另一甑料作为废糟。入窖发酵的四甑料，按加入新料的多少，分别被称为大渣、二渣和小渣，大渣和二渣所配的新料分别占新投原料总量的40%左右，剩下的20%左右原料拌入小渣。不加新料只加曲的称作回糟，回糟发酵蒸酒后即变成丢糟。老五甑法具有以下优点：原料经过多次发酵（一般两次以上），原料中淀粉利用率高，出酒率高；经多次发酵，有利于香味物质的积累，特别是以己酸乙酯为主的窖底香，有利于浓香型白酒的生产；采用混蒸混烧，热能利用率高，成本低；老五甑法的适用范围广，高粱、玉米和薯干等原料均可使用。

The process of distilling raw and fermented material together of old five-pot oder way. Distilling raw and fermented material together refers to the simultaneous distillation and steaming gelatinization of raw materials and the wine residue from the same distilling pot. The old five-pot oder way process is that each time the cellar is steamed for spirits, the alcoholic fermentative material from each fermentation pit is mixed with newly added raw materials, divided into five pots steamings, and after steaming, four pots alcoholic fermentative material are returned to the fermentation pit for fermentation, and the other is treated as spent grains. The four pots alcoholic fermentation for fermentation in the fermentative pit, according to the amount of new material added, are respectively called large, second, and small. The new materials mixed with large and second respectively account for about 40% of the total amount of newly added raw materials, and the remaining about 20% of the raw materials are mixed with small. The residue that does not add new materials but adds koji is called refermentation grains, and after refermentation grains fermentation and steaming, it becomes spent grains. The old five-pot oder way has the following advantages: the raw materials go through multiple fermentations (generally more than twice), resulting in high utilization of starch in the raw materials and high spirit yield; after multiple fermentations, it is conducive to the accumulation of aroma substances, especially the cellar bottom fragrance dominated by ethyl acetate, which is conducive to the production of Luzhou-flavor liquor; distilling raw and fermented material together are used, and the utilization rate of thermal energy is high, with low cost; the old five-pot oder way has a wide range of application, and raw materials such as sorghum, corn, and dried potatoes can all be used.

2. 酱香型（Moutai-flavor）

酱香型又被称为"茅香型"，其执行标准为 GB/T 26760—2011。所谓酱香，就是淀粉反复发酵时发出的一种味道。就香气而言，组成成分极为复杂，至今尚未有定论，但目前的观点普遍认为，酱香是由高沸点的酸性物质与低沸点的醇类物质组成的复合香气。

The Moutai-flavor, also known as "Mao-flavor", is implemented according to the standard GB/T 26760-2011. The so-called sauce-flavor is a taste emitted during the repeated fermentation of starch. In terms of aroma, the composition is extremely complex and there is no consensus to date. However, the current view generally believes that the sauce flavor is a composite aroma composed of high boiling point acidic substances and low boiling point alcohol substances.

高温酿造在酱香型白酒的工艺中，堪称继承传统工艺基础上最富有科技创新内涵的核心技术。它包括高温制作大曲、高温堆积发酵、高温蒸馏接酒三个至关重要的工艺环节。

High-temperature brewing is the core technology with the most scientific and technological innovation content, based on inheriting traditional craftsmanship, in the Moutai-flavor liquor process. It includes three crucial process links: high-temperature Daqu (fermentation starter) production, high-temperature stacking fermentation, and high-temperature distillation for picking up alcohol.

（1）高温制作大曲。高温曲糖化酶的糖化力都很低，几乎没有发酵力。它的主要作用是生香，把制曲过程中积累的酱香物质或酱香前体物质带入酒中，通过堆积和发酵，为最终生成酱香型酒主体香奠定基础。这是高温大曲与其他大曲最大的差异，这个差异决定了酱香型酒曲生产的两个显著工艺特点：一是大用曲量，二是窖外堆积。

High-temperature Daqu production. The saccharifying power of the high-temperature Daqu amylase is very low, and the fermenting power is almost negligible. Its main function is to generate aroma, bringing the sauce flavor substances or precursor substances accumulated during the Daqu production process into the alcohol. Through stacking and fermentation, it lays the foundation for the generation of the main aroma of the sauce-flavor type liquor. This is the biggest difference between high-temperature Daqu and other Daqu, and this difference determines the two significant process characteristics of sauce-flavor type wine production: one is the large amount of Daqu used, and the other is the stacking outside the cellar.

（2）高温堆积发酵。高温大曲几乎不具有发酵力，必须采用窖外堆积工艺。经过堆积，糟醅中微生物的数量增加，特别是酵母菌数量大大增多。一方面进行糖化；另一方面网罗空气中、场地上的酵母菌和其他微生物，起到二次制酒母的作用，使发酵过程正常进行。酱香型曲酒的大用曲量和窖外堆积工艺除了保证糖化、发酵正常进行外，更重要的是保证酱香型大曲酒的风格质量。

High-temperature stacking fermentation. Since the high-temperature Daqu almost has no fermenting power, it is necessary to adopt the process of stacking outside the fermentation pit. After stacking, the number of microorganisms in the fermented grains increases, especially the number of yeast increases greatly. On the one hand, saccharification is carried out, and on the other hand, it captures yeast and other microorganisms from the air and the site to serve the purpose of making a secondary yeast culture, ensuring that the fermentation process proceeds normally. The large amount of Daqu used and the process of stacking outside the fermentation pit in Moutai-flavor liquor not only ensure the normal progress of saccharification and fermentation but also more importantly, ensure the style and quality of Moutai-flavor liquor.

（3）高温蒸馏接酒。正所谓"生香靠发酵，提香靠蒸馏"，高温蒸馏是提取酱香物质的有效手段。高温将高沸点、水溶性的酱香物质最大程度地收于酒中，最大程度排除挥发性强的硫化物或其他刺激性的低沸点物质，形成酱香型白酒的典型香味特征：酸类、醇类物质多。高温馏酒，即提高蒸馏酒时冷却水的水温，按照不同轮次质量的酒精度数进行量质摘酒。接酒的过程要注意酒液的温度、浓度及口感，摘酒后要分开存放。分段取酒，能更好地保证取酒的质量，若发现有杂味，改用尾酒坛接酒，用于制作窖底醅或泼窖。

High-temperature distillation for picking up alcohol. As the saying goes, " fermentation generates aroma, and distillation enhances aroma ". High-temperature distillation is an effective means of extracting sauce-flavor substances. The high-temperature method maximizes the collection of high boiling point, water-soluble sauce-flavor substances into the alcohol and eliminates volatile sulfur-containing or other irritating low-boiling point substances as much as possible, forming the typical aroma characteristics of moutai-flavor liquor: a large amount of acidic and alcohol substances. High-temperature distillation means raising the water temperature of the cooling water during distillation, gathering distillate according to the quality of different batches, and paying attention to the temperature, concentration, and taste during gathering distillate process. After picking up alcohol, it should be stored separately. Gathering distillate in sections can better ensure the quality of alcohol. If there is a taste problem, use tail alcohol to pick up alcohol and store it in a tail alcohol vat for making pit bottom grains or splashing alcoholic fermentative material.

3. 清香型（Fen-flavor）

清香型白酒的执行标准为 GB/T 10781.2—2022，是以高粱等谷物为原料，以大麦或豌豆等制成的中温大曲为糖化发酵剂。采用清蒸清糟酿造工艺、固态地缸发酵、清蒸流酒，可大大减少杂菌的污染，发酵后的酒糟只能使用一次，强调"清蒸排杂、清洁卫生"，即在一个"清"字上下功夫，"一清到底"。工艺操作上有"四特殊"：即润料堆积，低温发酵，高度摘酒，适期贮存。

The execution standard for Fen-flavor liquor is GB/T 10781.2—2022. It is made from raw materials such as sorghum and other grains, the medium-temperature daqu, made from barley and peas, serves as the saccharifying and fermenting agent. The brewing process uses clear steaming and clear residue, solid state earthen jar fermentation, and clear steaming for wine extraction, which can greatly reduce the contamination of miscellaneous bacteria. The fermented wine residue can only be used once, emphasizing "clear steaming to remove miscellaneous flavors, clean and sanitary". That is, efforts are made on the word "clear" and "all the way clear". There are "four special" in the process operation: that is, moistening materials and stacking, low temperature fermentation, high degree of distillation, and appropriate period of storage.

（1）润料堆积。将粉碎后的高粱加入原料质量 55% ～ 62% 的热水，夏天水温 75 ～ 80 ℃，冬季为 80 ～ 90 ℃，拌匀后堆积润料 18 ～ 20 小时。要求润透、不落浆、无干糁、无异味、无疙瘩、手搓成面；在堆积过程中，倒糁 2 ～ 3 次，要倒彻底，放掉"窝气"，擦拦疙瘩，做到外倒里、里倒外、上倒下、下倒上。

Moistening materials and stacking. Add hot water at 55%–62% mass of the raw material to the crushed sorghum, with a water temperature of 75–80 ℃ in summer and 80–90 ℃ in winter. After mixing, stack the moistened material for 18–20 hours. The requirements are: thoroughly moistened, no falling pulp, no dry bran, no peculiar smell, no lumps, and can be rubbed into flour by hand; during the stacking process, turn the bran 2–3 times, thoroughly turn, release "the gas of construct", wipe off lumps, and make sure to turn inside out, outside in, top down, and bottom up.

（2）低温发酵。与其他的酒不一样，清香型白酒采用地缸低温发酵，发酵时间短，贮存时间短，口感醇厚、细腻、淡雅。这是因为高淀粉低酸度的条件，酒醅极易酸败，因此，更要坚持低温入缸，缓慢发酵。入缸温度常控制在 11 ～ 18 ℃，比其他类型的曲酒要低，以保证酿出的酒清香纯正。当然，入缸温度也应根据气温变化加以调整。

Low temperature fermentation. Different from other types of liquor, fen-flavor liquor uses low temperature fermentation in fermentation vat, with a short fermentation time and storage time, resulting in a mellow, delicate, and elegant taste. This is because under the conditions of high starch and low acidity, the alcoholic fermentation material is prone to acidification, so it is essential to maintain low temperature when entering the fermentation vat and slow fermentation. The temperature when entering the fermentation vat is usually controlled between 11 ～ 18 ℃, which is lower than that of other types of qu baijiu to ensure the fragrance and purity of the brewed wine. Of course, the temperature when entering the fermentation vat should also be adjusted according to the change of the weather.

（3）高度摘酒。清香型白酒需要摘酒主浊比例较高的头酒和尾酒。装甑后蒸馏，蒸汽经冷却后流出来的即为大茬酒，酒精含量在 75% 以上。含有较多的低沸点物质，口味冲辣，应单独接取存放，可回入窖中重新发酵。出完大茬酒的原料再重复加曲、发酵、蒸馏，生产出的酒为二茬酒，其酸、酯含量都较高，香味浓郁，二茬酒的酒精含量一般在 65% 左右。大茬酒、二茬酒按比例勾兑后放入酒库贮藏后即为成品。

High degree of distillation. Fen-flavor liquor requires the distillation of the main turbid part of the initial distillate and last distillate with a higher proportion. After loading the distilling pot, the steam is distilled, and the alcohol content is more than 75% after cooling. It contains more low-boiling point substances, with a strong and spicy taste, and should be collected separately for storage, which can be returned to the fermentation pit for re-fermentation. The amount of the initial distillate collected should be moderate. After the first wine is distilled, the raw material is repeatedly added with koji, fermented, and distilled to produce the second distiller wine, which has higher acid and ester content and a rich fragrance. The alcohol content of the second distiller wine is generally around 65%. The first distiller wine and the second distiller wine are blended in proportion and stored in the wine cellar before they become the finished product.

（4）适期贮存。蒸馏得到的大茬酒、二茬酒等，通常要分别贮存三年，在出厂前进行勾兑，然后灌装出厂。清香型白酒储存老熟是在陶缸里完成的，老熟比较快，但当酒体存储过了第四年之后，钠、钾元素上升速度明显增强，酯类下降比较明显，乙酸增长，导致乙酸乙酯的水果香没有了，酒体不再清香纯正。因此，清香型白酒的最佳储存期限为三到四年，储存时间过短，酒体老熟程度不够；储存时间过长，酒受损严重。

Appropriate period of storage. The distilled first distiller wine, second distiller wine, etc., usually need to be stored separately for three years, and then blended before going out of the factory. The aging of fen-flavor liquor is completed in a pottery jar, which ageing faster, but after the fourth year of storage, the rate of increase of sodium and potassium elements is significantly enhanced, the ester content decreases significantly, and acetic acid increases, resulting in the loss of the fruity fragrance of ethyl acetate, and the wine body is no longer fragrant and pure. Therefore, the best storage period for light fragrance type baijiu is three to four years. If the storage time is too short, the aging degree of the wine body is not enough; if the storage time is too long, the wine loss is severe.

4. 米香型（Rice-flavor）

米香型白酒的执行标准为 GB/T 10781.3—2006。米香型也称"蜜香型"，属小曲香型酒，一般以大米为原料。其典型风格是在"米酿香"及小曲香基础上，突出以乳酸乙酯、乙酸乙酯与β-苯乙醇为主体组成的幽雅清柔的香气。米香型白酒属半固态法发酵工艺，大米原料、小曲催化、补水发酵、釜式蒸馏，实现了从"黄酒"到"白酒"的转变，这在我国南方各省很常见。

The rice-flavor liquor follows the standard GB/T 10781.3-2006, also known as the "mi-flavor", which belongs to xiaoqu baijiu. It is generally made from rice as the main ingredient. Its typical style is based on "rice brewing fragrance" and xiaoqu baijiu, highlighting the elegant and gentle aroma composed mainly of lactate ethyl ester, acetic acid ethyl ester, and β-phenyl ethanol. Rice-flavor liquor belongs to the semi solid fermentation process, which includes rice raw material, small qu catalysis, water replenishment fermentation, and pot distillation, achieving the transformation from "yellow wine" to "white liquor." It is very common in various provinces in southern China.

米香型白酒以大米为主要原料，小曲也是用米糠、米粉制作而成，且酿造过程中不包含其他增香、增味环节，使米香型白酒闻之有淡淡的米饭香气，入口口感纯净，余味亦略有米香。米香型白酒制作时，先将大米蒸熟，拌好曲粉转移至开口容器（缸）内培养一天左右，待小曲微生物得到充分生长之后进行补水发酵。由于补水很好地稀释了发酵体系的淀粉浓度，使得大米原料经一次发酵之后便可利用干净，且发酵时间短（约7天），发酵醪液经蒸馏后即得米香型白酒。

Rice-flavor liquor mainly uses rice as the raw material, and the small qu is also made from rice bran and rice flour. The brewing process does not include other fragrance-enhancing and flavor-enhancing steps, which allows the rice-flavor liquor to have a faint rice aroma when smelled, a pure taste when sipped, and a slight rice fragrance in the aftertaste. In the production of rice-flavor liquor, the rice is first steamed, mixed with qu powder, and then transferred to an open container (jar) for cultivation for about a day. After the microorganisms in the small qu have grown sufficiently, water replenishment fermentation is carried out. Since water replenishment effectively dilutes the starch concentration of the fermentation system, the rice raw material can be used cleanly after a single fermentation, and the fermentation time is short (about 7 days). After distillation, the rice-flavor liquor is obtained.

(二)演变香型工艺（Evolution of Flavor Process）

1. 兼香型（Jian-flavor）

兼香型又称为"浓酱兼香型"，根据浓酱比例不同又可以分为以下两种：

Jian-flavor is also called "nongjiangjianxiang xing". According to the different proportions of flavor, it can be divided into the following two types:

（1）浓中兼酱：浓香酒和酱香酒各自酿造，分开存储，最后按 8∶2 比例混合；

Luzhou-flavor in the middle and Moutai-flavor on the side: Luzhou-flavor liquor and Moutai-flavor liquor are brewed separately and stored separately, and finally mixed in a ratio of 8∶2;

（2）酱中兼浓：多轮次固态发酵，前面轮次用酱香工艺，后面轮次用浓香的混蒸混烧工艺。

Moutai-flavor in the middle and Luzhou-flavor on the side: Multi-round solid state fermentation, the front rounds use Moutai-flavor process, the latter rounds use Luzhou-flavor distilling raw and fermented material together process.

2. 董香型（Dong-flavor）

董香型也称药香型。董香型白酒执行的是贵州地方标准 DB52/T 550—2013。因此，董香型白酒只针对贵州省境内，以高粱、小麦、大米为原料，制曲过程中加入中草药，经传统固态法大小曲发酵、串香蒸馏、长期陈酿、勾兑而成的，微带舒适药香，未添加食用酒精及非白酒发酵产生的呈香呈味物质，具有传统董香型风格的白酒。

Also known as the medicinal fragrance type, Dong-flavor liquor implements the local standard of Guizhou Province DB52/T 550-2013, so Dong-flavor liquor is only aimed at Guizhou Province. It is made from sorghum, wheat and rice as raw materials, and Chinese herbal medicine is added during the brewing process. After traditional solid state fermentation, distilling aroma of distilland, it is blended and aging. It has a slight comfortable medicinal fragrance, without adding edible alcohol and aroma substances produced by non-liquor fermentation, and has the traditional Dong-flavor liquor.

3. 特香型（Te-flavor）

特香型白酒的执行标准为 GB/T 20823—2017，其是以整粒大米为主要原料，以中高温大曲为糖化发酵剂，经传统固态法发酵、蒸馏、陈酿、勾兑而成。中国白酒泰斗周恒刚、沈怡方、曹述舜等全国知名专家在实地考察了江西省樟树四特酒后，将四特酒的工艺特点概括为"整粒大米为原料，大曲麦麸加酒糟，红褚条石垒酒窖，三型具备犹不靠"，将四特酒的香型定位一个全新的香型——"特香型"。四特酒自此成了"中国特香型白酒开创者"。

The implementation standard of Te-flavor liquor is GB/T 20823-2017, which mainly uses whole grain rice as raw material and medium-high temperature da qu as saccharification and fermentation agent. It is fermented, distilled, ageing and blended by traditional solid state fermentation. National famous experts such as Zhou Henggang, Shen Yifang and Cao Shushun, who are the authorities of Chinese baijiu, summarized the characteristics of SITIR in Jiangxi Province after on-the-spot investigation, as "whole grain rice as raw material, da qu bran added to the wine dregs, red Zhu Tiao stone to build the wine cellar, three types are available but not relied on". And the fragrance type of SITIR is positioned as a brand new fragrance type- "Te-flavor". SITIR has become the "pioneer of Chinese Te-flavor liquor" from then on.

4. 凤香型（Feng-flavor）

凤香型白酒的执行标准为 GB/T 14867—2007，其以粮谷为原料，经传统固态法发酵、蒸馏、酒海陈酿、勾兑而成，未添加食用酒精及非白酒发酵产生的呈香呈味物质，具有以乙酸乙酯和己酸乙酯为主体的复合香。凤香型属浓清兼香，由清香和浓香型白酒衍生而来，同时具有清香型白酒的主体香气（乙酸乙酯）和浓香型白酒的主体香气（己酸乙酯），在生产工艺上也兼有清香和浓香的工艺特征。

The implementation standard of Feng-flavor liquor is GB/T14867-2007, which is mainly made from grain as raw material and fermented, distilled, ageing in big Conservator for baijiu storage and blended by traditional solid state fermentation. It has not added edible alcohol and aroma substances produced by non-liquor fermentation, and has a composite fragrance with ethyl acetate and hexanoic acid ethyl ester as the main body. Feng-flavor belongs to the combination of Luzhou-flavor and Fen-flavor, derived from Fen-flavor and Luzhou-flavor liquor, and has both the main aroma of Fen-flavor liquor (ethyl acetate) and the main aroma of Luzhou-flavor liquor (hexanoic acid ethyl ester). In the production process, it also has the characteristics of both Fen-flavor and Luzhou-flavor processes.

5. 老白干香型（Laobaigan-flavor）

老白干香型白酒的执行标准为 GB/T 20825—2007。"老白干"这个名字背后隐藏着精湛的酿造水平，"老"指历史悠久；"白"指酒体纯净透明，不掺杂质；"干"指白酒燃烧后不留任何痕迹。其以高粱为原料，采用纯小麦制曲、地缸发酵、混蒸混烧、老五甑续糟法和"两排清"的生产工艺，决定了其主体香味物质是乳酸乙酯和乙酸乙酯，且两者之比 ≥ 0.8，香气清雅而不单一，带有浑厚一体的厚重感。

The standard for Laobaigan-flavored baijiu is GB/T 20825-2007. The name "LaoBaiGan" implies its exquisite brewing level, where "Lao" refers to its long history; "Bai" refers to the pure and transparent quality of the spirit without any impurities; and "Gan" refers to the absence of any residue after the spirit has been burned. It is made from sorghum, starter propagation made of pure wheat, fermented in fermentation vat, distilling raw and fermented material together, and produced using the old five-pot mixed distilland order way and double separating distilling raw and fermented material process. This determines its main fragrant substances to be ethyl lactate and ethyl acetate, with a lactate to acetate ratio of ≥ 0.8. The aroma is elegant but not monotonous, with a rich and integrated sense of heaviness.

6. 芝麻香型（Zhima-flavor）

芝麻香型白酒的执行标准为 GB/T 10781.9—2021。芝麻香型不是用芝麻酿造出来的白酒，而是用高粱、小麦为主酿造出酒体带有烤芝麻香气的白酒。它以芝麻香为主体，兼有浓、清、酱三种香型，故有"一品三味"之美誉。对于这种风味，其实现的核心在于高氮配料、高温堆积、高温发酵、长期贮存、多微共酵。原酒贮存是形成芝麻香风格的重要环节，贮存过程中，幽雅细腻焙炒芝麻的复合香气突出，陈香味较明显。

The standard for Zhima-flavor baijiu is GB/T 10781.9-2021. It is not a liquor made from sesame, but a liquor made from sorghum and wheat, which carries a roasted sesame aroma. It has a zhima-flavor as the main flavor profile, along with the Luzhou-flavor, Fen-flavor and Moutai-flavor, hence the reputation of "one product with three flavors." The core of achieving this flavor lies in high nitrogen ingredients, high temperature stacking, high temperature fermentation, long-term storage, and multi-microbiological fermentation. The storage of the original wine is an important part of forming the Zhima flavor. During the storage process, the elegant and delicate baked sesame aroma becomes prominent, and the aged flavor is more noticeable.

7. 馥郁香型（Fuyu-flavor）

馥郁香型白酒的执行标准为 GB/T 10781.11—2021，其采用五粮配方，以大曲和小曲为糖化发酵剂，发酵周期为 50～70 天，采用整粒原料，大小曲并用（小曲培菌糖化，大曲配糟发酵），泥窖固态发酵，清蒸清烧。分段摘酒，分级贮存，精心勾兑而成。馥郁是香气浓郁的意思，其中"馥"是"香"与"复"的组合，意思是这种香气不是普通的单体香气，也不是几种香气的简单叠加，而是由多种单香构成的完美协调的复合香气，蕴含了"海纳百川、中庸和谐"的传统文化精髓。

The standard for Fuyu-flavor baijiu is GB/T 10781.11-2021. It uses a five grains formula, with both daqu and xiaoqu as saccharification and fermentation agents, with a fermentation period of 50-70 days. It uses whole grains, with daqu and xiaoqu (xiaoqu for microbial saccharification, daqu for fermentation with residual grains). It is solid state fermentation in a clay pit, with distilling raw and ferment material apart and then fermenting apart. Gathering distillate in stages, stored in grades, and carefully blended. "Fu" means rich and fragrant. "Fu" is a combination of "fragrance" and "complex," which means this aroma is not an ordinary single aroma, nor is it a simple superposition of several aromas, but a perfectly coordinated composite aroma composed of multiple single fragrances, containing the essence of the traditional culture of "inclusiveness and harmony".

8. 豉香型（Chixiang-flavor）

豉香型白酒的执行标准为 GB/T 16289—2018，其以大米为原料，经蒸煮，用大酒饼作为主要糖化发酵剂，采用边糖化边发酵的工艺，釜式蒸馏，陈肉酝浸、勾兑而成，未添加食用酒精及非白酒发酵产生的呈香呈味物质，具有豉香的特点。豉香型白酒在陈酿过程中泡大肥肉，这在中国白酒中绝对是绝无仅有的。将蒸馏的新酒放入陈酿酒瓮中，加入肥肉，浸泡 3 个月左右，使脂肪慢慢溶解于酒中，给酒带来一股特有的豉香，有人也把这种香气称为"肉香"。

The standard for Chixiang-flavor baijiu is GB/T 16289-2018. It is made from rice, steamed, and the main saccharification and fermentation agent is da jiu bing. The process uses the technique of simultaneous saccharification and fermentation, with the distillation by pot, steeping process with Chen rou and blending, without the addition of edible alcohol and flavoring substances produced by non-baijiu fermentation, with the characteristics of Chixiang-flavor. In the aging process, the Chixiang-flavor baijiu soaks in large pieces of lard, which is absolutely unique in Chinese baijiu. The newly distilled wine is placed in an aging vat, and add fatty meat, soaked for about 3 months, allowing the fat to slowly dissolve in the wine, giving it a unique Chixiang-flavor. Some people also call this aroma "meat fragrance."

赛证直通 Competitions and Certificates

一、基础知识部分（Basic Knowledge Part）

二维码 1-2-2：
知识拓展

（一）专业名词解释（Explanation of Professional Terms）

1. 高度摘酒（Highly selective wine distillation）：
2. 原窖法工艺（Ferment in the same pit oder way）：

（二）思考题（Thinking Question）

谈谈不同香型白酒的特点。

Let's talk about the characteristics of different types of fragrant Chinese baijiu (alcohol).

二、技能操作部分（Skill Operation Part）

尝试品鉴不同香型的白酒，并描述其不同。

Try tasting different types of baijiu and describe their differences.

专题三

香飘万里：酱酒茅台

【微课】
香飘万里：
酱酒茅台

你认识图 1-2-5 中的作物吗？你知道它有什么用处吗？

_____

Do you know the crops in the Figure1-2-5? Do you know what it does?

_____

一、茅台概述（Summary of Moutai）

茅台酒独产于中国的贵州省遵义仁怀，是与英国苏

图 1-2-5　红缨子高粱（Red Tassel Sorghum）

格兰威士忌、法国科涅克白兰地齐名的世界三大名酒之一，也是我国著名的名优白酒。茅台酒的诞生，除了得益于茅台镇特有的自然环境外，更离不开其神秘的制造工艺。如果说环境和原料是天赋所致，那么茅台酒的生产工艺，便是人类上千年酿酒智慧的传承与发展。茅台酒酿造工艺被誉为我国白酒工艺的活化石。长达一年的生产周期，那么多个环节和细节，最后都凝聚到了人们品到的每一滴酒上。

Moutai is exclusively produced in Renhuai, Zunyi, Guizhou Province, China. It is one of the three famous liquors in the world, which is as famous as Scotch Whisky and Cognac Brandy, and is also a notable and excellent liquor in China. Apart from benefiting from the unique natural environment of Moutai Town, the birth of Moutai cannot go without its mysterious manufacturing process. If the environment and raw materials are the result of talent, then the production process of Moutai is the inheritance and development of thousands of years of human wine-making wisdom. Moutai, praised as the living fossil of liquor craft of China. Year-long production cycle with countless steps and details finally condenses into every drop of wine that people taste.

二、茅台制作工序（Manufacturing process of Moutai）

茅台酒的酿造遵循"12987"工艺，即端午制曲、重阳下沙、1年生产周期、2次投料、9次蒸煮、8次发酵、7次取酒，被誉为世界上最为复杂的蒸馏酒酿造工艺。

Moutai is brewed based on the "12987" process, that is starter propagation during Dragon Boat Festival, chong yang xia sha, 1 year production cycle, 2 times of feeding, 9 times steaming, 8 times fermentation, 7 times gathering distillate. And "12987" is known as the world's most complex distilling process.

首先是"端午制曲"。端午之后是一年中气温最高的时节，制曲车间里的温度经常高达40 ℃甚至以上，高温有利于微生物生长。曲是酒之骨。制曲是以小麦为原料，产出的为大曲。大曲联合本地优质糯高粱共同构成了茅台酒的原料。茅台至今还在坚持传统的人工制曲，用女性的双足踩出中间高、四边低，松紧适宜的"龟背型"大曲，这种形状有利于微生物的生长和后期发酵。据记载，茅台公司也曾经试用机械踩曲，但无法形成大曲的形状和密度，最终还是回归了人工踩曲。

First, "starter propagation during Dragon Boat Festival". After the Dragon Boat Festival, the temperature is the highest in a year. The temperature in the starter propagation workshop is often as high as 40 ℃, which is conducive to the growth of microorganisms. Starter is the bone of liquor. Starter propagation is based on wheat as raw material, and the product is daqu, which combines with local high-quality glutinous sorghum to form the raw material of Moutai. So far, Moutai has been using traditional hand-made starter propagation. Step out the "turtleback" daqu stack with the female feet, which is high in the middle and low on the four sides, with appropriate tightness. This shape is favorable for microbial growth and late fermentation. According to record, Moutai Company tried mechanical trampling, but failed to achieve the shape and density of daqu for making hard liquor. So they keep using the conventional ways of stepping on.

酒曲制好后需要入仓发酵40天，用高温对茅台酒中的微生物群进行筛选，弱者自然被淘汰，从而形成产香、产酒的独特微生物体系。茅台的大曲在发酵过程中温度高达60 ℃甚至以上，比其他白酒的制曲发酵温度都高 10 ～ 15 ℃。一般发酵后的曲块按颜色分为黄曲、白曲和黑曲，黄曲是一般适宜的曲块，白曲是发酵不够的，黑曲是发酵过头的，而茅台的黄曲占比达到了 80% 以上。

After making raw starter, it needs to be stored and fermented for 40 days. The microbiota in Moutai was screened by high temperature. The weak will be weeded out, thus forming a unique microbial system for aroma and liquor production. The fermentation temperature of Moutai daqu is over 60 ℃, which is 10-15 ℃ higher than starter propagation of any other baijiu. In general, raw starters after fermentation divided by color yellow, white, and black starters. Yellow starter is generally the suitable one, white starter is not fermented enough, black starter is fermented too much, and Moutai's yellow starter accounts for more than 80%.

在制曲和酒曲发酵的过程中，有一种神奇的小生物——曲蚊，起着举足轻重的作用。

There is a magical little creature—starter mosquitoes that play a vital role in in starter propagation and fermentation of raw starter.

国产酒制曲的地方一般都有曲蚊，种类多达二三十种。茅台镇有一种独一无二的曲蚊，外壳硬，小米粒般大小，在制曲过程中扮演了举足轻重的作用。这种曲蚊飞入制曲房内，钻进曲块中饱餐曲药，吃饱喝足后，在此繁衍后代产下幼虫，安居乐业。曲蚊幼虫含有大量蛋白质，这些蛋白质在酿酒过程中，对酱酒的风味起到了至关重要的作用。

There are generally 20 or 30 kinds of starter mosquitoes in the places where are used to starter propagation of domestic liquor. Moutai Town has a unique species of starter mosquito with a hard shell and the size of a small grain of rice, which plays an important role in the process of starter propagation. The kind of starter mosquitoes fly into the starter propagation chamber and burrow into the starters to eat them. After being full, they produce their offspring, give birth to their larvae, and live and work in peace and contentment. The larvaes of these starter mosquitoes contain a lot of protein. These proteins play a vital role in the flavor of sauce wine in the brewing process.

制好酒曲，到重阳节开始第二个关键步骤——"重阳下沙"。"沙"是指高粱。因为本地产的高粱细小而色红，所以称为"沙"。"下沙"就是指投放制酒的主料高粱。茅台镇夏季炎热，酒醅温度高，如果淀粉含量高，收堆、下窖升温过猛，生酸幅度过大，对酿酒极为不利。重阳下沙既避开了夏季高温期，又避开了夏季赤水河洪水期，这时期"潦水尽而寒潭清，烟山凝而暮山碧"，赤水河清澈见底，水质极佳，满足酿酒对水质的要求，且恰逢高粱成熟时期。

It would be Double Ninth Festival after the raw starters are made, we would welcome the second key step- "Chongyang xiasha". "Sha" here is sorghum. Given that local sorghum is small and red, it is called "sha". "Xiasha" means to put in main ingredient-sorghum. Moutai Town is very hot in summer, and the temperature of fermented grains is high. If the starch content is high, the temperature will rise too fast when the fermented grains are gathered and piled up and put into the fermentation pit, acid generation is too large, which is extremely adverse to brewing. Chongyang xiasha has avoided high temperature and Chishui River flood period in summer. As a saying goes "the water disappeared after the rain, cold pool clear water, the sky condensed light cloud smoke, a purple shows up in evening mist among the mountains." Chishui River is limpid, the water quality is excellent, which meets the requirements of brewing water quality, and this period coincides with the maturity of sorghum.

下沙投料分为两次，但两次都不取酒，只为增加发酵时间，尽可能多地培育微生物。一个月后，才开始第一次取酒，这时已到了岁末年初。之后周而复始，每月一次，直至第七次酒取完后，已经到了第二年的八月。每一次出来的酒，香味并不相同，茅台酒分三种酒体：酱香、醇甜和窖底。第三至五次出的酒最好，称为"大回酒"，第六次得到的酒为"小回酒"，第七次的酒为"追糟酒"。其中第三、四、五次出的酒最好喝，第一、二次酸涩、辛辣，最后一次发焦、发苦。但每一次的酒都有用处，最终出厂的酒就必须经过不同批次酒之间的勾兑。

As for feeding in both cases, no liquor preparation was taken, only to increase fermentation time and cultivate as many microbes as possible. One month later, we can start to take out the wine for the first time, this time has arrived at the end of the year. Later, running in cycle, do it once a month, until the seventh round of wine—the next year in August. Each time the wine comes out, it has different aroma. Moutai has three liquor bodies: sauce aroma, alcohol mellow sweetness and cellar flavor. The best wine from the third to the fifth time is called "da hui jiu", the sixth time is "xiao hui jiu", and the seventh time is "zhui zao jiu". Among them, three, four, five times taste the best; the first and the second tend to be sour and spicy, the last time of liquor preparation taste scorching bitter. Each one of them is useful, and the final product must be blended between different batches of wine.

勾兑一直是酿酒过程中比较神秘的工序，勾调师凭借自己的味觉进行搭配，把不同轮次的酒调在一起，寻找味道之间的平衡与层次感。勾兑的最后一项工作是调味，一般用年份老的酒进行味道的微调。勾兑调味后，酒还要分批次继续存放半年到一年，才能灌瓶出厂。也就是说，一瓶地道的茅台型酱香酒的生产至少五年。

Blending has always been a mysterious process in the wine making process. The person who mixes wine can blend different rounds of wine together to find balance and hierarchy between flavors. The final task of blending is seasoning, usually with older wines to fine-tune the flavor. After blending and seasoning, the liquor continues to be aged in batches for six months to a year before it is bottled. In short, a bottle of authentic type Moutai-flavor liquor production for at least five years.

三、茅台故事 (The Story of Moutai)

经过复杂而又漫长的过程酿造的茅台，酱香突出，香气扑鼻，满口生香，令人陶醉。茅台酒在国人心中有着特别的地位。1949 年开国大典的前一天夜里，周恩来总理在中南海怀仁堂召开会议，确定茅台酒为开国大典国宴用酒，并在北京饭店用茅台酒招待嘉宾。从此每年国庆招待会，均指定茅台酒。在日内瓦和谈、中美建交等历史性事件中，茅台酒都成为融化历史坚冰的特殊媒介。党和国家领导人无数次将茅台酒当作国礼，赠送给外国领导人。

After a complicated and lengthy process of brewing, Moutai has a prominent jiang-flavor, a strong aroma, and a mouthful of fragrance, which is intoxicating. Moutai occupies a special place in Chinese hearts. On the eve of the founding ceremony in 1949, Premier Zhou Enlai held a meeting in Huairen Hall, Zhong nanhai, and determined that Moutai was the wine for the state banquet of the founding ceremony. Mr. Premier entertained guests with Moutai in Beijing Hotel. Since then, Moutai has been designated for the National Day reception every year. In historic events, such as the Geneva Peace Negotiations and the establishment of diplomatic relations between China and the United States, Moutai has become a special medium to melt the historical ice. CPC and government leaders have always given Moutai as a national gift to foreign leaders for many times.

茅台酒已经超越了白酒本身，成为中国的一张文化名片。琼浆玉液有千种，而精致的茅台能香飘中国、香飘世界。

Moutai has surpassed baijiu to become a cultural name card in China. There are thousands of top-quality wine, only exquisite Moutai can make its fragrance spread throughout China and the world.

赛证直通 Competitions and Certificates

一、基础知识部分（Basic Knowledge Part）

二维码 1-2-3：
知识拓展

（一）专业名词解释（Explanation of Professional Terms）

重阳下沙（Chongyang xiasha）：

（二）综合分析题（Comprehensive Analysis Question）

中国白酒历史悠久，且种类繁多，其风味高度依赖于当地的自然环境。源远流长的中国酒文化以白酒为核心。白酒产地素有"中国白酒金三角"之称，即以四川省宜宾、泸州和贵州省遵义仁怀为基点，以长江上游及其支流赤水河为流域形成的白酒生产黄金区域。这个金三角诞生了浓香型白酒五粮液、酱香型白酒茅台这两大中国白酒品牌，并以这两座高峰引领四川和贵州白酒产业集群发展。

试比较浓香型白酒和酱香型白酒的不同点。

Chinese baijiu has a long history and various kinds, and the flavor of liquor is highly dependent on the local natural environment. Chinese liquor culture with a long history takes baijiu as the core, and its producing area is known as "Golden Triangle of Chinese baijiu", that is, Yibin and Luzhou in Sichuan Province, Zunyi Huairen in Guizhou Province, and the upper reaches of the Yangtze River and its tributaries Chishui River as the golden region for baijiu production.The golden triangle gave birth to two major Chinese baijiu brands, luzhou-flavor liquor Wuliangye and Moutai-flavor liquor Moutai, and led the development of liquor industry clusters in Sichuan and Guizhou with these two peaks.

Try to compare the difference between Luzhou-flavor liquor and Moutai-flavor liquor.

二、技能操作部分（Skill Operation Part）

介绍茅台酒制作的"12987"工艺。

Introducing the "12987" process of Moutai liquor production.

专题四
名人轶事：古代名人与酒

君不见黄河之水天上来，奔流到海不复回。君不见高堂明镜悲白发，朝如青丝暮成雪。人生得意须尽欢，莫使金樽空对月。

Do you not see the Yellow River's water coming from the sky, rushing into the sea and never returning?Do you not see the high hall bright mirror mourn the white hair, like green silk en-black in the morning and snow-white in the evening.When life is content, one should enjoy to the fullest, never let your wine cup empty in moonlight.

你知道上面的诗句出自何处吗？是谁的作品？

_____

Do you know where the poem above comes from? Whose work is it?

_____

一、古代名人与酒（Ancient Celebrities and Alcohol）

酒作为一种特殊的文化符号，对于古代名人的思想、行为和创作产生了深远影响。酒可以使人放松心情，激发灵感；有时被用作情感的出口，通过饮酒来表达人的快乐、悲伤或压力。

As a special cultural symbol, alcohol has had a profound influence on the thoughts, behaviors, and creative works of ancient celebrities. Alcohol can relax people's mood, stimulate inspiration; sometimes it is used as an outlet for emotions, expressing happiness, sadness, or stress through drinking.

（一）诗人与酒（Poets and Alcohol）

古代文学中，酒常常与诗人灵感迸发的时刻联系在一起，其中最为人所熟知的例子之一便是唐代诗人杜甫。杜甫在《饮中八仙歌》中，以酒为媒，描绘了八位文人雅士的饮酒风采和他们的文学成就。这首诗不仅反映了杜甫对这些文人饮酒风度的赞赏，也体现了他自己对于文人墨客饮酒作诗生活的向往和羡慕。

In ancient literature, alcohol is often associated with moments of inspiration for poets. One of the most well-known examples is the Tang Dynasty poet Du Fu. In "Song of Eight Drinking Immortals", Du Fu uses alcohol as a medium to depict the drinking style and literary achievements of eight literary and elegant gentlemen. This poem not only reflects Du Fu's admiration for these literati's drinking demeanor but also embodies his own yearning and envy for the life of literati drinking and composing poetry.

"饮中八仙"分别指的是：

贺知章：被称为"酒仙"，以豪饮著称。

李琎：以好饮酒出名。

李适之：曾任宰相，以饮酒和诗才闻名。

崔宗之：以风流才子著称，饮酒后风度翩翩。

苏晋：饮酒后常有妙语连珠。

李白：唐代著名诗人，以饮酒和诗歌齐名。

张旭：书法家，饮酒后书法如神助。

焦遂：饮酒五斗后，言谈方显卓绝。

"Eight Immortals in Drinking" refers to the following individuals:

He Zhizhang: known as the "Immortal of Wine," he is famous for his heavy drinking.

Li Jing: known for his fondness for alcohol.

Li Shizhi: he once served as a prime minister and was known for his drinking and poetic talent.

Cui Zongzhi: known as a romantic and talented scholar, he exhibited elegant demeanor after drinking.

Su Jin: often had witty remarks after drinking.

Li Bai: he was a renowned poet during the Tang Dynasty.

Zhang Xu: a calligrapher whose calligraphy seemed divinely inspired after drinking.

Jiao Sui: his conversation became outstanding after drinking five dou (a Chinese unit of volume).

知章骑马似乘船，眼花落井水底眠。

汝阳三斗始朝天，道逢麹车口流涎，恨不移封向酒泉。

左相日兴费万钱，饮如长鲸吸百川，衔杯乐圣称世贤。

宗之潇洒美少年，举觞白眼望青天，皎如玉树临风前。

苏晋长斋绣佛前，醉中往往爱逃禅。

李白一斗诗百篇，长安市长酒家眠，天子呼来不上船，自称臣是酒中仙。

张旭三杯草圣传，脱帽露顶王公前，挥毫落纸如云烟。

焦遂五斗方卓然，高谈雄辩惊四筵。

——《饮中八仙歌》

"诗仙"李白的许多诗作被认为是在饮酒后创作的。李白酷爱饮酒，他的诗歌充满了对酒的赞美和对酒后境界的描写，例如：

Many of the poems attributed to the "Immortal Poet" Li Bai are believed to have been composed after drinking. Li Bai loved drinking, and his poetry is filled with praise for wine and descriptions of the state of mind after consuming it, for example.

《月下独酌四首》之一中，"举杯邀明月，对影成三人。"

In one of the "Four Poems on Drinking Alone by Moonlight," it says, " raise the cup to invite the bright moon, and with my shadow, we become three."

《短歌行》中，"白日何短短，百年苦易满。苍穹浩茫茫，万劫太极长。"

In " A Short Song," it reads, " What is the daytime so short, a hundred years of suffering are easily fulfilled. The vast expanse of the heavens is boundless, and the cycles of eternity are immeasurably long."

李白的诗歌风格豪放、情感丰富，常表现出超脱世俗的意境，这与他饮酒后的心态有关。他的作品在中国文化中具有重要地位，影响了后世无数诗人。

Li Bai's poetic style is bold and unrestrained, with rich emotions, often expressing a transcendental realm that is related to his state of mind after drinking. His works hold an important place in Chinese culture and have influenced countless poets through the ages.

王羲之创作的《兰亭集序》与曲水流觞有关，或者说这个作品就是在曲水流觞的娱乐气氛推动下诞生的，其中："引以为流觞曲水，列坐其次。虽无管弦之乐，一觞一咏，亦足以畅叙幽情。"《兰亭集序》这一佳作的诞生也促使曲水流觞愈发成为一种具备文雅特征的游戏。

Wang Xizhi's creation of "The Orchid Pavillion" is related to drink water from a winding canal with one wine cup floating on it so as to wash away omionousness, or more precisely, this masterpiece was born under the influence of the entertaining atmosphere created by this game. It states: "together with a clear winding brook engirdled, which can thereby serve the guests by floating the wine glasses on top for their drinking.Seated by the bank of brook, people will still regale themselves right by poetizing their mixed feelings and emotions with wine and songs, never mind the absence of melody from string and wind instruments." The birth of this outstanding work also made the flowing cup in a meandering stream increasingly an elegant game.

另外，宋代诗人苏轼也以饮酒助兴而闻名。《赤壁赋》是他酒后乘舟游览赤壁时所作，文中不仅描写了赤壁的壮丽景色，还融入了对历史的深刻思考。

Additionally, the Song Dynasty poet Su Shi was also famous for using wine to enhance his creativity. "First Visit to the Red Cliff" was written while he was boating and visiting the Red Cliff after drinking. The text not only depicts the magnificent scenery of the Red Cliff but also incorporates profound reflections on history.

（二）书画家与酒（Calligraphy and Painting Artists and Alcohol）

历史上有一些喜好饮酒后作画写字的书画家，他们在酒后创作了不少著名的艺术作品。如"颠张醉素"指的是被誉为"草圣"的唐代著名书法家张旭和另一位草书大家怀素。两人常在饮酒后挥毫泼墨，这个称号反映了他们饮酒后作书的独特风格和狂放不羁的气势。

Historically, there have been some calligraphy and painting artists who preferred to paint and write after drinking alcohol. They have created many famous works of art after drinking. For example, "Dian Zhang and Zui Su" refers to the famous Tang Dynasty calligrapher Zhang Xu, known as the "Script-Sage," and another master of cursive script, Huai Su. They often created calligraphy after drinking alcohol, and this title reflects their unique style and unbridled momentum in creating calligraphy after drinking alcohol.

明代著名画家、文学家徐渭，其画风独特，擅长泼墨大写意画。徐渭酷爱饮酒，他的许多诗作中都提到了酒。徐渭的酒量很大，但他并不是沉溺于酒，而是通过饮酒激发自己的创作灵感。他喜欢酒后挥毫，其作品具有强烈的个性和情感表达。

The famous Ming Dynasty painter and writer, Xu Wei, had a unique painting style and was good at splash-ink large freehand brushwork. Xu Wei loved drinking alcohol, and many of his poems mentioned alcohol. Xu Wei had a large capacity for alcohol, but he was not addicted to it. Instead, he used drinking to inspire his creative inspiration. He liked to create after drinking, and his works have a strong personality and emotional expression.

二、名人轶事 (Anecdotes of Celebrities)

翻开中国文学史，常常能闻到扑鼻而来的酒香。多少古代的文人雅士与酒结下了不解之缘，他们中的许多人常自取或被人赋予与酒有关的雅号，如酒圣、酒仙、酒狂、醉翁、酒雄、酒鬼等，留下许多轶闻趣事。

Opening the pages of Chinese literary history, one can often sense the aroma of wine wafting through the air. Countless ancient literati and refined scholars have formed an inseparable bond with wine, and many of them have been given or have taken on elegant titles related to wine, such as the "Wine Sage," "Wine Immortal," "Wine Fiend," "Drunken Elder," "Wine Hero," and "Wine Ghost," leaving behind numerous anecdotes and interesting stories.

唐代诗人贺知章人称"酒仙"，与张旭、包融、张若虚并称"吴中四士"，个个都是嗜酒如命之人。有一次贺知章遇见李白，两人相见恨晚，遂成莫逆。一天，贺知章邀李白对酒共饮，正喝得尽兴，却发现兜里没钱，于是他毫不犹豫解下身上佩带的金龟换酒，与李白开怀畅饮，一醉方休。

The Tang Dynasty poet He Zhizhang was known as the "Wine Immortal" and was one of the "Four Scholars of Wu," along with Zhang Xu, Bao Rong, and Zhang Ruoxu, all of whom were fond of wine as if it were their life's blood. On one occasion, He Zhizhang met Li Bai, and they hit it off so well that they became close friends. One day, He Zhizhang invited Li Bai to drink wine together, and as they were enjoying themselves, they realized they had run out of money. Without hesitation, He Zhizhang took off the golden turtle he wore as a symbol of his official rank and exchanged it for more wine to continue their revelry.

李白更是爱喝酒，自称"酒仙"，一生喜酒、爱酒，写诗文时尤其离不开酒。无论得意、求仕期间，还是落魄之时，也不管何时何地，人多人少，有钱没钱，他都要想办法喝酒。暮年时，甚至将自己心爱的宝剑换了酒喝。他不但喜欢饮酒，而且几乎每饮必醉。

Li Bai, even more so, loved to drink. He called himself the "Wine Immortal," and he loved wine all his life, especially when writing poetry. Regardless of whether he was successful in his pursuits or down on his luck, whether he was in a good place or a bad one, surrounded by many or few people, with or without money, he always found a way to drink. In his later years, he even exchanged his beloved sword for wine. Not only did he enjoy drinking, but he also almost always got drunk.

有一个李白醉酒后广为流传的故事——高力士脱靴。唐玄宗时期，李白因其卓越的文学才华受到了唐玄宗的赏识，被召入宫中担任翰林待诏，这是一种文学侍从的职位。一日，李白在宫中与唐玄宗饮酒赋诗，酒醉之后，李白请求唐玄宗允许他自由吟诗。唐玄宗欣然同意，于是李白在金殿上大展诗才，吟咏了多首诗歌。李白在吟诗的兴头上，突然感到靴子有些紧，不便于行走。于是，他便命令站在一旁的高力士为他脱去靴子。高力士是唐玄宗身边的近侍，权势显赫，但面对李白的这一要求，高力士最终不得不在唐玄宗的默许下，为李白脱去了靴子。这个故事反映了李白不拘一格、豪放不羁的个性，以及他不畏权贵、自视甚高的态度。

There is a widely circulated story about Li Bai getting drunk-Gao Li Shi taking off his boots. During the Tang Xuanzong period, Li Bai was appreciated by Emperor Xuanzong for his outstanding literary talent and was summoned to the palace to serve as a Hanlin Academy official, a position for literary attendants. One day, Li Bai was drinking and composing poetry with Emperor Xuanzong in the palace. After getting drunk, Li Bai requested permission from Emperor Xuanzong to recite poetry freely. Emperor Xuanzong gladly agreed, and Li Bai showcased his poetic talent on the golden hall, reciting many poems. In the midst of his poetic recitation, he suddenly felt that his boots were too tight and not convenient for walking. So, he ordered Gao Li Shi, who was standing nearby, to take off his boots for him. Gao Li Shi was a trusted attendant to Emperor Xuanzong and held considerable power, but faced with Li Bai's request, he ultimately had to take off Li Bai's boots with Emperor Xuanzong's tacit approval. This story reflects Li Bai's unconventional and unrestrained personality, as well as his fearless attitude towards the powerful and his high self-esteem.

酒在古代文化中的重要性，不仅体现在饮酒习俗和酒宴文化上，更体现在对古代名人思想、行为和创作的影响上。总的来说，古代名人与酒的关系，是中国古代文化的重要组成部分。酒不仅是古代名人日常生活中的饮品，更是他们表达情感、展示才华、追求自由的重要方式。酒文化在古代社会中具有重要地位，对古代文化的发展和传承产生了深远影响。

The importance of wine in ancient culture is not only reflected in drinking customs and banquet culture but also in its impact on the thoughts, behaviors, and creations of ancient celebrities. In general, the relationship between ancient celebrities and wine is an important component of ancient Chinese culture. Wine was not only a daily drink for ancient celebrities but also an important way for them to express their emotions, display their talents, and pursue freedom. Wine culture held an important position in ancient society and had a profound impact on the development and inheritance of ancient culture.

赛证直通 Competitions and Certificates

一、基础知识部分（Basic Knowledge Part）

（一）专业名词解释（Explanation of Professional Terms）

曲水流觞（floating a wine cup along a winding stream）：

二维码 1-2-4：
知识拓展

（二）思考题（Thinking Question）

说说酒对古代名人思想、行为和创作的影响。

Discuss the impact of alcohol on the thoughts, behaviors, and creations of ancient celebrities.

二、技能操作部分（Skill Operation Part）

查阅资料，讲述其他古代中国的名人与酒的趣事。

Research materials and recount interesting anecdotes about other famous figures from ancient

China and their relationship with wine.

【单元二　反思与评价】Unit 2　Reflection and Evaluation

学会了：＿＿＿＿＿＿＿＿＿＿＿＿＿＿＿＿＿＿＿＿＿＿＿＿＿

Learned to: ＿＿＿＿＿＿＿＿＿＿＿＿＿＿＿＿＿＿＿＿＿＿＿

成功实践了：＿＿＿＿＿＿＿＿＿＿＿＿＿＿＿＿＿＿＿＿＿＿＿

Successful practice: ＿＿＿＿＿＿＿＿＿＿＿＿＿＿＿＿＿＿＿＿

最大的收获：＿＿＿＿＿＿＿＿＿＿＿＿＿＿＿＿＿＿＿＿＿＿＿

The biggest gain: ＿＿＿＿＿＿＿＿＿＿＿＿＿＿＿＿＿＿＿＿＿

遇到的困难：＿＿＿＿＿＿＿＿＿＿＿＿＿＿＿＿＿＿＿＿＿＿＿

Difficulties encountered: ＿＿＿＿＿＿＿＿＿＿＿＿＿＿＿＿＿＿

对教师的建议：＿＿＿＿＿＿＿＿＿＿＿＿＿＿＿＿＿＿＿＿＿＿

Suggestions for teachers: ＿＿＿＿＿＿＿＿＿＿＿＿＿＿＿＿＿

单元三　世界六大基酒

金酒、威士忌、白兰地、伏特加、朗姆酒和龙舌兰酒是六大基酒，它们是鸡尾酒的灵魂。

Gin, Whisky, Brandy, Vodka, Rum, and Tequila, they are the six major base liquors, they are the soul of cocktail.

单元导入 Unit Introduction

基酒在调酒中作为基础的酒精饮料，决定了鸡尾酒的主要口感和风格。那么世界六大基酒是什么？分别有什么特点？产于何处？工艺为何？怎么饮用？这个单元将带你走进基酒的世界。

The base liquor, as the fundamental alcoholic beverage in cocktail mixing, determines the main taste and style of the cocktail. So, what are the six major base liquors in the world? What are their characteristics? Where arc they produced? What is the production process? How to drink them? This unit will take you into the world of base liquors.

学习目标 Learning Objectives

知识目标（Knowledge Objectives）

（1）了解白兰地的概况、产区、制作工艺、特点、历史和饮用方法。

Gain an understanding of the general overview, production regions, manufacturing process, characteristics, history, and drinking methods of brandy.

（2）了解威士忌的概况、产区、制作工艺、特点、历史和饮用方法。

Gain an understanding of the general overview, production regions, manufacturing process, characteristics, history, and drinking methods of whiskey.

（3）了解金酒的概况、产区、制作工艺、特点、历史和饮用方法。

Gain an understanding of the general overview, production regions, manufacturing process, characteristics, history, and drinking methods of gin.

（4）了解伏特加的概况、产区、制作工艺、特点、历史和饮用方法。

Gain an understanding of the general overview, production regions, manufacturing process, characteristics, history, and drinking methods of vodka.

（5）了解朗姆酒的概况、产区、制作工艺、特点、历史和饮用方法。

Gain an understanding of the general overview, production regions, manufacturing process, characteristics, history, and drinking methods of rum.

（6）了解龙舌兰酒的概况、产区、制作工艺、特点、历史和饮用方法。

Gain an understanding of the general overview, production regions, manufacturing process, characteristics, history, and drinking methods of tequila.

能力目标（Ability Objectives）

（1）能介绍六大基酒的概况和饮用方法。

Can provide an overview of the six main base liquors and their methods of consumption.

（2）能分别用六大基酒调制鸡尾酒。

Can prepare a cocktail using each of the six main base liquors.

素质目标（Quality Objectives）

（1）让学生了解不同的酒文化。

Let students to understand different alcohol cultures;

（2）让学生感受不同生活环境、历史背景、传统习俗、价值观念、思维模式和社会规范等背景下，中外文化的不同。

Let students experience the differences between Chinese and foreign cultures in different living environments, historical backgrounds, traditional customs, values, thinking patterns and social norms.

（3）让学生感受酒文化是精神文化的一种重要体现。

Let students feel wine culture is an important embodiment of spiritual culture.

（4）培养学生具备良好的职业精神、专业精神和工匠精神。

Cultivate students with good professional spirit, professional spirit, and craftsmanship spirit;

（5）培养学生的创新精神。

Cultivate students' innovative spirit.

专题一
"葡萄酒灵魂"：白兰地

【微课】
"葡萄酒灵魂"：
白兰地

你知道 XO 吗？你是通过什么方式知道的？听到 XO，你们脑海中会联想到什么（图 1-3-1）？

_____

Do you know XO? How do you know that? What comes to mind when you hear XO（Figure1-3-1）?

_____

图 1-3-1　XO

一、白兰地概述（The Concept of Brandy）

白兰地，从狭义上讲，是指葡萄发酵后经蒸馏而得到的高度酒精，再经橡木桶储存而成的酒。白兰地是世界八大蒸馏酒之一，主要原料是葡萄，但也可以用其他水果如苹果、樱桃、梨等制作。如果是以其他水果原料酿成的白兰地，则应加上水果名称，如"苹果白兰地"。因为通过独特的工艺将葡萄酒的精髓升华，所以白兰地被誉为"葡萄酒的灵魂"。

In a narrow sense, brandy refers to a high-quality alcohol through distillation after fermentation of grapes, and then is stored in oak barrels. Brandy is one of the world's eight major distilled spirits, primarily made from grapes, but it can also be produced using other fruits such as apples, cherries, pears, and more. If the brandy is brewed from other fruit materials, then we should call it with the name of the fruit, such as " apple brandy ". Because the essence of wine is sublimated through a unique process, brandy is hailed as "the soul of wine."

二、白兰地产区及工艺（The Production Area and Craftsmanship of Brandy）

世界上有很多国家生产白兰地，但法国出品的最为有名，尤其是法国干邑地区生产的白兰地品质最佳。该地区位于法国西南部波尔多北部的一个小镇，当地土壤非常适宜葡萄的生长。干邑酿酒大多不会使用酿制红葡萄酒的葡萄，而是选用著名的白葡萄品种。这种葡萄酸度高、糖分低，能够有效防止酒液在发酵过程中变质，非常适合蒸馏。

There are many countries in the world that produce brandy, but the brandy produced in France is the most famous, especially the brandy produced in Cognac region of France is of the best quality. This area is located in a small town in the north of Bordeaux in southwestern France. The local soil is very suitable for the growth of grapes. Cognac winemaking mostly does not use grapes that make red wine, but chooses famous white grape. This type of grape has high acidity and low sugar content, which effectively prevents the wine from spoiling during the fermentation process, making it very suitable for distillation.

在白兰地的制作工艺中，特别讲究陈酿时间与勾兑的技术，其中陈酿的时间更是衡量白兰地酒质优劣的重要标准。干邑地区各个生产厂家储藏在橡木桶中的白兰地，有的甚至长达40～70年之久。制酒人选用不同年限的酒，按各自世代相传的秘方调配勾兑，创造出不同品质和风格的干邑白兰地。

In terms of the craftsmanship, the production of brandy needs to pay special attention to the aging time and blending technology. The aging time is an important criterion for judging the quality of brandy. The brandy stored in oak barrels by various manufacturers in the Cognac region, some may even be as long as 40–70 years. Winemakers select wines of different ages and blend them according to the recipe handed down from generation to generation to create cognacs with varieties of qualities and styles.

公元1909年，法国政府颁布法令明文规定，只有在干邑镇周围的36个县市所生产的白兰地方可命名为干邑（Cognac），除此以外的任何地区不能用"Cognac"一词命名。这一规定以法律条文的形式确立了干邑白兰地的生产地位，干邑就是"白兰地之王"。正如英语的一句话："All Cognac is brandy，but not all brandy is Cognac"。

In 1909, the French government issued a decree expressly stipulating that only the brandy produced in 36 counties and cities around the Cognac can be named Cognac, and brandy from other places cannot be called "Cognac". This regulation established cognac production status in the form of legal provisions, and Cognac is the "King of Brandy." Just like a saying in English: "All Cognac is brandy, but not all brandy is Cognac."

白兰地的香气主要来自葡萄原料、蒸馏工艺、橡木桶陈酿以及时间的沉淀。这些因素共同作用，赋予了白兰地复杂而迷人的香气层次。白兰地的基酒是由葡萄发酵而成的葡萄酒，葡萄品种本身的香气是白兰地香气的起点；蒸馏过程会浓缩葡萄基酒中的香气物质，同时产生新的香气；橡木桶陈酿是白兰地香气形成的关键环节，新桶通常香气更浓郁，而旧桶则更柔和。随着陈酿时间越长，白兰地的香气层次越丰富，顶级白兰地甚至可以展现出数十种不同的香气。

The aroma of brandy primarily stems from the grape raw materials, distillation process, oak barrel aging, and the passage of time. These factors work together to bestow upon brandy a complex and captivating array of aromatic layers. The base wine of brandy is made from fermented grapes, and the inherent aroma of the grape variety itself serves as the starting point for brandy's fragrance. The distillation process concentrates the aromatic compounds in the grape base wine while also generating new aromas. Oak barrel aging is a crucial stage in the formation of brandy's aroma; new barrels typically impart a more intense aroma, while older barrels lend a softer touch. As the aging period extends, the layers of aroma in brandy become richer, with top-tier brandies capable of exhibiting dozens of distinct fragrances.

三、白兰地特点（Characteristics of Brandy）

白兰地酒度在40度～43度，虽属烈性酒，但由于经过长时间的陈酿，具有优雅、细致的葡萄果香和浓郁的陈酿木香，其口感柔和、香味纯正，饮用后给人以高雅、舒畅的享受。白兰地呈美丽的琥珀色，富有吸引力，其悠久的历史也给它蒙上了一层神秘的色彩。白兰地有一个特点，即不怕稀释。在白兰地中放进白水，风味不变还可降低酒度。因此，人们饮用白兰地时往往放进冰块、矿泉水或苏打水，更有加茶水的，越是名贵的茶叶越好。白兰地的芳香加上茶香，具有浓郁的民族特色。

The degree of brandy is between 40 to 43 degrees. Although it is a strong wine, it has an elegant and delicate grape fruit aroma and a strong aging woody aroma after a long period of aging. It tastes soft with pure fragrance, making people feel elegant and comfortable. With beautiful amber color, brandy is so attractive and its long history had made it more mystery. One of the characteristics of brandy is that it still tastes good after diluted. Pouring water in the brandy will keep the flavor unchanged and reduce the alcohol content. Therefore, people often put ice cubes, mineral water or soda water when drinking brandy. Someone even adds tea in it, and the more precious the tea is, the better. The combination of the fragrance of brandy and tea makes it have strong national characteristics.

四、白兰地发展历史（The History of Brandy）

白兰地起源于法国。公元12世纪，干邑生产的葡萄酒就已经销往欧洲各国，外国商船也经常到滨海口岸购买葡萄酒。大约在16世纪中叶，为了方便葡萄酒的出口，减少占用空间，降低税金，也避免葡萄酒在长途运输过程中变质，干邑镇的酒商把葡萄酒进行蒸馏浓缩。

Brandy originated in France. In the 12 th century AD, the wine in Cognac has been sold to European countries, and foreign merchant ships often go to the coastal port to buy wine. Around the middle of the 16 th century, in order to facilitate the export of wine, reduce the occupation of space, lower taxes, and prevent the wine from spoiling during long-distance transportation, the wine merchants in Cognac began to distill and concentrate the wine.

到1701年，法国卷入西班牙战争，白兰地遭到禁运。酒商们只得将白兰地先暂时储存，等待时机。他们利用干邑当地盛产的橡木做成木桶，将白兰地储存其中。1704年战争结束，酒商们意外发现，酒不仅没有变质，本来无色的白兰地竟然变成了美丽的琥珀色，而且香味更加浓郁，口感也更加柔和。自此，橡木桶陈酿工艺就成为干邑白兰地的重要制作程序。

By 1701, France was involved in the Spanish War and brandy was also embargoed. The wine merchants had to temporarily store the brandy, waiting for the opportunity to sell. They made wooden barrels with the Cognac local rich oak and stored the brandy in them. At the end of the war in 1704, the wine merchants unexpectedly discovered that not only did the wine not deteriorate, the colorless brandy turned into a beautiful amber color with stronger aroma and softer taste. Since then, the oak barrel aging process has become an important process for cognac production.

　　白兰地在中国的发展，最早是作为一种高档的洋酒在上海、广州等沿海开放城市的租界和上层社会中流行，由于价格昂贵，它一度成为身份和地位的象征。然而直至中国第一个民族葡萄酒企业——张裕葡萄酿酒公司成立后，白兰地在国内才真正得以发展，张弼士先生对中国的葡萄酒发展功不可没。1915 年国产白兰地"张裕可雅"在万国博览会上获金奖，我国有了自己品牌的优质白兰地。

　　The development of brandy in China initially gained popularity as a high-end imported liquor in the concessions and upper-class societies of coastal open cities such as Shanghai and Guangzhou. Due to its high price, it once became a symbol of identity and status. However, it was not until the establishment of China's first national wine company—Changyu Pioneer Wine Company, that domestic brandy could truly develop, and Mr. Zhang Bishi contributed a lot to the development of wine in China. In 1915, the domestic brandy "Zhangyu Keya" won the gold medal at the World Exposition, and China has its own brand of high-quality brandy.

五、白兰地饮用方法（The Drinking Method of Brandy）

　　白兰地是一种典雅、庄重的美酒，饮一杯白兰地，仿佛置身于高雅的殿堂，品味人生的美好。白兰地的饮用方法多种多样，可作消食酒，可作开胃酒，可以不掺兑任何东西"净饮"，也可以加冰块饮、掺兑矿泉水饮或掺兑茶水饮。对于具有绝妙香味的白兰地来说，无论怎样饮用都可以。一般来说，不同档次的白兰地，采用不同的饮用方法，可以收到更好的效果。

　　As an elegant and solemn wine, brandy will put yourself in an elegant palace and make you enjoy the beauty of life. There are many ways to drink brandy. It can be used as a digestive wine or an aperitif. It can be " straight up " without mixing anything, or drunk with ice, mixed with mineral water or tea. For a brandy with a wonderful aroma, you can drink it any way you like. Generally speaking, different drinking methods for different grades of brandy can get better results.

　　例如，XO 级白兰地，是在橡木桶里经过十几个春夏秋冬的贮藏陈酿而成，是酒中的珍品和极品。这种白兰地最好的饮用方法是什么都不掺，这样原浆原味使人更能体会到这种酒艺术的精髓和灵魂。

　　For example, XO-grade brandy is aged in oak barrels after more than a dozen years, so it is a treasure and the best type in wine. The best way to drink this brandy is to mix it with nothing, so that it has the original flavor to make people enjoy the essence and soul of this kind of wine art.

　　V.S 级白兰地，只有 2～4 年的酒龄。如果直接饮用，难免有酒精的刺口辣喉感，而掺兑矿泉水或夏季加入冰块饮用，既能使酒精浓度得到充分稀释、减轻刺激，又能保持白兰地的风味不变。

　　In terms of VS-grade brandy, which is only 2-4 years old. If you drink it directly, you will inevitably feel thorny and spicy. While drinking it with mineral water or ice cubes in summer will not only fully dilute the alcohol strength, reduce irritation, but also can keep the brandy's flavor unchanged.

赛证直通 Competitions and Certificates

一、基础知识部分（Basic Knowledge Part）

（一）专业名词解释（Explanation of Professional Terms）

1. 白兰地（brandy）:
2. 天然老熟（natural aging）:

（二）思考题（Thinking Question）

白兰地的制作工艺包括哪些程序？

What are the procedures for making brandy?

二维码 1-3-1：
知识拓展

二、技能操作部分（Skill Operation Part）

实践白兰地常见的饮用方式，说说你最喜欢的品饮方式及原因。

Common ways to enjoy brandy in practice, tell us about your favorite way to drink it and the reasons why.

【微课】
"液体黄金"：
威士忌

专题二
"液体黄金"：威士忌

你在影视剧中听过"来一杯 whisky"这句台词吗？你知道"whisky"是什么酒吗（图 1-3-2）？

————————————

Have you heard this line " Have a whisky" in the movie？Do you know what whisky is（Figure1-3-2）？

————————————

图 1-3-2　调酒（bartending）

一、威士忌的概念及产区（Concept and Region of Whisky）

威士忌是一种由大麦、玉米等谷物作为原料，发酵蒸馏后放入橡木桶中陈酿多年的烈性蒸馏酒。它的酒精度为 43 度左右。英国是威士忌的主要生产国。高品质的白兰地，尤其是陈年佳酿，因其稀有和昂贵，常被称为"液体黄金"。

Whisky is a kind of strong distilled liquor made of barley, corn and other grains as raw materials, after fermentation and distillation the materials will be put in oak barrels for many years until the brewing is finished. The alcohol level is about 43 degrees. Britain is the main producer of whisky. High-quality brandy, especially aged vintage, is often referred to as "liquid gold" due to its rarity and high cost.

按照产地威士忌可以分为：苏格兰威士忌、爱尔兰威士忌、美国威士忌和加拿大威士忌四大类。

According to the place of origin, whisky can be divided into the four major categories: Scotch whisky, Irish whisky, American whisky and Canadian whisky.

二、威士忌的制作工艺（Making Process of Whisky）

威士忌的酿制工艺过程可分为七个步骤：

The brewing process of whisky can be divided into seven steps:

第一步：发芽。将麦类或谷类浸泡在温水中使其发芽，一般需要一至两周的时间，待其发芽后再将其烘干或使用泥煤熏干，等冷却后再储放大约一个月的时间，发芽的过程即算完成。值得一提的是，所有威士忌中，只有苏格兰地区所生产的是使用泥煤熏干，因此，苏格兰威士忌有一种独特的泥煤烟熏味。

Step 1: Malting. Soaking malt or grain in warm water to germinate, usually it takes one or two weeks. After germination, drying or smoke drying with peat. Store it for about a month after cooling, then the germination process is completed. It is worth mentioning that among all whiskies, only those produced in Scotland are dried with peat, so Scotch Whisky has a unique taste of peat smoke.

第二步：磨碎。将发芽的麦类或谷类捣碎并煮熟成汁。在磨碎的过程中，温度及时间的控制是相当重要的环节，过高的温度或过长的时间都会影响品质。

Step 2: Mashing. Grind and cook sprouted wheat or grains into juice. In the grinding process, the control of temperature and time is very important, too high temperature or the too long time will affect the quality.

第三步：发酵。将冷却后的麦芽汁加入酵母菌进行发酵。由于酵母能将麦芽汁转化成酒精，因此，在完成发酵过程后会产生酒精浓度为5%～6%的液体，此时的液体被称为"wash"或"beer"。

Step 3: Fermentation. The cooled wort is added to yeast for fermentation, as yeast can convert wort into alcohol, a liquid with an alcohol concentration of about 5%-6% will be produced after the fermentation process. At this time, the liquid is called wash or beer.

第四步：蒸馏。蒸馏具有浓缩的作用。发酵形成低酒精度的酒液后，还需经过蒸馏才能形成威士忌酒，这时的威士忌酒精浓度为60%～70%，被称为"新酒"。

Step 4: Distillation. Distillation has the function of concentration. The liquid with low alcohol formed after fermentation needs distillation to be the whisky. At this time, the alcohol concentration of whisky is about 60%-70%, which is called "new wine".

第五步：陈年。蒸馏过后的新酒必须经过陈年的过程，使其经过橡木桶的陈酿吸收植物的天然香气，产生漂亮的琥珀色，同时也能降低高浓度酒精的强烈刺激感。在苏格兰地区有相关的法令规定陈年的酒龄时间，即每一种酒所标示的酒龄都必须是真实无误的。

Step 5: Maturing. After distillation, the new wine must go through the maturing process, so that it can absorb the natural aroma of plants by oak barrel maturing to make itself be the beautiful amber, and reduce the strong stimulation of high concentration alcohol. In Scotland, there are relevant laws to regulate the maturing time of wine, that is, the wine age marked by each wine must be true.

第六步：混配。由于麦类及谷类原料的品种众多，因此，所制造的威士忌酒也存在着各不相同的风味。这时就靠各个酒厂的调酒大师依其经验的不同和本品牌对酒质的要求，按照一定的比例调配勾兑出与众不同口味的威士忌酒。因此，各个品牌的混配过程及方法都被视为绝对的机密。

Step 6: Blending. Because there are many varieties of malt and grain raw materials, the whisky produced also has different flavors. At this time, it depends on the masters bartender of each winery to get their own whisky with different flavors by a certain proportion blending according to their different experience and the requirements of the wine quality of their own brands. Therefore, the blending process and methods of each brand are regarded as absolute secrets.

第七步：装瓶。调配后，则是装瓶。调配好的威士忌再过滤一次除掉杂质后装瓶，贴上标签后就可以进入市场和消费者见面了。

Step 7: Boltting. After blending the next step is bottling. The blended whisky needs to be filtered again to remove the impurities before the bottling, and then can enter the market to meet consumers after labeling.

三、不同威士忌的特点（Characteristics of Whisky）

由于不同地区的原料、酿造工艺、气候条件以及文化传统有所不同，不同地区的威士忌各有其独特的风味和特点。

Due to differences in raw materials, brewing techniques, climatic conditions, and cultural traditions across various regions, each region's whisky possesses its unique flavor and characteristics.

苏格兰威士忌：原料主要使用大麦，部分使用其他谷物。必须在橡木桶中熟成至少 3 年，风味复杂多样，以独特的烟熏味和焦香味著称。

Scotch Whisky: Primarily made from barley, with some using other grains. Must be matured in oak barrels for at least 3 years, known for its complex and diverse flavors, particularly its distinctive smoky and toasted notes.

爱尔兰威士忌：原料主要使用大麦，部分使用其他谷物。必须在橡木桶中熟成至少 3 年，爱尔兰威士忌口感较为柔和，带有一定的甜味。

Irish Whiskey: Primarily made from barley, with some using other grains. Must be matured in oak barrels for at least 3 years. Irish whiskey is known for its relatively smooth taste, with a hint of sweetness.

美国威士忌：原料主要使用玉米，部分使用黑麦、大麦等。通常在新的橡木桶中陈酿，带有浓郁的香草、焦糖、橡木和香料味，口感较为甜美。

American Whiskey: Primarily made from corn, with some using rye, barley, etc. Typically aged in new oak barrels, it features rich flavors of vanilla, caramel, oak, and spices, with a relatively sweet taste.

加拿大威士忌：原料主要使用玉米，部分使用黑麦、大麦等。通常在经过烤制的橡木桶中陈酿，口感较为清淡、顺滑，甜味较少，带有轻微的香草、蜂蜜和谷物香气。

Canadian Whiskey: Primarily made from corn, with some using rye, barley, etc. Usually aged in charred oak barrels, it has a lighter and smoother taste, with less sweetness and subtle notes of vanilla, honey, and grain.

四、苏格兰威士忌（Scotch Whisky）

苏格兰威士忌酒闻名世界主要有以下几个原因：

Scottish whisky is famous in the world for the following reasons:

第一，苏格兰产酒地的气候与地理条件适宜这种大麦农作物的生长。

First, the climate and geographical conditions of place of origin are suitable for the growth of crop barley.

第二，在这些地方蕴藏着一种称为泥煤的煤炭，这种煤炭在燃烧时会发出特有的烟熏气味，在苏格兰制作威士忌酒的传统工艺中要求必须使用这种泥煤来烘烤麦芽。

Second, there is a kind of coal called peat in these places, which gives off a unique smoky smell when burned. In the traditional process of making whisky in Scotland, this peat must be used to bake malt.

第三，苏格兰蕴藏着丰富的优质矿泉水，为酒液的稀释、勾兑提供了优质的材料。

Third, Scotland is rich in high-quality mineral water, which provides high-quality materials for the dilution and blending of liquor.

第四，苏格兰人有着传统的酿造工艺和严谨的质量管理方法。

Fourth, Scots have traditional brewing technology and rigorous quality management methods.

五、威士忌饮用方式（Drinking Method of Whisky）

根据个人习惯和爱好，威士忌有多样的饮用方式。

According to personal habits and hobbies, whisky can be drunk in a variety of ways.

（一）纯饮（Pure Drink）

纯饮可以获取单一麦芽威士忌的真谛。将威士忌直接倒入酒杯，静静感受琥珀色的液体流过身体，芳香瞬间弥漫。

Pure drink can get the true meaning of single malt whisky. Pour the whisky directly into the glass and quietly feel the amber liquid sliding down your throat. The aroma is instantly diffuse.

（二）加水（With Water）

加水堪称是全世界最"普及"的威士忌饮用方式，即使在苏格兰，加水饮用仍大行其道。许多人认为加水会破坏威士忌的原味，其实加适量的水并不至于让威士忌失去原味，相反地，此举可能让酒精味变淡，引出威士忌潜藏的香气。依据理论，将威士忌加水稀释到20%的酒精度，是表现出威士忌所有香气的最佳状态。

Adding water is the most "popular" way to drink whisky in the world. Even in Scotland, drinking with water is still popular. Many people think that adding water will destroy the original flavor of whisky. In fact, adding an appropriate amount of water will not make whisky lose its original flavor. On the contrary, it may dilute the alcohol flavor and bring out the latent aroma of whisky. According to the theory, diluting whisky with water to 20% alcohol level is the best state that can best show all the aroma of whisky.

低年份酒较高年份酒有更强的刺激性。因此，要达到最佳释放香气的状态，低年份威士忌所需稀释用水的量，便会高于高年份威士忌。一般而言，1∶1 的比例最适用于 12 年威士忌，低于 12 年的威士忌水量要增加，高于 12 年的威士忌水量要减少。如果是高于 25 年的威士忌，建议是加一点水或不需要加水。

New vintage wine compares to old vintage wine, has stronger irritation. Therefore, in order to achieve the best release of aroma, the amount of dilution water required for new vintage whisky will be higher than that for old vintage whisky. In General, the proportion of 1:1 is most suitable for 12-year whisky. If it is lower than 12 years, the water volume should be increased, and if it is higher than 12 years, the water volume should be reduced. If it is higher than 25 years, it is recommended to add a little water or no water.

（三）加冰（On the Rock）

加冰又称"on the rock"，主要是给想降低酒精的刺激，又不想稀释威士忌的酒客们另一种选择。然而，威士忌加冰块虽能抑制酒精味，但也连带因降温而让部分香气闭锁，难以品尝出威士忌原有的风味特色。

Drink with Ice, also known as "on the rock", is mainly an alternative for drinkers who want to reduce alcohol stimulation and do not want to dilute whisky. However, although whisky with ice can inhibit the smell of alcohol, it also locks part of the aroma due to cooling, so it is difficult to taste the original flavor characteristics of whisky.

（四）加汽水（Highball）

以烈酒为基酒，再加上汽水的调酒称为加汽水（Highball）。加可乐普遍用于美国威士忌，至于其他种类威士忌，大多是用姜汁汽水等其他的苏打水调制。

Adding soda is based on liquor, and blending with soda is called "highball". Adding coke is widely used in American whisky. As for other kinds of whisky, it is mostly prepared with ginger soda and other soda.

（五）火热托迪（Hot Toddy）

在寒冷的苏格兰有一种叫作火热托迪（Hot Toddy）的传统威士忌酒谱，不但可祛寒，还可治愈小感冒。这种酒多以苏格兰威士忌为基酒，调入柠檬汁、蜂蜜等，最后加入热水，即成御寒又好喝的调制酒。

In freezing Scotland, there is a traditional whisky recipe called hot toddy, which can not only dispel the cold, but also cure a small common cold. This Wine usually takes Scotch Whisky as the base wine, mixed with lemon juice, honey, etc., and finally blended with hot water to become a cold resistant and delicious wine.

赛证直通 Competitions and Certificates

一、基础知识部分（Basic Knowledge Part）

（一）选择题（Multiple Choice）

苏格兰威士忌的主要原料是（ ）。

A. 大麦　　　　B. 玉米　　　　C. 小麦　　　　D. 燕麦

The main ingredient of Scotch whisky is（ ）.

A. barley　　　　B. corn　　　　C. wheat　　　　D. oats

（二）思考题（Thinking Question）

影响威士忌风味的因素有哪些？

What are the factors that affect the flavor of whisky?

二、技能操作部分（Skill Operation Part）

品鉴一款威士忌，描述其颜色、香气、滋味等特征。

To taste a whiskey, describe its color, aroma, flavor, and other characteristics.

专题三
"鸡尾酒心脏"：金酒

你知道金酒（图 1-3-3）吗？说说你所了解的金酒知识或故事。

Do you know Gin（Figure 1-3-3）? Tell me about your knowledge or stories of Gin.

图 1-3-3　金酒（Gin）

一、金酒发展历史（The History of Gin）

金酒，又称杜松子酒或琴酒，是一种以谷物为原料，通过发酵和蒸馏过程制作的烈酒。金酒具有芬芳诱人的香气，无色透明，味道清新爽口，可单独饮用，也可调配鸡尾酒。金酒是鸡尾酒中使用最多的基酒之一，有"鸡尾酒心脏"的美誉，其酒精含量通常在35% ～ 55%。

Gin, also known as Jack or Genever, is a distilled spirit made from grains through fermentation and distillation. It has a fragrant and enticing aroma, is colorless and transparent, and has a fresh and refreshing taste. It can be consumed on its own or mixed with other ingredients. Gin is one of the most widely used spirits in cocktails and is often referred to as the "heart" of the cocktail, with an alcohol content typically ranging from 35% to 55%.

金酒的起源可以追溯到中世纪的荷兰。像许多烈酒一样，金酒最初也常被当作药品使用，据说杜松子有治疗痛风、解热、利尿的功效。荷兰商人路卡斯·博斯看中了金酒的商业价值，通过改良药用金酒，酿造出了口味更甜、更容易被消费者接受的金酒，并于 1575 年在荷兰斯奇丹建立了博斯酒厂，这家酒厂至今都是荷兰金酒最主要的生产商。到了 17 世纪，这种酒开始在英国流行，并逐渐演变成为今天人们所知的金酒。金酒在英国特别受欢迎，成了英国文化的一个重要组成部分。

The origins of gin can be traced back to medieval Holland. Like many spirits, gin was initially used as a medicine, with juniper berries believed to have the benefits of treating gout, reducing fever, and promoting diuresis. The Dutch merchant Lucas Bols recognized the commercial potential of gin, and by improving the medicinal variety, created a sweeter and more palatable version that was more easily accepted by consumers. He established the Bols Distillery in Schiedam, Holland, in 1575, which remains the main producer of Dutch gin to this day. By the 17th century, the spirit began to gain popularity in England and gradually evolved into the gin known today. Gin is particularly popular in the UK and has become an important part of British culture.

二、金酒的分类（The Classification of Gin）

金酒可以从口味、地域、原料、风格等不同维度进行分类。按照口味，金酒可以分为四类：干金酒、加甜金酒、果味金酒、荷兰金酒。

Gin can be classified from various perspectives, including taste, region, ingredients, and style. According to taste, gin can be divided into four categories: Dry Gin, Sweet Gin, Fruit Gin, and Dutch Gin.

（一）干金酒（Dry Gin）

干金酒又叫英式干金酒、伦敦干金酒。"干"是指不含糖，辛辣，喝起来没有甜味，与葡萄酒的"干"是一个概念。金酒上如果有标示"LONDON DRY GIN"就是干金酒。通常用于调酒的金酒主要是干金酒，它的风味相对清淡柔和，是调酒的首选金酒类型。

Dry gin, also known as London Dry Gin or London Gin. "Dry" means it contains no sugar, resulting in a spicy taste without sweetness, similar to the concept of "dry" in wine. If a gin bottle is marked with "LONDON DRY GIN," it is a dry gin. Dry gin is the primary type of gin used for mixing in cocktails, with a relatively light and mild flavor, making it the preferred choice for cocktail recipes.

（二）加甜金酒（Sweet Gin）

加甜金酒是指在干金酒中加入了糖分，使其带有怡人的甜辣味。18 世纪，有精明的商人用木质的猫形状的容器卖金酒，买主把钱放进猫的嘴里，金酒就从藏在猫爪里的管子中流出来。所以加甜金酒也有"老汤姆金酒"之称。"汤姆柯林斯"这款经典鸡尾酒就是用老汤姆金酒调制而成的。

Sweet gin is made by adding sugar to dry gin, giving it a pleasant sweet and spicy flavor. In the 18 th century, a clever merchant sold gin in a wooden container shaped like a cat. Customers would put money into the cat's mouth, and the gin would flow out of a tube hidden in the cat's paw. As a result, sweet gin is also known as "Old Tom Gin," and the classic cocktail "Tom Collins" is made using Old Tom Gin.

（三）果味金酒（Fruit Gin）

果味金酒是在干金酒中加入了成熟的水果和香料，如柑橘金酒、柠檬金酒、姜汁金酒等，此类金酒多用串香的工艺和勾兑工艺制成。现在一些金酒品牌都有推出这一类型的金酒，让金酒有多样的风味，满足更多人群的需求。

Fruit gin is made by adding ripe fruits and spices to dry gin, such as orange gin, lemon gin, and ginger gin. These gins are often produced using the infusion and blending processes. Some gin brands now offer this type of gin, adding more variety to the flavor profile and catering to a wider audience.

（四）荷兰金酒（Dutch Gin）

荷兰金酒除了具有浓烈的杜松子气味外，还有麦芽的芬芳，带有明显的谷物风味。带着相对明显的药味芳香，这可能就是原产地的一种传承。虽然不是当年的药物，但是相对干金酒而言它的杜松子和药味会更重一些。

In addition to its strong juniper aroma, Dutch gin also has a malty fragrance, with a distinct gain flavor. It has a relatively distinct medicinal aroma, which may be a legacy of its origins, although it is no longer used as a medicine. Compared to dry gin, Dutch gin has a stronger juniper and medicinal flavor.

三、金酒的制作流程（The Production Process of Gin）

金酒发展至今，荷兰、英国、美国、德国、法国、比利时等很多国家和地区都出产金酒，主要有两类生产方式。

Gin has developed to the point where many countries and regions, including the Netherlands, the United Kingdom, the United States, Germany, France, and Belgium, produce it. There are mainly two production methods.

（一）传统法（Traditional Method）

传统法是以大麦、黑麦、谷物为原料，经粉碎、糖化、发酵、蒸馏、调配而成的方法。其中加入杜松子的方法又有三种。第一种是在酿酒原料中加入压碎成小片的杜松子，加入整个糖化发酵蒸馏环节，比较类似于中国白酒中的草药入曲环节；第二种是用酒精浸泡杜松子，让杜松子的有效物质与风味都溶于乙醇，浸泡一周后重新蒸馏获得杜松子风味的乙醇；第三种是直接把杜松子用过滤网布包好放置于蒸馏器出口位置，酒体流出时浸出杜松子的有效物质与风味。

This method uses barley, rye, and other grains as raw materials, and involves crushing, saccharification, fermentation, distillation, and blending. There are three ways to add juniper berries. The first is to add crushed juniper berries to the brewing raw materials and include them in the entire saccharification, fermentation, and distillation process, which is similar to the process of adding herbs to the brewing process in Chinese baijiu. The second is to soak juniper berries in alcohol, allowing the effective substances and flavors of the juniper berries to dissolve in ethanol. After soaking for a week, the alcohol is redistilled to obtain ethanol with the flavor of juniper berries. The third is to wrap the juniper berries in a filter cloth and place them at the outlet of the still. When the liquid flows out, the effective substances and flavors of the juniper berries are extracted.

（二）合成法（Synthetic Method）

合成法也被称为化学法或添加法。在合成法中，酒精基通常是中性的，不含有特殊的香味，然后通过添加预先蒸馏或提取的植物香料来赋予金酒特有的风味。首先，需要制备酒精基，这通常是通过发酵和蒸馏谷物或水果得到的。酒精基需要是中性的，没有太多的自身风味，以便不会干扰最终产品的味道；然后选择用于制作金酒的香料，如杜松子、桂皮、柑橘皮等，将香料与酒精基混合，让酒精提取香料中的香气成分。这个过程可以通过冷浸或热浸的方式进行。冷浸是将香料和酒精混合后放置一段时间，热浸则是将混合物加热以加速提取过程。提取后，将香料残渣从酒精中分离过滤，以获得清澈的液体；然后根据需要确定是否还需添加额外的香料或酒精，以调整金酒的风味和酒精度。

The synthetic method for making gin, also known as the chemical method or addition method, involves using a neutral alcohol base that has no special fragrance. The characteristic flavor of gin is then imparted by adding pre-distilled or extracted botanical spices. First, an alcohol base needs to be prepared, which is usually obtained through the fermentation and distillation of grains or fruits. The alcohol base must be neutral and not have too much of its own flavor, so as not to interfere with the taste of the final product. Then, spices used for making gin, such as juniper berries, cinnamon, citrus peels, etc., are selected and mixed with the alcohol base, allowing the alcohol to extract the aromatic components from the spices. This process can be carried out either by cold maceration or hot maceration. Cold maceration involves mixing the spices with the alcohol and allowing it to stand for a period of time, while hot maceration involves heating the mixture to accelerate the extraction process. After extraction, the spice residue is separated and filtered from the alcohol to obtain a clear liquid. Depending on the requirements, additional spices or alcohol may be added to adjust the flavor and alcohol content of the gin.

合成法制作的金酒通常成本较低，生产速度较快，但在风味上可能不如传统蒸馏法制作的金酒复杂和丰富。

Gin made by the synthetic method usually has a lower cost and faster production speed, but may not be as complex and rich in flavor as gin made by the traditional distillation method.

四、金酒的饮用方式（Drinking Method of Gin）

金酒的饮用方式多种多样，如净饮、调酒、加冰、配餐、温热、混合饮用等，可以根据个人喜好和场合选择。

Gin can be consumed in a variety of ways, such as neat, mixed, on the rock, with a meal, warm, or blended, depending on personal preference and the occasion.

（一）净饮（Neat）

直接品尝金酒的原始风味，适合喜欢浓郁草本香气的人。将金酒倒入酒杯中，稍微摇晃杯身，让酒液与空气接触，释放香气。慢慢啜饮，感受金酒的复杂层次。这种品饮方式适合高品质的金酒，如伦敦干金酒或陈年金酒。

Directly savoring the original flavor of gin, this method is suitable for those who enjoy rich herbal aromas. Pour the gin into a glass, gently swirl the glass to allow the liquor to come into contact with the air, releasing its aromas. Sip slowly to appreciate the complex layers of the gin. This tasting approach is ideal for high-quality gins, such as London Dry Gin or aged gins.

（二）加冰（On the rock）

加入冰块能稍微稀释酒液，降低酒精的刺激感，同时让金酒的香气更加柔和。先在酒杯中加入大块冰块，倒入适量金酒，稍等片刻让酒液冷却并稀释，然后慢慢饮用。此种方法适合口感较重的金酒。

Adding ice cubes can slightly dilute the liquor, reducing the sharpness of the alcohol while making the aroma of the gin more mellow. First, place large ice cubes in the glass, pour in an appropriate amount of gin, and wait a moment for the liquor to cool and dilute. Then, sip slowly. This method is suitable for gins with a heavier flavor profile.

（三）调酒（Cocktail）

金酒是调制鸡尾酒的常用基酒，与各种果汁、糖浆、碳酸饮料或其他烈酒混合，可以制作出各种风味的鸡尾酒，像金汤力、马提尼、内格罗尼、长岛冰茶等。虽然品尝纯净的或加冰的金酒受到很多人的追捧，但是金酒的最主要用途还是作鸡尾酒的基酒，它是调酒的优质原料，芳香柔软，可以调出众多款美酒。

Gin is a common base spirit for mixed drinks, mixed with various juices, syrups, carbonated beverages, or other spirits, it can create a variety of flavors, such as gin and tonic, martini, negroni, long island iced tea, etc. Although enjoying pure or on the rocks gin is popular with many people, but the main use of gin is still as the base spirit for cocktails. It is a high-quality ingredient for mixing drinks, with a fragrant and soft aroma, capable of mixing a variety of beautiful drinks.

五、常见的金酒品牌（Common Brands of Gin）

（一）必富达（Beefeater）

必富达来自英国伦敦，是世界各地金酒爱好者的首选品牌，其在伦敦生产、灌装，具有地道的伦敦金酒风格。必富达创始于 19 世纪，传统的金酒酿造配方由一代又一代的酿酒大师传承，其经典酿造风味沿用至今。必富达伦敦干金酒已成为调酒师的必备酒种，也是世界上获奖最多的金酒。

Beefeater originates from London and is the preferred brand for gin enthusiasts worldwide. It is produced and bottled in London, embodying the authentic London dry gin style. Established in the

19 th century, the traditional gin recipe has been passed down from one generation of master distillers to another, with its classic flavor still in use today. Beefeater London Dry Gin has become a must-have for bartenders and is the most awarded gin in the world.

（二）老汤姆金（Old Tom Gin）

老汤姆金是非常经典的金酒配方，在 18 世纪的英国广受欢迎。最初的汤姆柯林斯是用英国老汤姆金酒制作的。当时正是杜松子酒风头无两的热潮期，随着其在民间广为流传，"老汤姆"这个名字便有了很多起源。这个名字已经成了一种传奇性的称谓，但被人们所认同最多的，应该是"老汤姆猫"的标志被用来传达金酒的甜美风格这一说法。

Old Tom Gin is a very classic gin recipe that was popular in 18 th century England. The original Tom Collins was made with British Old Tom Gin. At that time, it was the heyday of the gin craze, and as it spread widely among the people, the name Old Tom became associated with many origins. This name has become a legendary title, but the most widely accepted is probably the " Old Tom Cat " logo, which is used to convey the sweet style of the gin.

（三）猴王 47（Monkey 47）

尽人皆知的"猴王"来自德国黑森林，这里有丰富的蒸馏技术和数百年的经验。采用最高质量的新鲜原料，和取用来自深砂岩泉的黑森林水，德国黑森林具备了诞生最优质金酒的先决条件。猴王 47 采用 47 种不同的植物成分，这些异国风味的香料，包含六种不同的有机胡椒，浸泡在法国甘蔗糖蜜制成的烈酒中 36 个小时。这是一款大师级的金酒，带有丰富的草本、花香、柑橘调、辛香以及微妙的蔓越莓气息，柔顺的口感，即便酒精浓度高达 47%，也丝毫不过于刺激。

The well-known " Monkey King " comes from the Black Forest in Germany, where there is a wealth of distillation technology and centuries of experience. Using the highest quality fresh ingredients and water from deep sandstone springs in the Black Forest, Germany's Black Forest has the prerequisites for the birth of the finest gin. Monkey 47 uses 47 different botanical ingredients, including six different types of organic pepper, soaked in a spirit made from French cane sugar molasses for 36 hours. This is a master-class gin with a rich herbal, floral, citrus, spicy, and subtle cranberry aroma, a smooth mouthfeel, and even with an alcohol concentration of up to 47%, it is not overly stimulating.

赛证直通 Competitions and Certificates

一、基础知识部分（Basic Knowledge Part）

（一）专业名词解释（Explanation of Professional Terms）

1. 干金酒（Dry Gin）：

2. 加甜金酒（Sweet Gin）：

二维码 1-3-3：
知识拓展

（二）思考题（Thinking Question）

金酒从口味上可以分为哪几类酒？分别有什么特点？

What are the types of gin based on taste? What are their characteristics?

二、技能操作部分（Skill Operation Part）

品鉴一款金酒，描述其颜色、香气、滋味等特征。

To taste a gin, describe its color, aroma, flavor, and other characteristics.

【微课】
"生命之水"：
伏特加

专题四
"生命之水"：伏特加

你在影视剧作品中听过伏特加（图 1-3-4）吗？哪些国家的人最喜欢伏特加？

_____

Have you heard about vodka (Figure 1-3-4) in film and television works? Which countries' people like vodka the most?

_____

图 1-3-4　伏特加（Vodka）

一、伏特加的历史起源（The History and Origin of Vodka）

（一）伏特加的起源（The Origins of Vodka）

关于伏特加的确切起源有些争议，但普遍认为它最早是在 14 世纪末或 15 世纪初在波兰或俄罗斯出现的。波兰和俄罗斯都声称自己是伏特加的发源地，而且两者都有足够的历史文献和考古证据来支持自己的主张。在波兰，最早的书面记录提到了一种名为"gorzałka"的酒精饮料，这个词在波兰语中意为"燃烧"，这被认为是伏特加的前身。而俄罗斯的文献则提到了一种名为"водка"的饮料，这个词在俄语中有"水"的意思，因为伏特加的酒精度很高，在制作过程中经过多次蒸馏，以达到纯净无异味的效果，所以被称为"生命之水"。

There is some controversy about the exact origin of vodka, but it is widely believed to have first appeared at the end of the 14th century or the beginning of the 15th century in Poland or Russia. Both Poland and Russia claim to be the birthplace of vodka, and both have sufficient historical and archaeological evidence to support their claims. In Poland, the earliest written records mention an alcoholic beverage named "gorzałka," which means "burning" in Polish and is considered the predecessor of vodka. Russian documents, on the other hand, mention a beverage called "водка", which means "water" in Russian. Because vodka has a high alcohol content and undergoes multiple distillations during the production process to achieve a pure and odorless effect, it is also known as "water of life."

（二）伏特加的发展历程（The Development of Vodka）

15 至 16 世纪，伏特加在俄罗斯和波兰等地开始流行，并逐渐成为俄罗斯的国酒。17 至 18 世纪，随着俄罗斯帝国和波兰—立陶宛联邦的扩张，伏特加开始向周边国家传播。同时，贸易和外交往来也促进了伏特加在欧洲的传播。19 世纪，随着工业化和现代化进程的推进，伏特加的生产变得更加规模化和标准化，这也为其更广泛的传播奠定了基础。在 20 世纪的两次世界大战中，伏特加作为俄国士兵的补给品之一，随着军队的移动被带到了更多的国家和地区。20 世纪后半叶，随着全球化和国际贸易的快速发展，伏特加开始进入更多国家的市场，成为国际知名的烈酒品牌。21 世纪，伏特加已经成为全球烈酒市场的重要组成部分，各种品牌和口味的伏特加在全球范围内销售，其地位与威士忌、朗姆酒、白兰地等其他烈酒相提并论。

Between the 15th and 16th centuries, vodka began to gain popularity in Russia and Poland, gradually becoming the national drink of Russia. During the 17th and 18th centuries, as the Russian Empire and the Polish–Lithuanian Commonwealth expanded, vodka started to spread to neighboring countries. At the same time, trade and diplomatic exchanges also facilitated the dissemination of vodka across Europe. In the 19th century, with the advancement of industrialization and modernization, vodka production became more scaled and standardized, laying the foundation for its broader spread. In the two World Wars of the 20th century, vodka, as one of the supplies for Russian soldiers, was brought to more countries and regions along with the movement of the army. In the latter half of the 20th century, with the rapid development of globalization and international trade, vodka began to enter the markets of more countries, becoming an internationally renowned brand of spirits. In the 21st century, vodka has become an important part of the global spirits market, with various brands and flavors being sold worldwide, and its status is on par with other spirits such as whiskey, rum, and brandy.

伏特加在传播过程中，与各国的文化和饮酒习惯相融合，出现了许多具有地方特色的伏特加品牌和饮用方式。伏特加的全球传播不仅促进了国际贸易和文化交流，也丰富了全球烈酒市场的多样性。如今，伏特加已经成为全球酒精饮料文化的一个重要组成部分。

In the process of spreading, vodka has also integrated with the cultures and drinking habits of various countries, resulting in many vodka brands and drinking methods with local characteristics. The global spread of vodka has not only promoted international trade and cultural exchanges but also enriched the diversity of the global liquor market. Nowadays, vodka has become an important part of the global alcoholic beverage culture.

二、伏特加的制作工艺（The Production Process of Vodka）

根据 GB/T 11858—2008《伏特加（俄得克）》规定，伏特加是以谷物、薯类、糖蜜及其他可食用农作物等为原料，经发酵、蒸馏制成食用酒精，再经过特殊工艺精制加工制成的蒸馏酒。国家标准中要求甲醇含量 ≤ 50 mg/L。以原味伏特加为基酒，加入的食品用香料味道可制成风味伏特加酒。

According to GB/T 11858−2008 Vodka (Russian style), vodka is made from raw materials such as grains, tubers, molasses, and other edible crops. It is produced through fermentation, distillation to create edible alcohol, and then refined and processed through special techniques to create a distilled spirit. The national standard requires that the methanol content should be ≤ 50 mg/L. Based on the plain vodka, flavored vodka can be produced by adding food−grade flavoring essences.

伏特加的制作工艺虽然简单，但对原料的选择和蒸馏工艺有很高的要求。伏特加可以使用多种原料制作，包括谷物（如小麦、黑麦、大麦）、土豆以及各种水果。在选择原料时，需要确保原料的质量和纯度。伏特加的蒸馏至少进行两次，但有些高品质的伏特加会进行三到四次蒸馏。在多次蒸馏的过程中，会分阶段收集酒精蒸气，这些蒸气冷却后变成液体，被称为"酒心"。酒心既没有酒头的刺鼻气味，也没有酒尾的杂味，口感清爽、纯净，杂质含量极低，酒精浓度适中，是制作高品质伏特加的关键。

Although the production process of vodka is relatively simple, it has high requirements for the selection of raw materials and distillation techniques. Vodka can be made from a variety of raw materials, including grains (such as wheat, rye, and barley), potatoes, and various fruits. When selecting raw materials, it is necessary to ensure the quality and purity of the ingredients. Vodka distillation is typically carried out at least twice, but some high−quality vodkas undergo three to four distillations. During the multiple distillation process, alcohol vapor is collected in stages. After cooling, these vapors turn into liquids, known as " central liquor of the spirit." The central liquor of the spirit, neither carries the pungent aroma of the initial distillate nor the miscellaneous flavors of the tail end. It boasts a refreshing and pure taste with an extremely low level of impurities and a moderate alcohol concentration, making it the key to producing high−quality vodka.

伏特加的制作工艺虽然大致相同，但不同品牌和国家可能会有自己的独特工艺和配方。不管哪一种伏特加，它的酒液一定呈现清澈如水、丝滑圆润的观感，这是伏特加的基本特征。

Although the production process of vodka is roughly the same, different brands and countries may have their own unique processes and recipes. However, no matter which type of vodka, its liquid must present a clear appearance like water and a smooth and mellow texture, which is the basic feature of vodka.

三、伏特加的分类（The Classification of Vodka）

伏特加的产区，毋庸置疑是以俄罗斯和波兰为典型代表，除此之外的瑞典、法国以及美国也有大量的伏特加生产，市面上常见的伏特加大部分来自这几个产区。每种伏特加都有其独特的风味和制作工艺，消费者可以根据个人喜好选择不同的品牌和风格。

Undoubtedly, Russia and Poland are the typical representatives of vodka−producing regions. In addition, countries like Sweden, France, and the United States also have a large production of vodka, and most of the vodka commonly seen on the market comes from these regions. Each type of vodka has its unique flavor and production process, allowing consumers to choose different brands and styles based on their preferences.

（一）俄罗斯伏特加（Russian Vodka）

原料：通常使用小麦、黑麦或玉米等谷物。

Ingredients: grains such as wheat, rye, or corn are usually used.

工艺：俄罗斯伏特加的生产工艺强调纯净，会进行多次蒸馏以确保高酒精度和口感纯净。

Process: the production process of Russian vodka emphasizes purity, and the distillation process may be carried out multiple times to ensure a high alcohol content and a pure taste.

口感：典型的俄罗斯伏特加口感较为强烈，有明显的酒精感，但同时也追求纯净和顺滑。

Taste: typical Russian vodka has a strong taste, with a clear feeling of alcohol, but it also pursues purity and smoothness.

代表品牌：百加得、绿标等。

Representative Brands: Bacardi, Green Label, etc.

（二）波兰伏特加（Polish Vodka）

原料：波兰伏特加通常使用谷物（如小麦、黑麦）或马铃薯。

Ingredients: Polish vodka usually uses grains such as wheat, rye, or potatoes.

工艺：波兰伏特加的制作工艺精细，特别是其独特的蒸馏和过滤技术，如使用木炭过滤以提高酒液的纯净度。

Process: the production process of Polish vodka is exquisite, especially its unique distillation and filtration techniques, such as using charcoal filtration to improve the purity of the liquor.

口感：波兰伏特加口感通常较为柔和、顺滑，具有较好的平衡感和细腻度。

Taste: Polish vodka usually has a soft and smooth taste, with a good sense of balance and delicacy.

代表品牌：雪树、维波罗、瓦斯米诺等。

Representative Brands: Belvedere, Wyborowa, Smirnoff etc.

（三）瑞典伏特加（Swedish Vodka）

原料：瑞典伏特加可能使用谷物或马铃薯，有时还会使用当地的特色原料。

Ingredients: Swedish vodka may use grains or potatoes, sometimes with local specialty ingredients.

工艺：瑞典伏特加的制作工艺注重创新，会采用特殊的蒸馏技术或过滤方法。

Process: the production process of Swedish vodka focuses on innovation, and it may adopt special distillation techniques or filtration methods.

口感：瑞典伏特加以其清新、圆润的口感而闻名，有时还会带有淡淡的果香或花香。

Taste: Swedish vodka is known for its fresh and mellow taste, sometimes with a light fruit or floral fragrance.

代表品牌：绝对等。

Representative Brands: Absolute, etc.

四、伏特加饮用方式（The Drinking Methods of Vodka）

早在 20 世纪 50—60 年代，伏特加就已经占据了酒柜的中心位置，而在 20 世纪 80 年代，更是成了调酒师心中"完美的调酒基酒"。

As early as the 1950s to 1960s, vodka had already occupied the central position in the liquor cabinet. In the 1980 s, it became the "perfect cocktail base" in the hearts of bartenders.

（一）纯饮（Neat）

虽然六大基酒都属于高度烈酒，但伏特加因为其绝对的酒精纯度，纯饮会感觉尤其猛烈。入口有强烈的辛辣感在口腔炸开，吞下的时候辛辣感顺着食道往下直到胃部，能感到胃部微微发热。等这股辛辣感过去后，才能感受到往回返上来的酒精香气。

Although the six major base liquors are all high-alcohol spirits, vodka, due to its absolute alcohol purity, tastes particularly strong when drunk neat. Upon entry, there will be a strong pungent sensation that explodes in the mouth. When you swallow, the pungent sensation will follow the esophagus down to the stomach, and you can feel the stomach warming up slightly. After this pungent sensation passes, you can feel the alcohol aroma coming back up.

（二）调制（Mixed）

伏特加是一种非常流行的基酒。一方面，它在生产过程中经过多次蒸馏和过滤，以达到非常纯净的口感，在调制鸡尾酒时，可以很好地与其他成分的味道融合，而不会盖过其他风味。另一方面，伏特加的口感较为中性、平滑，酒精度相对较高，它可以与多种风味的配料相搭配，并在调制的鸡尾酒中保持一定的力度，确保饮品的整体平衡。同时作为一种国际性烈酒，在全球有众多品牌和风格，这为调酒师和消费者在选择基酒时提供了丰富的选项。

Vodka is a very popular base liquor. On the one hand, it is usually distilled and filtered multiple times during production to achieve a very pure taste. When mixed, it can blend well with the flavors of other ingredients without overpowering other flavors. On the other hand, vodka has a neutral and smooth taste with a relatively high alcohol content. It can match with a variety of flavored ingredients and maintain a certain strength in the mixture, ensuring the overall balance of the drink. As a global spirit, there are many brands and styles worldwide, providing rich options for bartenders and consumers when choosing a base liquor.

赛证直通 Competitions and Certificates

一、基础知识部分（Basic Knowledge Part）

二维码 1-3-4：
知识拓展

（一）专业名词解释（Explanation of Professional Terms）

酒心（Central Liquor of the Spirit）：

（二）思考题（Thinking Question）

按照生产国家不同，伏特加主要分为哪些类别？

According to the producing countries, what are the main categories of vodka?

二、技能操作部分（Skill Operation Part）

品鉴一款伏特加，描述其颜色、香气、滋味等特征。

To taste a vodka, describe its color, aroma, flavor, and other characteristics.

专题五
"加勒比传奇"：朗姆酒

【微课】
"加勒比传奇"：
朗姆酒

一、朗姆酒的起源（The origin of rum）

15 世纪，葡萄牙和西班牙航海家们开始了大西洋航海冒险，哥伦布第二次航行美洲时来到古巴。他从加纳利群岛带来了制糖甘蔗的根茎，并且这些根茎代替了人们在土著人称作 Cipango 的这个岛上想要寻找的金子。

In 15th century, the Portuguese and Spanish navigators began their Atlantic adventure, and Columbus came to Cuba on his second voyage to the Americas. He brought the rhizome of sugar cane from Canary Islands. It was unexpected that these rhizomes replaced the gold that people were looking for on the island called Cipango by the natives.

而古巴的气候条件、肥沃的土壤，水质和阳光正好适合作为制糖原材料的甘蔗的生长，由此，古巴制糖工业也得以飞速发展。

Cuba's climate conditions, fertile soil, water quality and sunlight are just right for the growth of sugar cane as a raw material for sugar production, and the Cuban sugar industry has also been able to develop rapidly.

甘蔗榨糖后会产生两种叫作糖蜜（molasses）和糖稀的副产品，这两种副产品中都含有大量未结晶的糖分和其他杂质，其黏度很高、味道刺鼻、苦味较重，同时，也不能再继续进行加工，庄园主们觉得这些副产品没有任何价值，于是便将其都丢给奴隶们处理。

Two by-products, called molasses and dilute sugar, are produced by the milling of sugar cane. Both of these by-products contain large amounts of uncrystallized sugars and other impurities that are viscous, punchy and bitter. They can no longer be processed. So they left it all to the slaves.

生活资源匮乏的奴隶们通过不断实践用糖蜜制作出一种含有轻微酒精的饮料，这种饮料最初味道古怪，异常难喝，但后来，奴隶们学会使用一种极其简陋的装置，把这种自酿饮料进行蒸馏。蒸馏后的饮料烈度大大加强，更富刺激性，有使人醉酒的效果，喝后使人兴奋并消除疲劳。这种呛辣烈酒当时被称为"塔菲亚"，也就是朗姆酒的雏形。

The slaves produced a slightly alcoholic drink from molasses. The drink was strange in taste at first and it was unpleasant to drink, but the slaves clumsily distilled the self-made drinks with the extremely simple device. The drink intensity after distillation was largely enhanced. It became more irritant and could make people drunk. The slaves felt excited after drinking and the fatigue may be alleviated. Such spicy spirit was called "Tafia", which was the embryonic form of rum.

由于品质粗糙、价格低廉，朗姆酒最初的酒客大多是古巴的奴隶和底层人民，尤其与水手和海军"如影随形"。造成这种现象的原因有很多，其中一个重要原因是在欧洲大航海时代，为了长时间携带宝贵的淡水，海员们会将淡水放在大木桶中储存，有条件的还会把淡水烧开后再装桶，然后在水桶中置放一银板杀菌（银离子可以杀菌），但即使是这样，淡水在船上往往坚持不到三个星期就会长满绿苔、滋生水虫，饮用这种发臭的淡水，还很容易胃疼、胃泻或是痢疾，饮用这种"淡水"对于水手们来说是一种艰难的考验，必须忍受极大的生理不适才能喝下这种散发着下水道恶臭的水。

Initially, the rum was popular among the slaves and people at the bottom of Cuba, especially the sailors, pirates and navies. There are many reasons for this phenomenon, mainly in the era of European navigation, influenced by the lagging scientific and technological means. The sailors may store the fresh water in the wooden cask, or boil the fresh water and then pack it into barrels if conditions permit. Then, they would place the silver plate in it for sterilization (the silver ion could be used for sterilization). But even so, the green algae and water bugs may appear in fresh water within three weeks.

为了确保自己可以喝下这种带有恶臭的淡水，海员们不得不进行加工处理，一般的做法都是往淡水中添加大量酒液，以延长淡水保质期，而朗姆酒因为酒精含量高，香味浓烈，价格低廉，容易获取，受到海员们的一致追捧。当水手们获得朗姆酒后，除了将少部分朗姆酒直接饮用外，剩余的大部分都会兑入到饮水中混合饮用，因此，朗姆酒被称为"海员之酒"和"传奇之酒"。

To ensure that they could drink the foul fresh water, the sailors had to process water. In general, they would add much wine to extend the guarantee period. Due to the high alcohol content, strong aroma, low price and easy access, rum is unanimously sought after by sailors. When the sailors got rum, they would drink a small part and mix the remaining part in the fresh water, which was why the rum is called "the sailor's wine" or "the wine of legends".

二、朗姆酒的发展（The development of rum）

当意识到朗姆酒的作用和商业价值后，种植园主们开始主动酿造和销售朗姆酒，由于价格低廉，朗姆酒受到了很多人的热烈追捧，并逐渐向海外拓展。而随着酿造技术的提高，尤其是工业设备的陆续引入，生产者们开始对糖蜜进行多次过滤和蒸馏，朗姆酒也不再是劣酒的代名词，开始逐渐被更多人接受。为了满足上层人士对质量的要求，人们也不再只用蔗糖的副产品酿造朗姆酒，一些酿酒师开始精选优质甘蔗压榨酿酒，采用精密的铜质蒸馏器提取酒精，然后把酒放在木桶里储存。一段时间后，木桶里竟然飘出醉人的香气，口感细致，香味馥郁，就这样，今天大家所熟悉的朗姆酒诞生了。

After being aware the functions and commercial values of rum, plantation owners started to proactively brew and sell the rum. The low price attracted more people and the rum was gradually exported overseas. With improvement of brewing techniques in the colony, especially the introduction of the industrial equipment, the manufacturers conduct the multiple filtering and distillation for the molasses. The rum was no longer synonymous with inferior wine and it would be accepted by more people. To satisfy the quality requirements of the people from the upper class, people no longer brew the rum with the by-products of saccharose and some winemakers chose the high-quality sugarcane and press it for brewing. They adopted the precision copper distiller to extract the alcohol and stored the win in the wooden cask. After a period of time, the intoxicating aroma arose with the fine taste and fragrant scent. Hence, the rum that everyone is familiar with today was born.

三、朗姆酒的生产工艺（Production process of rum）

朗姆酒是一种以甘蔗汁或糖蜜为原料，通过发酵和蒸馏而制成的烈性酒，其生产过程主要包括：

Rum is a kind of cane juice or molasses as raw materials, through fermentation and distillation made of spirits, its production process mainly includes:

（一）原料准备（Raw material preparation）

朗姆酒的主要原料是甘蔗汁或糖蜜。首先，将甘蔗剥皮并压榨，得到甘蔗汁，然后，将甘蔗汁进行过滤和澄清，去除杂质和固体颗粒，接着，将甘蔗汁进行浓缩，使其含糖量满足酿造要求。如果使用糖蜜作为原料，则可以直接使用。

The main ingredient of rum is sugarcane juice or molasses. First, the sugar cane is peeled and pressed to obtain cane juice. The juice is then filtered and clarified to remove impurities and solid particles. Next, the sugarcane juice is concentrated so that its sugar content reaches a certain standard. If molasses is used as a raw material, it can be used directly.

（二）发酵（Fermentation）

将处理好的甘蔗汁或糖蜜倒入发酵罐中，加入酵母和其他辅助发酵剂。酵母会将糖分解成酒精和二氧化碳，这个过程称为发酵。发酵过程一般需要几天到几周的时间，具体时间取决于所使用的酵母菌株和发酵条件。

Pour the treated cane juice or molasses into the fermenter and add the yeast and other auxiliary starter. Yeast breaks down sugar into alcohol and carbon dioxide in a process called fermentation. The fermentation process generally takes a few days to a few weeks, depending on the strain of yeast used and the fermentation conditions.

（三）蒸馏（Distillation）

发酵完成后，得到的液体称为"发酵液"或"酒糟"。然后，需要对"发酵液"或"酒糟"进行蒸馏，以提取酒精。蒸馏过程分为两次，分别称为"粗蒸馏"和"精细蒸馏"。

粗蒸馏：将发酵液倒入蒸馏锅中，加热使其沸腾。由于酒精的沸点较低，它会先蒸发出来，然后通过冷凝器冷却成液体，这个液体称为"粗酒"。粗酒中含有酒精、水和其他杂质。

精细蒸馏：将粗酒倒入蒸馏锅中再次加热，这次只提取酒精。由于酒精的沸点较低，它会先蒸发出来，通过冷凝器冷却成液体，这个液体就是"朗姆酒"，其酒精含量通常在40%至60%之间。

After fermentation is complete, the resulting liquid is called "fermentation liquor" or "lees". At this point, a distillation process is required to extract the alcohol. The distillation process is divided into two stages, called "crude distillation" and "fine distillation".

Crude distillation: Pour the fermentation liquid into a still pan and heat it until it boils. Because the alcohol has a lower boiling point, it will first evaporate out and then cool through the condenser into a liquid, which is called "coarse wine". Coarse wine contains alcohol, water and other impurities.

Fine distillation: Pour the coarse wine into the still and heat it again, this time extracting only the alcohol. Since alcohol has a lower boiling point, it evaporates first and is cooled by a condenser into a liquid, which is called "rum". Rum is usually between 40% and 60% alcohol by volume.

（四）陈酿（Aging）

蒸馏后的朗姆酒需要进行陈酿，以提高口感和香气。陈酿过程中，朗姆酒会在橡木桶中存放一段时间，让酒精和橡木桶中的单宁发生化学反应，使朗姆酒的口感更加醇厚，香气更加浓郁。陈酿时间可以从几个月到几十年不等，但通常在 1~3 年之间，具体取决于朗姆酒的品质要求。

The distilled rum needs to be aged to improve the taste and aroma. During the aging process, rum will be stored in oak barrels for a period of time, so that the alcohol and the tannin in the oak barrel chemical reaction, making the rum taste more mellow, more intense aroma. Aging time can range from a few months to decades, but is usually between 1~3 years, depending on the quality requirements of the rum.

（五）调配（Deployment）

为了提高朗姆酒的品质和口感，酿酒师会对不同年份、不同品种的朗姆酒进行调配。调配是一种将多种酒混合在一起，以达到理想的口感和香气的方法，调配后的朗姆酒品质会更加稳定，口感更加丰富。

In order to improve the quality and taste of rum, winemakers will mix different years and different varieties of rum. Blending is a method of mixing a variety of wines together to achieve the desired taste and aroma. The resulting rum is more stable and has a richer taste.

（六）瓶装和包装（Bottling and packaging）

经过陈酿和调配后，朗姆酒就可进行瓶装和包装，以便后期销售和运输。

After aging and blending, rum can be bottled and packaged. First, pour the rum into the bottle and close the cap to ensure a tight seal. The bottles are then labeled and packaged for sale and transportation.

四、主要产地（Main places of origin）

朗姆酒最初起源于古巴，现今主要的产地有西半球的西印度群岛，以及美国、墨西哥、牙买加、海地、多米尼加、特立尼达和多巴哥、圭亚那、巴西等国家。另外，非洲岛国马达加斯加也出产朗姆酒。

Rum originates from Cuba and the current places of origin include the West Indies in the western hemisphere, USA, Mexico, Jamaica, Haiti, Dominica, Trinidad and Tobago, Guyana, Brazil and other countries. In addition, Madagascar, an island country in Africa, also produces Rum.

五、认知朗姆酒（Know about Rum）

（一）颜色（Color）

按颜色可分为白朗姆酒、金朗姆和黑朗姆（图1-3-5）。

According to the color, it could be divided into white rum, golden rum and dark rum (Figure 1-3-5).

图1-3-5　不同颜色的朗姆酒
Different colors of rum

（二）口感（Taste）

具有细致、甜润的口感，芬芳馥郁的酒精香味。

Rum has a delicate, sweet taste and fragrant alcohol aroma.

（三）酒感（Sense of wine）

1. 酒体轻盈、酒味极干的朗姆酒（Light-bodied and extremely dry rum）

这类朗姆酒主要由西印度群岛属西班牙语系的国家生产，如古巴、波多黎各、维尔京群岛（VIRGIN ISLANGS）、多米尼加、墨西哥、委内瑞拉等，其中以古巴朗姆酒最负盛名。

Such rum is mainly produced by Spanish-speaking countries in West Indies, such as Cuba, Puerto Rico, Virgin Islands, Dominica, Mexico and Venezuela. Among them, the Cuba rum is the most popular.

2. 酒体丰厚、酒味浓烈的朗姆酒（Full-bodied and strong rum）

这类朗姆酒多为古巴、牙买加和马提尼克的产品。酒在木桶中陈年的时间长达5～7年，甚至15年，有的要在酒液中加焦糖调色剂（如古巴朗姆酒），因此其色泽金黄、深红。

Such rum is mostly the product of Cuba, Jamaica and Martinique. Wine is aged in wooden barrels for 5-7 years, or even 15 years. Sometimes, the caramel toner shall be added to the wine (such as Cuban rum), so its color is golden yellow and crimson.

3. 酒体轻盈、酒味芳香的朗姆酒（Light-bodied and fragrant rum）

这类朗姆酒主要是古巴、爪哇群岛的产品，其酒香气味是由芳香类药材所致。

Such rum is mainly the products of Cuba and Java Islands, its long-lasting aroma is caused by aromatic medicinal materials.

（四）风格（Style）

朗姆酒分为清淡型和浓烈型两种风格。

Rum is divided into the light and strong style.

清淡型朗姆酒是用甘蔗糖蜜、甘蔗汁加酵母进行发酵后蒸馏，在橡木桶中陈酿多年，再勾兑配制而成。酒液呈浅黄到金黄色，酒精度为 45% ～ 50%vol。清淡型朗姆酒主要产自波多黎各和古巴。

The light rum is made from sugarcane molasses, sugarcane juice and yeast after fermentation, distillation, aging in oak barrels for many years and then blending. The wine is light-yellow to golden yellow in color and the alcohol content is 45% to 50%vol. The light rums are mainly produced in Puerto Rico and Cuba.

浓烈型朗姆酒是由掺入榨糖残渣的糖蜜在天然酵母菌的作用下缓慢发酵制成的。酿成的酒经过两次蒸馏，为无色的透明液体，然后在橡木桶中熟化 5 年以上。浓烈朗姆酒呈金黄色，酒香和糖蜜香浓郁，味辛而醇厚，酒精含量 45% ～ 50% vol。浓烈型朗姆酒以牙买加所产的为代表。

The strong rum is manufactured by slow fermentation of molasses mixed with the residue of sugar extraction under the action of natural yeast. After two distillations, it becomes a colorless transparent liquid, and then it is aged in oak barrels for more than 5 years. The strong rum is golden-yellow in color, rich in wine and molasses, pungent and mellow in taste, with the alcohol content of 45%–50% vol. Strong rum is represented by Jamaica.

（五）风味（Flavor）

根据风味特征，可将朗姆酒分为浓香型和清香型。

According to the flavor features, the rum is divided as strong flavor and mild flavor.

浓香型：首先将甘蔗糖澄清，再加入能产丁酸的细菌和产酒精的酵母菌，发酵 10 天以上，用壶式锅间歇蒸馏，得 86% 左右的无色原朗姆酒，在木桶中贮存多年后勾兑成金黄色或淡棕色的成品酒。

Strong flavor: Firstly, the saccharose is clarified. Then, the butyric acid producing bacteria and alcohol producing yeast are added, and it is fermented for more than 10 days. Later, the batch distillation is carried out by the kettle type pot to obtain about 86% colorless original rum. After aging in the wooden cask for many years, the finished wine blended into a golden or light brown color.

清香型：甘蔗糖只加酵母，发酵期短，塔式连续蒸馏，产出 95% 度的原酒，贮存勾兑成浅黄色到金黄色的成品酒，以古巴朗姆为代表。

Mild flavor: The yeast is only added to sucrose with the short fermentation period. With the tower-type continuous distillation, the 95% raw wine is obtained, which is stored and blended into the light yellow and golden yellow finished wine, represented by Cuban rum.

六、饮用方法（Drinking Method）

朗姆酒名列世界八大烈酒之一，其口感非常强劲，它的喝法也很有讲究。现下比较流行的喝法主要有：

Rum is one of the top eight spirits in the world. the rum's taste is strong and its drinking method is also very particular. Here are some of the popular methods to drink:

加冰：酒精度较高（大概 70 度以上）的朗姆酒加冰饮用最佳。冰块能压制一些酒精，让口感更容易被大多数人接受。

Add ice: Rum with high alcohol content (above 70 degrees) is best to drink with ice. Ice can suppress the alcohol and make the taste more acceptable to most people.

加青柠汁：将青柠汁、百加得白朗姆酒、白糖放入容器中搅拌至糖融化。然后，加入一半冰块一半碎冰，摇匀并过滤到冰镇过的酒杯中，口感醇厚又清冽。

Add lime juice: There are many methods to add the lime juice, and here is a bartending method I saw on the Internet: Add lime juice and Bacardi to the white rum and put the white sugar to mix for melting. Then, add half the ice and half the crushed ice, shake up and filter to the iced cup.

加可乐：把可乐倒入放冰块的朗姆酒中，可乐本身就是一种能让人快乐的饮料，两者混合，增添了汽水感，口感更丰富。

Add cola: Pour the cola into the rum with ice. The cola itself is a happy drink. When the two are mixed, the soda feels stronger and tastes better. Some people prefer to add the orange juice or lemon slices.

加马天尼：将20毫升单糖浆、15毫升青柠汁、2滴安高天娜苦酒等配料与朗姆酒放入容器中摇匀，然后用网倒入冰过的酒杯，最后再倒满杯马天尼，口感浓烈、香味独特。

Add martini: 20 mL simple syrup, 15 mL lime juice, 2 drops of Angostura bitters and mint leaves. Shake up all ingredients and rum in a container. Then, pour it to the ice cup via filter screen, and finally fill the martini.

加苏打水（莫吉托 Mojito）：杯中加入冰块，加入大约60毫升苏打水，5-10毫升柠檬汁，30毫升白朗姆酒，所有配料一起搅拌、再加6～8片新鲜薄荷叶，最后放入一两片青柠檬，即可饮用，口感非常清爽。

Add soda water (Mojito): Add ice and 60ml soda water, 5- 10 mL lemon juice, 30 mL white rum, 6–8 pieces of fresh mint leaves. Then, add one or two pcs of lemon or white sugar. Mix all ingredients together and shake well

赛证直通 Competitions and Certificates

二维码 1-3-5
知识拓展

一、基础知识部分（Basic knowledge part）

（一）简答题（Short answer questions）

1. 简述朗姆酒的主要产区。
Briefly describe the main regions of Rum.
2. 简述朗姆酒的分类。
Briefly describe the classification of Rum.

（二）思考题（Thinking Question）

谈谈你对朗姆酒发展历史的看法。
Talk about your thoughts on the history of rum.

二、技能操作部分（Skill operation part）

请品鉴一款浓郁型朗姆酒，并对其进行详细描述。

Taste a rich rum and describe it in detai.

专题六
"墨西哥灵魂"：龙舌兰酒

【微课】
"墨西哥灵魂"：
龙舌兰酒

一、龙舌兰酒的历史起源（Historical origin）

西元三世纪，居住于中美洲地区的印第安人已经发现了发酵酿酒的技术，他们取用生活里面任何可以得到的糖分来源来造酒，除了主要作物玉米与当地常见的棕榈汁之外，含糖量不低且多汁的龙舌兰也成为他们造酒的原料。

As early as the third century AD, the Indian civilization living in Central America had already discovered the technology of brewing wine by fermentation. They used any available source of sugar in their lives to make the wine. In addition to the main crop corn and the local common palm juice, the agave which is rich in sugar and juicy has naturally become their raw material for making wine.

以龙舌兰汁发酵后制造出来的被称为——Pulque 的纯发酵龙舌兰酒，经常被用来作为宗教信仰用途，除了饮用，还可以帮助祭司们与神明的沟通（这其实是醉酒后产生幻觉现象），他们在活人祭献之前会先让牺牲者饮用，使其失去意识或减弱反抗能力，方便开展仪式。

Pulque wine is made by the fermentation of agave juice. It is often used for religious purposes. It can help priests communicate with gods (Actually, it's the drunkenness or hallucinations after drinking). Besides that, they will ask the victims to drink Pulque to make them unconscious or at least reduce their resistance before the human sacrifice, so as to facilitate the ceremony.

在西班牙的征服者们将蒸馏术带来新大陆之前，龙舌兰酒一直保持着纯发酵酒的身份。后来，由于西班牙人从家乡带来的酒水消耗量大却补给困难，他们想在当地寻找新的酒水来取代。于是，他们看上了有着奇特植物香味的 Pulque，但又嫌这种发酵酒的酒精度远比葡萄酒低，因此尝试使用蒸馏的方式提升 Pulque 的酒精度，于是，以龙舌兰制造的蒸馏酒得以产生。

Tequila remained a pure fermented wine until the conquistador from Spain who came across the Atlantic brought distillation to the new lands. Because the Spaniards wanted to find a suitable raw material to replace their wine or other European spirits that they had brought from home at great expense and were not enough to meet their huge consumption. So they took a fancy to the Pulque with a peculiar botanical fragrance. However, they thought the alcohol content of this fermented wine was much lower than that of wine. Therefore, they tried to use the method of distillation to improve the alcohol purity of the Pulque, and a distilled liquor made from agave was produced.

由于这种新产品的用途是用来取代葡萄酒，于是获得了 Mezcal wine 的名称。Mezcal wine 经过长久的尝试与改良，逐渐演变成为今天被称为墨西哥国酒的 Mezcal/Tequila。

It got the name of Mezcal wine because this new product was intended to replace the use of wine. The prototype of Mezcal wine had been tried and improved for a long time before it gradually evolved into the national wine Mezcal/Tequila of Mexico as we can see today.

二、制作工艺 Production process

龙舌兰酒由以成熟的龙舌兰为原料酿而成，此种植物至少需要 6 年的时间才能成熟，那些种植在高地的龙舌兰酒往往需要更长的时间才能成熟，最长可达 12 年。生产龙舌兰酒需要投入大量的时间和精力。通常，龙舌兰酒的制作需要历经收割——烹煮——发酵——蒸馏——陈年等几个阶段，且每个阶段都需要耗费大量的人力物力，这也是龙舌兰酒价格不菲的原因。

Tequila is made from mature agave plants, which take at least six years to mature. Those grown in the highlands tend to take longer to mature, up to 12 years. It takes a lot of time and effort to produce tequila. Typically, tequila needs to go through several stages—harvesting, cooking, fermenting, distilling and aging—and each stage requires a lot of labor and resources, which is why tequila is so expensive.

三、酒品分类 The wine classification

（一）普利克（Pulque）

以龙舌兰草心为原料，经发酵而成的酒类，也是所有龙舌兰酒的基础原型。
A fermented wine made from the heart of agave, it is also the basis of all tequila.

（二）特其拉（Tequila）

龙舌兰酒中的顶级，只有在特定地区、使用一种称为蓝色龙舌兰草（Blue Agave）的植物作为原料所制造的此类酒品，才能冠以 Tequila 之名。

The top of the Tequila, only in certain regions, using a plant called Blue Agave as the raw material of this type of wine, can be called Tequila.

（三）梅兹卡尔（Mezcal）

所有以龙舌兰草心为原料制造出的蒸馏酒的总称。请注意，Tequila 可以是 Mezcal 的一种，但并不是所有的 Mezcal 都能称作 Tequila。

Mezcal is the general name of all distilled wines made from tequila tender leaves. In a few words, Tequila is a kind of Mezcal. But not all Mezcal can be called Tequila. Tequila can also be divided into different grades according to the purity and age of the brew .

四、等级标准 Grade standard

（一）白色 / 银色龙舌兰（Blanco 或 Plata）

Blanco 与 Plata，西班牙文意思是"白色"与"银色"的意思，在龙舌兰酒的领域里，它被视为一种短暂陈酿或未经陈酿的酒款。若放入橡木桶，不可超过 30 日。需要注意的是，这种等级标示只说明产品的陈年特性，无关成分。Blanco/Plata 等级的龙舌兰酒通常比较辛辣、植物香气直接，对龙舌兰酒钟爱者而言，唯有这个等级方能尽显其独特风味与非凡魅力。

Blanco and Plata, which means "white" and "silver" in Spanish, are considered a short or unaged wine in the tequila world. If placed in oak barrels, not more than 30 days. It should be noted that this grade label only describes the aging characteristics of the product, not the ingredients. The Blanco/Plata grade of tequila is usually more spicy and plant-based, and for tequila lovers, only this grade can fully reveal its unique flavor and extraordinary charm.

（二）年轻龙舌兰（Joven abocado）

Joven abocado 西班牙文意思是"年轻且顺口的"，此等级的酒也常被称为 Oro（金色的）。金色版本的龙舌兰加了少量的酒用焦糖与橡木萃取液（其重量比不能超过 1%），从而看起来有点像陈年酒品。理论上，金色龙舌兰属于 Mixto，没有 100% 龙舌兰制造的那么高级，但由于价格实惠，仍然是销售主力。

Joven abocado means "young and smooth" in Spanish. This grade of wine is often called Oro (golden). Basically, the golden tequila and white tequila can be the same thing. The golden tequila has the local color modulation and seasoning (including caramel for wine and oak extract liquor, and its weight ratio should not exceed 1%). These make them look like the aging products. In terms of classification, all these wines belong to Mixto. Although this kind of wine is less advanced than those made of 100% agave products in theory, this grade of wine is still the main products in sales in the export market because of its affordable price.

（三）微陈龙舌兰（Reposado）

Reposado，西班牙文意思是"休息过的"，此等级的酒要经过 2 个月到 1 年之间的橡木桶陈酿，因为酒吸收了橡木桶的风味和颜色，不仅口味会变得浓厚、复杂，颜色也较深。目前，是墨西哥本土 Tequila 销售量的六成，市场占有率最大。

Reposado, which means "rested" in Spanish, is aged between two months and a year in oak barrels, which make it rich, complex and dark, as the wine absorbs the flavors and colors of the barrels. At present, it accounts for 60% of the sales of Tequila in Mexico and has the largest market share.

（四）陈年龙舌兰（Añejo）

在西班牙文里面原意是指"陈年过的"，其橡木桶陈放时间超过 1 年且没有上限。与前三种等级相比，陈年龙舌兰酒受政府管制要严格许多，它们必须使用容量不超过 350 升的橡木桶封存，由政府官员上封条，虽然规定上只要超过 1 年的都可称为 Añejo，但有少数非常稀有的高价产品，例如 Tequila Herradura 著名的顶级款 "Selección Suprema"，就是陈年超过四年的超高价产品之一。一般来说，龙舌兰最适合的陈年期限是 4 ～ 5 年。

Añejo originally means "aged" in Spanish. In a few words, wines aged in oak barrels for more than one year belong to this grade, and there is no upper limit. However, aged tequila wine is different from the previous three grades. It is much more strictly regulated by the government. They must be sealed in oak barrels with a capacity of no more than 350 liters. It is sealed by government officials. Although it is stipulated that it can be called Añejo as long as it lasts for more than one year. But there are a few rare high-priced products. For example, the famous top one "selección Suprema" of Tequila Herradura is one of the super-high-priced products that have been aged for more than four years. Its market is not even inferior to a bottle of Scotch Whisky aged 30 years. Generally speaking, experts agree that the most suitable aging period for Tequila is four to five years. After that, the alcohol in the barrel will volatilize too much.

除了少数陈年有 8 ～ 10 年的特殊酒款外，大部分的 Añejo 都是在陈年时间满了后，直接移到无陈酿效用的不锈钢桶中保存等待装瓶。

五、饮用方法（Drinking Method）

（一）传统饮法（Traditional drinking method）

在墨西哥，龙舌兰酒的传统饮法是，首先把盐撒在手的虎口处，用拇指和食指握一小杯纯龙舌兰酒，再用无名指和中指夹一片柠檬片，迅速舔一口虎口上的盐，接着把酒一饮而尽，再咬一口柠檬片，整个过程一气呵成。食用少许盐和柠檬，是因为盐可以适当中和酸味以平衡口感，而柠檬可消减酒中的涩味（图 1-3-6）。

In Mexico, the traditional way to drink tequila is to sprinkle salt on the mouth of the tiger, hold a small glass of pure tequila between the thumb and index finger, then hold a lemon slice between the ring and middle fingers, quickly lick the salt on the mouth of the tiger, then drink the wine, then take a bite of the lemon slice, and the whole process is complete. Use a little salt and lemon because salt can properly neutralize the acidity to balance the taste, while lemon can reduce the astringency of the wine (Figure 1-3-6).

图 1-3-6　龙舌兰酒的传统饮用方式
The traditional way of drinking tequila

（二）常见饮法（Common drinking methods）

龙舌兰酒也适宜冰镇后纯饮，或是加冰块饮用（图 1-3-7）。

Tequila is also suitable for drinking after freezing or drinking with ice (Figure 1-3-7).

（三）调制鸡尾酒（Make cocktails）

由于龙舌兰特有的风味以及接近 40% vol 的较高酒精

图 1-3-7　加冰的龙舌兰
Tequila with ice

度，它也很适合调制各种鸡尾酒。著名的玛格丽塔、马提尼酒、龙舌兰日出、血腥玛丽等鸡尾酒，里面都融入了龙舌兰酒。

Due to the unique flavor of Tequila and the high alcohol content of nearly 40% vol, it is also suitable for making all kinds of cocktails. The famous Margarita, Martini, Tequila Sunrise, Bloody Mary and other cocktails are mixed with tequila.

赛证直通 Competitions and Certificates

二维码 1-3-6
知识拓展

一、基础知识部分（Basic knowledge part）

（一）专业名词解释（Explanation of professional terms）

1. Tequila：
2. Blanco 与 Plata：

（二）简答题（Short answer questions）

龙舌兰酒的传统饮用方式里，为什么一定要用到盐和柠檬？

In the traditional way of drinking tequila, why does salt and lemon have to be used?

二、技能操作部分（Skill operation part）

请分别品鉴年轻龙舌兰酒和陈年龙舌兰酒，并作出详细评价

Please taste the Joven abocado and the Anejo separately and make a detailed evaluation

【单元三　反思与评价】Unit 3　Reflection and Evaluation

学会了：_____

Learned to: _____

成功实践了：_____

Successful practice: _____

最大的收获：_____

The biggest gain: _____

遇到的困难：_____

Difficulties encountered: _____

对教师的建议：_____

Suggestions for teachers: _____

单元四 配制酒

桌上这碗酒宛如太阳，粉红色的酒是其光芒，如果没有酒，仿佛环绕在太阳四周的行星般的我们就无法发光。

——理查·B.谢瑞敦（爱尔兰作家）

单元导入 Unit Introduction

配制酒是一个比较复杂的酒品系列，它的诞生晚于其他单一酒品，但发展很快。配制酒的品种繁多，风格也各有不同。配制酒是酒类里面一个特殊的品种，不能专属于哪个酒的类别，是混合的酒品。本单元将学习——中外配制酒文化。

Integrated alcoholic beverages is a more complex wine series, its birth is later than other single wine, but the development is fast. There are many varieties and styles of integrated alcoholic beverages. Integrated alcoholic beverages is a special variety of wine, can not be exclusive to which category of wine, is a mixed wine. In this unit, we will study the culture of integrated alcoholic beverages at home and abroad.

学习目标 Learning Objectives

➤ 知识目标（Knowledge Objectives）

（1）了解中外配制酒的概念。

Understand the concept of integrated alcoholic beverages at home and abroad.

（2）了解中外配制酒的起源发展。

Understand the origin and development of integrated alcoholic beverages at home and abroad.

（3）认知开胃酒的分类。

Recognize the classification of aperitif.

（4）认知利口酒的分类。

Recognize the classification of liqueur

（5）中外配制酒的饮用方法。

Methods of drinking integrated alcoholic beverages at home and abroad.

➤ 能力目标（Ability Objective）

能根据所学知识认知并选择适合的开胃酒、利口酒。

You can recognize and choose suitable aperitif and liqueur according to the knowledge learned.

➤ **素质目标（Quality Objectives）**

（1）学习和认识西方开胃酒的起源历史，发展及技艺，中国配制酒的发展及功效。

To learn and understand the origin history, development and skills of Western aperitif, the development and efficacy of Chinese integrated alcoholic beverage.

（2）提升学生的文化素养、专业素养，培育并提升学生的文化自信。

To enhance students' cultural literacy, professional literacy, and cultivate students' cultural confidence.

【微课】
触碰味蕾：
开胃酒

专题一
触碰味蕾：开胃酒

一、开胃酒的概念（The Concept of Aperitif）

开胃酒，泛指在餐前饮用能刺激胃口、增加食欲的所有酒精饮料，俗称餐前酒；也专指以葡萄酒基或蒸馏酒基为主酒，并加入植物的根、茎、叶等配制而成的有开胃功能的酒精饮料。

Aperitif, generally refers to all alcoholic beverages that can stimulate appetite and increase appetite before a meal, commonly known as pre-meal wine; Also refers to wine based or distilled wine based wine, and add plant roots, stems, leaves, herbs, spices and other preparations of the appetizing function of the alcoholic beverage.

二、开胃酒的起源（The Origin of Aperitif）

开胃酒，源自拉丁文 aperare，意为"打开"，旨在餐前打开食欲。其历史起源说法多样，有人追溯至古罗马帝国的甜酒，有人认为起源于中世纪的药酒或香料葡萄酒，还有人认为其背景与中产阶级的兴起相关。不过，"开胃酒"一词作为名词使用的历史，据现有资料最早可追溯到 1888 年，它指代多种酒类及含酒精饮料，如味美斯酒、奎纳皮酒、茴香酒、苦味酒等，也包括水果白兰地和利口酒。在不同的国家和文化背景下，饮用开胃酒的习俗和方式也各有特色。

Aperitif, from the Latin aperare, "to open", is intended to stimulate the appetite before a meal. Its historical origins vary from the sweet wine of the ancient Roman Empire to the medicinal or spiced wine of the Middle Ages, and to the rise of the middle class. However, the use of the term "aperitif" as a noun dates back to 1888, and it refers to a variety of alcoholic and alcoholic beverages, such as Vermouth, quinard, anisette, bitters, etc., as well as fruit brandy and liqueurs. In different countries and cultural backgrounds, the customs and ways of drinking aperitif also have their own characteristics.

三、开胃酒的特点（The characteristics of aperitif）

开胃酒为什么具有开胃的功效呢？原因是：

Why does aperitif have an appetizing effect? The reason is:

（1）开胃酒的酒精度数通常不高，在餐前饮用时不会过于强烈地刺激感官，同时又能保持清醒，有助于享受后续的用餐过程。

Aperitif usually has a low alcohol content so it does not stimulate the senses too strongly before a meal, keeps you awake and helps you enjoy the rest of the meal.

（2）开胃酒有着轻盈的酒体，有助于避免给胃部带来负担，同时又更容易被人体吸收，从而更快地发挥开胃作用。

Aperitif has light body, which helps to avoid putting a burden on the stomach, is easier to absorb, thus acting as an appetizer more quickly.

（3）开胃酒适合的酸度，非常有助于刺激味蕾，增加食欲。

Aperitif suitable acidity, very helpful to stimulate the taste buds, increase appetite.

（4）开胃酒偏干的口感可以避免酒中的糖分影响食欲，保持口感的清爽。

The dry taste of aperitif can avoid the sugar in the wine to affect the appetite, to keep the taste fresh.

（5）开胃酒风味普遍细腻且清淡，香气馥郁但不浓烈，能增加食欲，却不会喧宾夺主，不会掩盖食物的原味。以上这些特质，很好地解释了餐前开胃酒的功效原因。

Aperitif flavor is generally delicate and light, aroma is rich but not strong, can increase appetite, but will not dominate, will not cover up the taste of food. These characteristics explain the effectiveness of aperitif before dinner.

四、开胃酒的类型（Type of Aperitif）

现代开胃酒主要分为三类：味美思、比特酒和茴香酒。

Modern aperitif is divided into three main categories: Vermouth, bitters and anisette.

（一）味美思（Vermouth）

味美思，也叫"苦艾酒"，以葡萄酒为基酒，加入植物、药材等物质浸制而成；起源于希腊，有强烈的草本植物味道；酒精含量在17%至20%，根据酒的颜色和含糖量分为干性味美思、白色味美思、红色味美思和玫瑰红味美思四种。该酒最好的生产国是法国、意大利。常见品牌有：法国产乐华里、马天尼（白）、意大利产马天尼（红）、意大利产仙山露（白）、意大利产仙山露（红）。

Vermouth, also known as "Absinthe", is a wine based liquor made by adding plants, medicinal materials and other substances. It originated in Greece and has a strong herbal taste. The alcohol content is between 17% and 20%, and is often divided into dry Vermouth, white Vermouth, red Vermouth, and rose Vermouth depending on the color and sugar content of the wine. The best producing countries are France and Italy. Common brands are: French Noilly Prat; Italian Martini (white); Italian Martini (red); Italian Cinzano (white); Italian Cinzano (red).

（二）比特酒（Bitters）

比特酒，又称必打士，或苦酒，是在葡萄酒或蒸馏酒中加入树皮、草根、香料及药材浸制而成的酒精饮料。该酒酒味苦涩，酒精度在 16% 至 40%。"苦酒" 即有一定苦涩味和药味，其酒精含量在 18% 至 49%。该酒产自意大利、法国等国，常见品牌有金巴利、杜本纳等。

Bitters are alcoholic beverages made from wine or distilled spirits with bark, roots, spices, and herbs. The wine has a bitter taste and is between 16% and 40% proof. The meaning of " bitter wine ", has a certain bitter and medicinal taste, and the alcohol content is between 18% and 49%. The wine has produced in Italy, France and other countries, the common brand products are: Campari, Dubena and so on.

（三）茴香酒（Anisette）

茴香酒是用蒸馏酒与茴香油配制而成，有浓郁的茴香气味，口味浓重且刺激性强，通常光泽明亮。酒精含量在 25% 左右。著名产地是法国波尔多地区，常见产品有潘诺、巴斯特 51 等。

Anisette is prepared with distilled wine and anise oil, with a strong anise smell, strong taste and pungent, usually bright. The alcohol content is around 25%. The famous origin is the Bordeaux region of France, the common products are Panot, Buster 51 and so on.

五、开胃酒的饮用方法（Drinking Method of Aperitif）

（一）净饮（Clean Drink）

使用工具：鸡尾酒杯、量杯。

操作步骤：直接倒入酒杯中，每次饮用 1 ～ 2 盎司或 40 毫升左右，在餐前饮用，以感受其芳香和味道，刺激食欲。

Tools to use: cocktail glass, measuring cup.

Operation steps: Directly into the glass, each drink 1–2 ounces or 40 mL or so, before the meal to feel its aroma and taste, stimulate the appetite.

（二）加冰（Drinking with Ice）

使用工具：平底杯、量杯、酒吧匙。

操作步骤：先在平底杯加进半杯冰块，然后加入 1.5 量杯开胃酒，再用酒吧匙搅拌 10 秒，最后加入一片柠檬便可饮用，口感冰爽，令人愉悦。

Use tools: flat cup, measuring cup, bar spoon.

How to do: Fill a flat glass with half a cup of ice, then add 1.5 cups of aperitif, stir with a bar spoon for 10 seconds, and finally add a slice of lemon.

（三）混饮（Mixed drinking）

可用苏打水、果汁来搭配开胃酒。以金巴利酒举例说明：先将半杯冰块、一片柠檬放入柯林杯，然后再倒入 42 毫升金巴利酒和 68 毫升苏打水，再用小勺搅拌 5 秒即可饮用，非常爽口清新。

Use soda water and juice to accompany the aperitif. Take campari as an example: First put half a cup of ice cubes and a slice of lemon into a corin glass, then pour 42 mL campari wine and 68 mL soda water, and then stir with a small spoon for 5 seconds to drink, very refreshing.

赛证直通 Competitions and Certificates

一、基础知识部分（Basic Knowledge Part）

二维码 1-4-1：
知识拓展

（一）专业名词解释（Explanation of Professional Terms）

1. 开胃酒（Aperitif）：
2. 比特酒（Bitters）：

（二）思考题（Thinking Question）

谈谈对开胃酒特点的理解。
Tell me about your understanding of the characteristics of aperitif.

二、技能操作部分（Skill Operation Part）

某酒店将举行母亲节晚宴。现场女士居多，但也有部分男士。请为这个宴会选择一款适合的餐前开胃酒，并说明原因。

A hotel is having a Mother's Day dinner. Most of the guests are women, but there are also some men. Please choose a suitable aperitif for this banquet and explain why.

专题二

甜蜜加倍：甜食酒

【微课】
甜蜜加倍：
甜食酒

一、甜食酒的概念（The Concept of Dessert Wine）

甜食酒是西餐中，搭配餐后甜点饮用的酒品，以葡萄酒为基酒，通过添加白兰地提高酒精浓度来终止发酵，从而保留较高的糖分，因此也被称为强化葡萄酒，其糖度和酒精含量均高于一般葡萄酒，酒精含量通常达到 25% 左右。常见的甜食酒有雪莉酒、波特酒、马德拉酒、冰酒等。

Sweet wine is a wine that is drunk with dessert in Western food. It is based on wine and terminates fermentation by adding brandy to improve the alcohol concentration, so as to retain higher sugar, so it is also known as fortified wine. Its sugar and alcohol content are higher than that of general wine, and the alcohol content usually reaches about 25%. Common sweet wine is sherry, port, Madeira wine, ice wine ,etc.

二、甜食酒的生产（Sweet wine production）

甜食酒的生产独具匠心，通常是在葡萄酒发酵过程中，通过加进酒精，及时阻止发酵，有效保留糖分，从而保持了葡萄酒的甜味，又提高了酒精度数，使其不仅口感丰富，更添一份醇厚。

The production of sweet wine is unique, usually in the process of wine fermentation, by adding alcohol to prevent fermentation in time, effectively retain sugar, so as to maintain the sweetness of the wine, but also improve the alcohol degree, so that it is not only rich taste, but also add a mellow.

三、甜食酒特点（Characteristics of sweet wine）

（1）含糖量高：由于加入蒸馏酒阻止发酵，甜食酒的糖分得以保留，糖度较高。

High sugar content: Due to the addition of distilled wine to prevent fermentation, the sugar of sweet wine is retained and the sugar content is high.

（2）酒精度高：由于以葡萄酒为酒基，又加入蒸馏酒或白兰地，酒精含量提高到 19% 或以上。

High alcohol content: Due to the wine as the wine base, and the addition of distilled wine or brandy, the alcohol content is increased to 19% or more.

（3）口感香醇浓郁：由于糖分的保留以及酒精度数得提高，使其口感丰富且醇厚。

Mellow and rich taste: due to the retention of sugar and the improvement of alcohol content, the taste is rich and mellow.

四、甜食酒的分类（The Classification of Dessert Wine）

（一）波特酒（PORTO）

波特酒产于葡萄牙，是世界上最优秀的甜食酒之一。其生产历史悠久，已有 300 多年的历史，酿造工艺精湛独特。该酒原名"PORT"，取自葡萄牙北部第二大港口城市波尔图。

PORTO produced in Portugal, it is the world's best sweet wine. The production has a long history of more than 300 years, and the brewing technology is exquisite and unique. The wine formerly known as "PORT", from porto the second largest port city in northern Portugal.

1. 成为波特酒的条件（The Conditions for Becoming PORTO）

（1）用杜罗河上游的奥特·杜罗地域所种植的葡萄酿造，用来掺兑的白兰地也要使用本地区葡萄酿造。

Brewing grapes which grown in the Alto Douro region on the upper reaches of the Douro River, and the brandy used for blending is also made from grapes in the region.

（2）必须在杜罗河口的维拉·诺瓦·盖亚酒库内陈化储存，并从对岸的波特港口运出。

It must be aged and stored in the Vila Nova Gaia Winery at the mouth of the Douro River and shipped out from Port on the other side.

（3）产品酒精度在 16.5° 以上。

The alcohol content of the product is above 16.5°.

2. 波特酒的分类（The classification of PORTO）

根据色泽和陈酿时间不同，波特酒分为以下种类：①宝石红波特酒；②茶红波特酒；③白色波特酒；④年份波特酒。

The PORTO is divided into the following categories according to color and aging time: ① Ruby port; ② Tawny port; ③ White port wine; ④ Vintage port.

3. 波特酒品牌（The Brand of PORTO）

（1）科伯恩。科伯恩酒庄由罗伯特·科伯恩创建于 1815 年，产品销量世界第一。

Cockburn's. Cockburn's created by Rort Cockburn in 1815, and it is the world's best-selling product.

（2）桑德曼。（Sandeman.）

（3）达尔瓦。（Dalva.）

（4）高乐福。高乐福酒庄创立于 1678 年，拥有葡萄牙最大的葡萄园，以精选的葡萄为原料酿制波特酒。

Croft. Croft Port founded in 1678, Croft owns the largest vineyard in Portugal and produces port wine from selected grapes.

（5）泰勒。泰勒是葡萄牙名酒。泰勒公司创建于 1692 年。

Taylor's. Talor's is a Portuguese wine, the company was founded in 1692.

（二）雪利酒（Sherry）

雪利酒，又称雪莉酒、些厘酒，是由西班牙语"Jerez"音译而来，在西班牙，它的名字是"赫雷斯"酒。和很多的欧洲名酒的命名规律一样，也以产地得名。赫雷斯是位于西班牙南部海岸的一个小镇，小镇附近富含石灰质的土壤，适于品种葡萄巴洛米诺的生长。这种白葡萄酿成的酒再添加葡萄蒸馏酒进行酿制，就成为雪利酒，所以国外一般称它为强化葡萄酒。虽然雪利酒是西班牙的国酒，但是英国人对雪利酒的喜爱程度却超过了世界上所有国家，就连"雪利酒"这一现用名都是英国人所命名的（过去西班牙人习惯上称它为"xeres"，后来英国人则以其谐音命名为"Sherry"）。

Sherry, is a transliteration of the Spanish word "Jerez", and in Spain, its name should be Jerez wine. Like many famous European wines, it is named after the place where it is produced. Jerez is a small town on the southern coast of Spain, near the town rich in lime soil, suitable for the growth of the variety of grape Palomino (Palomino), this white grape wine with distilled wine added to make sherry, so it is generally called Fortified wine (Fortified Wine). Although sherry is the national wine of Spain, but the British love sherry more than all countries in the world, so even the current name of sherry is named by the British (The Spanish used to call it "xeres", and later the British named it "Sherry").

1. 雪利酒的特点（The Characteristics of Sherry）

雪利酒的酒精含量较高，为 15～20 度。雪莉酒有着丰富而复杂的口感和香气，通常呈现出干果、巧克力、咖啡、橡木和香料的味道。其中，干果的甜味和坚果的香气是其特色元素之一。雪莉酒还有着深红琥珀色的外观，给人以视觉上的享受。

Sherry has a higher alcohol content, about 15 to 20 degrees. Sherry has a rich and complex taste and aroma, it is often presented with dry fruit, chocolate, coffee, oak and spices. Among them, the sweetness of dried fruits and the aroma of nuts are one of its characteristic elements. Sherry also has a deep red amber appearance, giving a visual pleasure.

2. 葡萄品种（Grape Variety）

在 1894 年的葡萄根瘤蚜虫害之前，西班牙估计有超过 100 种葡萄用于生产雪利酒，但现在法定用来酿造雪利酒的葡萄品种只有三个，都是白葡萄品种，分别是帕洛米诺、佩德罗·希梅内斯、亚历山大麝香。其中，帕洛米诺葡萄在赫雷斯镇的种植比例高达 95%，是最主要的酿造雪利的葡萄品种。佩德罗·希梅内斯和亚历山大麝香主要用来酿造甜型雪利酒。

Before the 1894 root nodular aphid attack, it was estimated that more than 100 varieties of grapes were used in the production of sherry in Spain, but now there are only three varieties legally used to make sherry, all white varieties. They are Palomino, PX(Pedro Ximenez) and Muscat of Alexandria. Among them, the Palomino grape in the town of Jerez, the proportion of up to 95%, is the most important grape variety for sherry. PX(Pedro Ximenez) and Muscat of Alexandria are mainly used to make sweet sherry.

3. 雪莉酒的风格分类（Styles of Sherry）

（1）干型雪莉酒（Dry Styles of Sherry）。

①菲奴：颜色淡黄而明亮的干型雪利酒，是雪利酒中颜色最淡的酒品。此酒香气悦人，口味甘洌，清新爽口，酒精含量为 15.5%～17%，不宜贮藏，即买即喝。

Fino: a dry sherry with a light yellow and bright color. It is the lightest of the sherry wines. This wine has a pleasant aroma, dry taste, fresh and refreshing, alcohol content of 15.5%−17%, should not be used to hide, that is, buy and drink.

②曼萨尼亚：在西班牙圣卢卡尔·巴拉梅德镇酿造的菲奴雪利酒称为曼萨尼亚。原因是这个镇的气候条件更为冷凉和潮湿，这让陈酿的葡萄酒菲奴风味更浓郁。这种酒色泽微红，透亮晶莹，口感清淡，甘洌，略带苦味，酒精含量为 15.5%～17%。

Mainzanilla: Fino sherry made in the town of Sanlucar de Barrameda Spain is called Manzanilla. The reason is that the climate in this town is cooler and wetter, which gives the aged wine a stronger flavor of Fino. This wine has reddish color, translucent crystal, light taste, dry, slightly bitter, alcohol content of 15.5%−17%.

③阿蒙提那多：陈酿至少八年的雪利酒，酒液呈琥珀色，口味甘洌而清淡，带有坚果香气，酒精含量为 16%～18%。

Amontillado: Sherry aged at least 8 years, the liquor is amber in color, dry and light in taste, nutty aromas, 16%−18% alcohol content.

④奥罗露索：加强型酒品，酒液呈金黄、棕红色，透明度极好；香气浓郁扑鼻，并有坚果香气、越陈越香；口味浓烈、绵柔、甘洌，但有甘甜之感，酒体丰硕圆润，酒精含量为18% ～ 20%。

Cloroso: enhanced wine, the liquor is golden brown red, excellent transparency. The aroma is rich and tangy, and has a nutty aroma, which becomes more fragrant with ages. The taste is strong, soft, dry, but sweet feeling, rich and round body, alcohol content of 18%–20%.

⑤帕罗卡特多：非常稀少，是有名的雪利酒之一，酒体丰满圆润，色泽金黄。这种酒是经过陈酿的干型成品酒，具有菲奴和奥罗露索两种类型的特点，被称为"具有菲奴酒香的奥罗露索"，是雪利酒中的珍品。

Palo Cortado: very rare, is one of the famous sherry, full body, golden color. This wine is after aging dry finished wine, with Fino and Oloroso two types of characteristics, known as " with Fino aroma Oloroso", is a rare sherry.

⑥阿莫露索：又叫"爱情酒"，是用奥罗露索与甜酒勾兑配制而成的甜雪利酒。酒液深红色，酒香与奥罗露索相近，但不那么突出，口味凶烈，劲足力大，甘甜圆正，是英国人喜爱的品种。

Amoroso: also called " love wine ", it is a sweet sherry mixed with Oloroso and sweet wine. Deep red wine, similar to Oloroso, but not so prominent, fierce taste, strong foot, sweet round. A breed loved by the British.

（2）天然甜型雪利酒：天然甜型雪利酒在当地叫作" Jerez Dulce Naturales"，分为 PX 甜雪利酒和麝香雪利酒两种。

Naturally sweet sherry, locally called " Jerez Dulce Naturales ", is divided into two types: PX Sweet Sherry and Muscat Sherry.

①PX 甜雪利酒：由佩德罗·希梅内斯葡萄酿造，颜色为深棕色，酒中残糖含量非常高，经常达到 500 g/L，具有水果干、咖啡、甘草等香气。

PX Sweet Sherry: made from PX(Pedro Ximenez) grapes, the color is dark brown, the residual sugar in the wine is very high, often reaching 500 g/L, with dried fruit, coffee, licorice and other aromas.

②麝香雪利酒：由亚历山大麝香（Muscat of Alexandria）葡萄酿造，其风格与 PX 甜雪利酒相似，但多了一些干柠檬皮的香气。

Muscat Sherry: made from the Muscat of Alexandria grape, it is similar in style to PX Sweet Sherry, but with some dry lemon peel aromas.

（3）混合型雪利酒（Blended Sherries）。

①白奶油雪利：将菲奴雪利酒用浓缩的葡萄汁加甜后得到的雪利酒就是白奶油雪利。

Pale Cream: Fino sherry is sweetened with concentrated grape juice and the resulting sherry is Pale Cream.

②中等甜度雪利：将阿蒙提拉多雪利酒与自然甜型雪利酒混合，形成的半甜型雪利酒。

Medium Cream: Amontillado sherry is mixed with natural sweet sherry to form semi-sweet sherry.

③奶油雪利：将奥罗露索雪利酒与自然甜型雪利酒混合，形成的半甜型或甜型雪利酒。

Cream Sherry: Mix Oloroso sherry with natural sweet sherry to form semi-sweet or sweet sherry.

（4）著名品牌（Famous brand）。

①当佐伊罗：当佐伊罗公司创建于 1851 年，是西班牙 120 个同行业企业中最大的集团公

司，各公司以创业者的名字命名。Don 表示"首领"的意思。

Don Zoilo: Dan Zoilo founded in 1851, the Don Zoilo Company is Spain's largest group of 120 companies in the same industry, each named after the entrepreneur. Don means leader.

②克罗夫特：该公司建立于 1678 年，目前在葡萄牙生产 Croft 品牌的波特酒，而在西班牙生产雪利酒，酿酒技术独特，产品声誉很高。

Croft: it founded in 1678, the company currently produces Croft brand porto in Portugal and Sherry in Spain, with unique winemaking techniques and a high reputation for its products. it was created by Mr. Dee Terry in 1833.

③哈维斯：哈维斯公司创建于 1796 年，其生产的雪利酒先在西班牙酿造出原酒，再运到英国调配装瓶而成。哈维斯味道香醇，迎合英国人的口味，产品畅销英伦三岛及世界 100 多个国家。

Har Veys: Har vegs founded in 1796, the Harveys Company produces sherry that is first brewed in Spain and then shipped to England for blending and bottling. Harevys is delicious and caters to the taste of the British people, and its products are sold well in the British Isles and more than 100 countries around the world.

④桑德曼（Sandeman）：桑德曼公司于 1790 年由桑德曼创立，总公司设在英国伦敦。该公司在葡萄牙生产波特酒，在西班牙则生产雪利酒，产品销往世界各国。

Sandeman: the Sandman Company was founded in 1790 by Mr. Sandman and is headquartered in London, England. The company produces port wine in Portugal and sherry in Spain, and its products are sold around the world.

（三）马德拉酒（Madeira）

马德拉酒产于葡萄牙属地马德拉岛，该岛位于非洲西海岸。马德拉酒是用当地生产的葡萄酒与白兰地勾兑而成的一种强化葡萄酒。

Madeira wine is produced on the Portuguese island of Madeira, off the west coast of Africa. Madeira is a fortified wine made from locally produced wine mixed with brandy.

1. 马德拉酒的特点（Characteristics of Madeira）

马德拉酒酒精含量为 16%～18%，其干型强化葡萄酒是优质的开胃酒，甜型强化葡萄酒是著名的甜食酒。其中，长年陈酿的酒是世界上长寿的酒品之一。马德拉酒饮用初期需要稍烫一下，越干越好喝。

Madeira alcohol content of 16%–18%, its dry fortified wine is a high-quality aperitif, sweet fortified wine is a famous dessert wine. Among them, the long-aged wine is one of the longest-lived wines in the world. Madeira wine needs to be slightly ironed at the beginning of drinking, and the more dry the better to drink.

2. 马德拉酒的分类（Classification of Madeira）

（1）舍西亚尔：酒液呈金黄色或淡黄色，色泽艳丽，香气卓绝，带有清香的杏仁味，人称"香魂"，属干型酒，口味醇厚、浓正，西方厨师常用之作为料酒。

Sercial: the liquor is golden yellow or light yellow, beautiful color, excellent aroma, with fragrance almond taste, known as " fragrant soul", is a dry wine, taste mellow, strong, Western chefs commonly used as cooking wine.

（2）弗德罗：干型酒，但比舍西亚尔稍甜，适合多数人的口味。酒液呈金黄色，光泽动人，香气优雅，口味甘洌、醇厚、纯正，是马德拉酒中的精品。

Verdelho: dry wine, but sweeter than Sercial, suitable for most tastes. The wine is golden yellow, luster moving, elegant aroma, dry taste, mellow, pure, is the fine Madeira wine.

（3）布阿尔：属半干型，色泽栗黄或棕黄，香气浓郁，富有个性，口味甘洌、浓醇，甜而不腻，最适合作甜食酒。

Bual: semi-dry type, chestnut yellow or brownish yellow color, rich aroma, full of personality, the taste is dry and rich, sweet but not greasy, the most suitable for making sweet wine.

（4）马尔姆赛：甜型酒，在马德拉酒中享誉最高。该酒呈褐黄或棕黄色，香气悦人，口味极佳，甜适润爽，比其他同类酒更醇厚浓正，风格和酒体给人以富贵豪华的感受，是较好的甜食酒之一。

Malmsey: sweet wine, the highest reputation among Madeira wines. The wine is brown or brownish yellow, pleasant aroma, excellent taste, sweet and moist, more mellow than other similar wines, style and wine body give people a rich and luxurious feeling, is one of the best sweet wine.

五、甜食酒的饮用方法（Sweet wine drinking method）

甜食酒的饮用方法灵活多样，关键在于根据个人喜好和场合选择合适的饮用方式。主要有：

Sweet wine drinking methods are flexible and varied, the key is to choose the right way to drink according to personal preferences and occasions. The main ones are:

（1）纯饮：不兑水也不加入其他饮料混合，直接饮用，口感纯粹。

Pure drink: do not mix with water or other drinks, drink directly, taste pure.

（2）加冰饮用：在葡萄酒杯或鸡尾酒杯中加入半块碎冰块，再加入甜食酒，可以降低酒的温度，使口感更加清爽。

Serve over ice: Fill a wine glass or cocktail glass with half a crushed ice cube, add sweet wine. lowers the temperature and adds a refreshing taste.

（3）混饮：甜食酒也可以与其他饮料混合饮用，如牛奶、咖啡、茶等，独具创意、风味新颖。

Mixing: dessert wine can also be mixed with other drinks, such as milk, coffee, tea, etc., unique creativity, novel flavor.

赛证直通 Competitions and Certificates

一、基础知识部分（Basic Knowledge Part）

（一）专业名词解释（Explanation of Professional Terms）

1. 甜食酒（Dessert Wine）：

2. 帕罗卡特多（Palo Cortado）：

二维码 1-4-2：
知识拓展

（二）思考题（Thinking Question）

简述甜食酒的分类。

Briefly describe the classification of dessert wine.

二、技能操作部分（Skill Operation Part）

某晚宴的甜点是苹果派和香蕉布丁，请选择适合的甜食酒进行搭配，并说明选择原因。

The dessert of a dinner party is apple pie and banana pudding, please choose a suitable dessert wine to match, and explain the reason for choosing.

专题三
锦上添花：利口酒

【微课】
锦上添花：
利口酒

一、利口酒的概念（The Concept of Liqueur）

利口酒（Liqueur）也称为餐后甜酒，是由法语"Liqueur"音译而来，美国人称其为"Cordial"（拉丁文），与"Liqueur"同义，意为"心脏"，指酒对心脏有刺激作用。在法国，人们称其为"Digestifs"，指这种酒有助于消化。我国港澳台地区称为力娇酒。它是以蒸馏酒（白兰地、威士忌、朗姆酒、金酒、伏特加、龙舌兰）为基酒配制各种调香物品并经过甜化处理的酒精饮料。利口酒香味浓郁，含糖量高，故又叫"香甜酒"。

Liqueur, also known as dessert wine after dinner, is transliterated from the French "Liqueur", Americans call it "Cordial" (Latin), synonymous with "Liqueur", meaning "heart", referring to the wine has a stimulating effect on the heart. In France, they're called "Digestifs", which means this wine contribute to digestion. In Hong Kong, Macao and Taiwan, it is called Lijiao wine. It is distilled liquor (brandy, whiskey, rum, gin, vodka, tequila) as the base liquor to prepare a variety of flavoring items, and through the sweetening treatment of alcoholic beverages. Liqueur has a strong flavor and high sugar content, so it is also called "sweet wine".

二、利口酒的起源（The Origin of Liqueur）

利口酒最初源于医用目的，由修道院的僧侣用烈酒加药草或水果提取物制成，用于治疗各种疾病。后来，人们发现加入蜂蜜、香草等材料可以调和葡萄酒的酸味，经过过滤，便制成了最初的利口酒。随着新大陆和亚洲植物的引进，利口酒的原料日益丰富。公元1314年，西班牙学者首创新技术，用酒精析出柠檬、橘子花等香味并配上颜色，进一步创新了利口酒的制造方式。16世纪，意大利人通过稀释蒸馏过的葡萄酒并加入多种配料，使利口酒再次创新并声誉鹊起。18世纪后，随着科学进步，利口酒逐渐转变为以水果香味为主的美味型香甜酒。

Originally originating for medical purposes, liqueur was made by monks in monasteries who used spirits with herbs or fruit extracts to treat various ailments. Later, it was discovered that adding honey, vanilla and other materials could balance the sour taste of the wine, and after filtration, the original liqueur was made. With the introduction of New World and Asian plants, the raw materials of liqueur became increasingly abundant. In 1314 AD, Spanish scholars first innovated technology, using alcohol to separate out lemon, orange flower and other flavors and colors, and further innovated the manufacturing method of liqueur. In the 16th century, the Italians made liqueurs innovative and famous again by diluting the distilled wine and adding a variety of ingredients. After the 18th century, with the progress of science, liqueur gradually changed into a delicious sweet wine with fruit aroma.

三、利口酒的特点（The Characteristics of Liqueur）

（1）颜色娇美、气味芬芳独特、酒味甜蜜。

The color is delicate, the smell is unique, and the taste of wine sweet

（2）具有舒筋活血、促进消化的功效适合餐后饮用。

It has the effect of relaxing muscles and promoting blood circulation and digestion Suitable for drinking after meals.

（3）含糖量高、色彩鲜艳，可用性广，常被用于调制鸡尾酒，增添色彩和风味，突出个性。同时，也是烹调、烘烤、制作甜点的理想调料。

High sugar content, bright colors, wide availability, often used to make cocktails, add color and flavor, highlight personality. At the same time, it is also an ideal seasoning for cooking, baking and making desserts.

四、利口酒的生产方法（Production Method of Liqueur）

（一）蒸馏法（Distillation Method）

蒸馏法主要分为两类：一类是将原料直接浸渍在基酒中后一同蒸馏；另一类则是浸渍后取出原料，仅用基酒浸泡过的汁液蒸馏。两种方法所得酒液均为无色透明，之后通过添加甜浆和植物色素，增加酒液的甜度和色彩。此方法主要用于提取香草类、柑橘类等原料的香气和风味。

Distillation method mainly divided into two categories: one is the raw material directly impregnated in the base wine after distillation; The other is to remove the raw material after maceration and only distillation with the juice soaked in the base wine. The liquor obtained by the two methods is colorless and transparent, and then the sweetness and color of the liquor are increased by adding sweet pulp and vegetable pigment. This method is mainly used to extract the aroma and flavor of vanilla, citrus and other raw materials.

无论用哪种方法，所得的酒液都是无色透明的，再加入甜浆和植物色素，使酒液变甜、变色。蒸馏法主要用于香草类、柑橘类的干皮等原料的提香、提味上。

Whatever method is used, the resulting liquor is colorless and transparent, and then sweet pulp and vegetable pigments are added to make the liquor sweet and discolored. Distillation method is mainly used for vanilla, citrus and other raw materials such as dry peel, flavor enhancement.

（二）浸渍法（Impregnation Method）

浸渍法通过将新鲜原料浸泡在基酒中，使酒液充分吸收其味道和色泽，随后滤出原料，并加入甜浆和植物色素以增加甜度和色彩，最后滤清装瓶，也可选择使用橡木桶进行陈酿。

Impregnation method by soaking fresh raw materials in the base wine, so that the wine fully absorbs its taste and color, then filter out the raw materials, and add sweet pulp and plant pigments to increase sweetness and color, and finally filter and bottle, you can also choose to use oak barrels for aging.

（三）渗透过滤法（Osmosis Filtration Method）

渗透过滤法也称蒸汽蒸馏法，常用于草药、香料利口酒的生产。此法将原料置于上方容器，基酒则放于下方容器，加热后水汽或酒精上升，穿过原料层摄取风味，反复循环直至酒液获得足够风味。

Osmosis filtration method. Also known as steam distillation, it is often used in the production of herbal and spice liqueur. In this method, the raw material is placed in the upper container, and the base wine is placed in the lower container. After heating, the water vapor or alcohol rises, and the flavor is absorbed through the raw material layer, and the cycle is repeated until the wine gets enough flavor.

（四）香精法（Flavor Method）

将植物性的天然香精加入基酒中，进行甜度和颜色的调和。但此法酿造的利口酒一般品质较差。

The natural vegetable flavor is added to the base wine to harmonize the sweetness and color. However, the liqueur produced by this method is generally of poor quality.

五、利口酒的分类（Classification of Liqueur）

（一）根据酒精含量分类（Classified by Alcohol Content）

按照酒精度数分类，利口酒可以分为低度利口酒（酒精度数在 15°～20°）、中度利口酒（酒精度数在 20°～25°）和高度利口酒（酒精度数在 25° 以上）。

According to the alcohol classification, liqueur can be divided into low liqueur (alcohol content of 15° to 20°), moderate liqueur (alcohol content of 20° to 25°) and high liqueur (alcohol content of more than 25°).

（二）根据香料物质分类（Classified by Alcohol Content）

1. 果料利口酒（Liqueurs de Fruits）

果料利口酒以水果，包括苹果、樱桃、柠檬、草莓等的果皮及其肉质为辅料与基酒配制而成。该酒主要采用浸泡法配制，具有天然的水果色泽，风格明显，口味清新，适宜新鲜时饮用。

Fruit liqueur is made from the skin and meat of fruit, including apple, cherry, lemon, strawberry, etc., as auxiliary materials and wine base. The wine is mainly prepared by soaking method, it has natural fruit color, obvious style, fresh taste, suitable enjoyed fresh.

2. 草料利口酒（Liqueurs de Plants）

草料利口酒以草本植物，包括金鸡纳树皮、樟树皮、当归、龙胆根、甘草、姜黄及各种花类等为辅料，与酒基配制而成。该酒一般是无色的，如果有颜色也是外加的。这类酒是利口酒中的高级品。

Plant liqueur is prepared from plant and wine base, including cinchona bark, camphor bark, angelica, gentian root, licorice, turmeric and various flowers, etc. The wine is generally colorless, if any color is added. This kind of wine is the highest grade of liqueur.

（三）根据所用基酒分类（Classified by the base Wine used）

（1）以威士忌为基酒的利口酒：如杜林标、爱尔兰之雾等。

Whiskey based liqueur such as Drambuie, Irish Mist and so on.

（2）以白兰地为基酒的利口酒：如夏朗德百诺、蛋黄酒等。

Liqueur based on brandy such as Pineau des Charentes, Advocaat.

（3）以金酒为基酒的利口酒：如黑刺李酒等。

Liqueur based on gin such as Sloe Gin.

（4）以朗姆酒为基酒的利口酒：如甘露咖啡酒、添万利。

Liqueur based on Run such as Kahlua, Tia Maria.

（5）以中性谷物酒精为酒基的利口酒：这类酒占利口酒的大多数，如薄荷酒、顾美露等。

Liqueur based on neutral grain alcohol: this type of liquor accounts for the majority of liqueurs, such as Peppermint and Kummel.

六、利口酒的饮用方法（How to Drink Liqueur）

（一）纯饮（Straight drink）

选用纯度高的利口酒，倒在专用杯里，用嘴一点点慢慢地啜饮，细细品饮，但多数人认为这样喝过于甜腻。

Choose a high-purity liqueur, pour it in a special cup, sip it slowly with your mouth, and savor, but most people think that this drink is too sweet and greasy.

（二）兑饮（Mixed）

兑饮即饮用时加苏打水、矿泉水、果汁。喝前先将酒倒入平底杯中，约为杯子容量的60%，再加满苏打水即可。如觉得水分过多，可添加一些柠檬汁，半个柠檬较合适，在上面再加碎冰。

Add soda water, mineral water, juice. Before drinking, pour the wine into a flat glass, which holds about 60% of the cup's capacity, and fill it with soda water. If you feel that the water is too much, you can add some lemon juice, half a lemon is suitable, and add crushed ice on the top.

（三）冰饮（Ice-cold Drink）

将碎冰放入杯中，再倒入利口酒，就可直接享用。

Place crushed ice in a glass, then pour in the liqueur and enjoy.

 赛证直通 Competitions and Certificates

一、基础知识部分（Basic Knowledge Part）

（一）专业名词解释（Explanation of Professional Terms）

1. 利口酒（Liqueur）：
2. 君度酒（Cointreau）：

（二）思考题（Thinking Question）

结合所学知识，谈谈利口酒与甜食酒的区别。

Combined with the knowledge, talk about the difference between liqueur and sweet wine.

二、技能操作部分（Skill Operation Part）

请为某酒店自助圣诞晚餐搭配适合的利口酒并说明原因。

Please serve a suitable liqueur for the buffet Christmas dinner at a hotel and explain why.

专题四

酒中之秘：中国配制酒

一、中国配制酒的概念（Chinese Concept of Wine Preparation）

我国 2021 年新修订的国家标准《饮料酒术语和分类》（GB/T 17204—2021）中，对配制酒的解释是："以发酵酒、蒸馏酒、食用酒精等为酒基，加入可食用的原辅料或食品添加剂，进行调配或再加工制成的饮料酒。"

In the newly revised national standard " Terminology and Classification of beverage wine " (GB/T 17204–2021), the interpretation of prepared wine is: " beverage wine made from fermented wine, distilled wine, edible alcohol, etc., adding edible raw materials or food additives, and being blended or reprocessed."

二、中国配制酒的起源（The Origin of Chinese Wine Preparation）

我国最早的配制酒是作为药用的，甲骨文中就有"鬯其酒"的记载，"药酒"一词始见于西汉司马迁的《史记·留侯世家》中："夫药酒苦于口，利于病。忠言逆耳，有利于行。药酒，病之利也。正言，治之病也"。班固在《汉书·食货志》中也称酒为"百药之长"。到了唐朝，大医学家孙思邈的医学专著中，也记载了很多药酒方。

China's earliest preparation of wine is used for medicinal purposes, there is a record of "Chang its wine" in the oracle bone inscriptions, the word "medicinal wine" began to appear in the Western Han Dynasty Sima Qian's "Shi Ji Liu Hou Family": "Husband medicinal wine bitter in the mouth, conducive to disease. Honest advice is bitter to the ear, and good to the deed. Medicine wine, disease benefit also. Correct speech, cure the disease also". Ban Gu also called wine "the length of a hundred medicines" in the Book of Han·Food Annals. In the Tang Dynasty, the great medical scholar Sun Simiao's medical monograph also recorded a lot of medicinal wine prescriptions.

三、中国配制酒的发展（The Development of Chinese Wine Preparation）

（一）秦汉时期（Qin and Han Dynasties）

我国现存最早的医学理论专著为《黄帝内经》，一般认为它是经战国时期到汉代一个相当长的时期汇编而成的，后来又分成《素问》及《灵枢》两部书。《黄帝内经》虽是医学理论专书，但也录有方剂 13 个，如治尸厥的左角发酒、治鼓胀的鸡矢醴，均为早期药酒的代表。

The earliest existing monograph on medical theory in China is "Yellow Emperor's Canon of Internal Medicine", which is generally believed to have been compiled during a long period from the Warring States Period to the Han Dynasty, and later divided into two books: "Plain Questions" and "Miraculous Pivot". Although the "Yellow Emperor's Canon of Internal Medicine" is a special book on medical theories, it also contains 13 prescriptions, such as Zuo Jiao Fa Jiu for treating dead syncope and the Ji shi li for treating meteorism, which are the representatives of early medicinal wine.

（二）魏晋南北朝时期（Wei Jin and Southern and Northern Dynasties）

南朝齐梁时期的著名本草学家陶弘景，总结了前人采用冷浸法制药酒的经验，在《神农本草集经注》一书中，提出了采用冷浸法制作药酒的一套常规方法："凡渍药酒，皆须细切，生绢袋盛之，乃入酒密封，随寒暑日数，视其浓烈，便可漉出，不必待至酒尽也。滓可暴燥微捣，更渍饮之，亦可散服。"

Tao Hongjing, a famous herbologist in the Qiliang period of the Southern Dynasties, summarized the experience of predecessors in using cold soaking method to make medicinal wine, and put forward a set of conventional methods of using cold soaking method to make medicinal wine in "Shennong Becao Jing Jizhu": "Where the stained medicinal wine, must be fine cut, put in the raw silk bag, is sealed into the wine, with the number of cold and summer days, depending on its strong, it can be filtered, do not have to wait until the wine is done. The dregs can be violently dry and slightly mashed, and even more stained and drunk, and can also be dispersed."

（三）唐宋时期（Tang and Song Dynasties）

该时期的药酒制法有继承前代的酿造法、冷浸法及热浸法，以前两者为主。如《外台秘要》"古今诸家酒"的 11 种药酒配方中，就有 9 种是采用加药酿造法制取的，其生产工艺颇为详尽。在《圣济总录》中，则有多种药酒采用隔水加热的水浴法"煮酒"的，如"腰痛门"中的狗脊酒，要求将药浸于酒中、封固容器，"重汤煮"（即隔水加热）后，方能取出、放凉饮用。这种热浸法对后世有重要影响。

The method of making medicinal wine in this period has inherited the brewing method, cold soaking method and hot soaking method of the previous generation, and the former two are mainly. For example, among the 11 kinds of medicinal wine formulations of "Ancient and modern family wines" in "Wai Tai Mi Yao", 9 kinds are made by adding medicine and brewing, and the production process is quite detailed. In the "Sheng Ji Zong Lu", there are a variety of medicinal wine using the water bath method of water heating to "boil wine", such as the Gouji wine in the prescript "Yao Tong Men", which requires the medicine to be soaked in the wine, sealed the container, and "heavy soup cooking" (that is, heated on water), then it can be taken out, cooled and drunk. This method of hot immersion has an important influence on later generations.

（四）元明清时期（Yuan Ming and Qing Dynasties）

该时期的养生保健酒不断发展，如元代的《饮膳正要》是我国第一部营养学专著，此书从食疗的角度，选辑了 10 多种药酒，其用药少而精，且多有保养作用；明代的《扶寿精方》，集方极精，其中有著名的延龄聚宝酒及史国公药酒等，在《万病回春》《寿世保元》两书中，载有近 40 种配伍较好的、以补益作用为主的药酒，如八珍酒、延寿酒、长春酒、红颜酒、延寿瓮头春、扶衰仙凤酒、长生固本酒等；清代更盛行养生保健酒，如乾隆饮用的益寿药酒"松龄太平春酒"，对老者的诸虚百损、关节酸痛、纳食乏味、夜不成眠等症，都有较明显的辅助疗效。清代对上述这类酒的服用方法、作用机理及其疗效，均有详细的研究和记载。

我国的医药学曾长期居于世界先进行列，并对朝鲜、日本等国产生重大影响。后人理应在前人的基础上，不断深入地对中药材成分及其作用进行研究，以推动配制酒等工业的发展。

The period of health care wine continued to develop, such as the Yuan Dynasty "drink food is about" is China's first monograph of nutrition, this book from the perspective of food therapy, selected more than 10 kinds of medicinal wine, it use less medicine but fine, and more maintenance effects; Ming Dynasty's "Fu Shou Jing Fang", "set the Square very fine", which has the famous Yanling Jubao wine and Shi Guogong medicine wine, etc., in the "Wan Bing Hui Chun", "Shou Shi Bao Yuan" two books, containing nearly 40 kinds of medicine wine with better, to tonic effect, such as Ba Zhen Jiu, Yan Shou Jiu, Changchun Jiu, Hongyan Jiu, Yanshou Weng Tou Chun, Fushuai Xian Feng Jiu, Chang Sheng Gu Ben Jiu; The Qing Dynasty more popular health wine, such as emperor Qianlong drinking life-prolonging medicine wine "SongLing TaiPing Chun Jiu", on the old people's various deficiency hundred damage, joint pain, anorexia, insomnia and other diseases, have a more obvious auxiliary effect. In the Qing Dynasty, the administration method, mechanism of action and curative effect of these wines were also studied and recorded in detail.

China's medicine has long been among the world's advanced ranks, and has a significant impact on Korea, Japan and other countries. On the basis of the predecessors, the future generations should continue to deeply study the ingredients of Chinese medicinal materials and their functions in order to promote the development of wine industry.

（五）现代配制酒发展（The Development of Modern Intergrated Alcoholic Beverages）

20 世纪 50 年代至 20 世纪 60 年代，我国配制酒无论在品种数量和产量，以及产品质量和生产技术上，均有很大的增加、提高和发展。但在 20 世纪 60 年代，由于众所周知的原因，大量"三精一水"（酒精加糖精、合成香精及水）式的配制酒（包括配制型汽酒）充斥市场，这种"多、快、'好'、省"式的生产方式，给以后的配制酒工业及市场，长期蒙上了一层难以抹去的阴影，可以说，在某种程度上，至今仍未摆脱这种不良影响。20 世纪 70 年代末期至 20 世纪 80 年代末，配制酒的产量又较快地增长。在 1988 年，果露酒的产量达到了历史上最高水平；但随后又呈下降趋势，至今仍未出现明显转机。据国家有关权威部门统计，1989 年我国饮料酒的总产量为 2 787.96 万 t，其中啤酒为 2 098.7 万 t，白酒为 502.26 万 t，黄酒为 140 万 t，葡萄酒为 27 万 t，果露酒为 20 万 t（未说明果酒和露酒两者的比例）。在 2001 年春节之前的北京市场上，连配制型的山楂酒也难以找到，只有一种加强型的苹果酒、葡萄汽酒及配制型的蜜枣酒；另有一种杯装或瓶装的梅酒，虽然均为配制酒，但价格也不菲。有的瓶装梅酒，采用酒精、糖、蜂蜜及青梅制成，在容量为 720 mL 的瓶中，装有总质量为 80 g 的几颗青梅果。由此可见，目前市场上配制酒的品种，实在是少得可怜，但也表明其广阔的发展空间。

进入 21 世纪，随着人们生活水平的提高，以及环境污染和各种压力增加导致的"现代病"威胁日益加剧，人们对健康更加关注，各种"保健酒"开始大行其道。需要说明的是，尽管许多配制酒的发明是治疗疾病的需要，但是配制酒并不是药物，并不能当作治疗疾病的主要手段，它更多的是作为一种辅助手段。如茅台生产的大汉玉液酒、五粮液的苦荞酒，都是一种健康配制酒，但是并不能保证就一定能治疗疾病，只能是辅助调理。

From the 1950s to the 1960s, Chinese integrated alcoholic beverages has greatly increased, improved and developed in terms of the number of varieties and output, as well as product quality and production technology. However, in the 1960s, due to well-known reasons, a large number of " three refined and one water " (alcohol and saccharin, synthetic flavor and water) type of prepared wine (including prepared sparkling wine) flooded the market, this " more, faster, ' good ', provincial " type of production mode, to the future integrated alcoholic beverages industry and market, long-term cast an indelible shadow, it can be said, to some extent, it is still not rid of this bad influence. From the late 1970s to the late 1980s, the production of integrated alcoholic beverages increased rapidly, and in 1988, the output of " fruit wine " reached the highest level in history. But then there was a downward trend, and so far, there has been no obvious turnaround. According to the statistics of the relevant authorities of the state, the total output of beverage wine in China in 1989 was 27.879 6 million t, of which beer was 20.987 million t, baijiu was 5.022 6 million t, rice wine was 1.4 million t, wine was 270 000 t, fruit wine was 200 000 t. It is not stated that the proportion of fruit wine and liqueur is roughly half. In the Beijing market before the Spring Festival in early 2001, even the formulated hawthorn wine was difficult to find, only one kind of enhanced apple wine, gasified wine and formulated Jujube wine. There is also a cup or bottle of plum wine, although they are prepared wine, but the price is not cheap. Some bottled plum wine, made of alcohol, sugar, honey and green plums, in a bottle with a capacity of 720 mL, there are several green plums with a total weight of 80 g. It can be seen that the varieties of wine prepared on the market at present are really very few, but it also shows that there is a broad space for development.

In the 21st century, with the improvement of people's living standards, as well as the increasing threat of so-called "modern diseases" due to environmental pollution and various pressures, people are more concerned about health, and various "health wine" has become popular. It should be noted that although many of the invention of integrated alcoholic beverages is the need to treat diseases, but the preparation of wine is not medicine, and can not be used as the main means of treating diseases, it is more as an auxiliary means to use, like Moutai production of Dahan Yuye wine, Wuliangye Tartary buckwheat wine, are a kind of healthy preparation of wine, but they can not guarantee that they will be able to treat diseases, only say assisted conditioning.

四、中国配制酒的生产（The Development of Chinese Wine Preparation）

一种是在酒和酒之间进行勾兑配制；另一种是以酒与非酒精物质，包括液体、药材、香料和植物等浸泡而成的或直接添加食品添加剂进行调配。

One is blending between wine and wine; The other is to blend wine with non-alcoholic substances, including liquids, herbs, spices and plants, or directly add food additives.

五、中国配制酒的特点（Characteristics of Chinese Integrated Alcoholic Beverage）

（1）基酒为白酒和黄酒：中国配制酒的制作方法与外国配制酒大致相同。不同之处是中国配制酒所用的基酒为白酒和黄酒，绝大多数配制酒均采用中药材为辅料，具有较高的医疗价值。

The liquor base is baijiu and yellow rice wine: the production method of Chinese integrated alcoholic beverage is roughly the same as that of foreign integrated alcoholic beverage. The difference is that the liquor base used in China is baijiu and yellow wine, and most of the liquor is made of Chinese medicinal materials, which has high medical value.

（2）采用动物性原料：外国配制酒一般不采用动物性原料，而我国加入乌鸡、鹿茸、蛇等动物性原料制成滋补型、疗效型配制酒。

The use of animal raw materials: foreign integrated alcoholic beverage generally does not use animal raw materials, but in China adds black chicken, deer antler, snake and other animal raw materials to make a tonic, therapeutic integrated alcoholic beverage.

（3）多药材配制：我国最初采用"一酒一药"，即一种酒只用一种药材制成，后来逐渐发展到用多种药材配制成一种酒。

Preparation of multi-medicinal materials: China initially adopted "one wine and one medicine", that is, a wine made of only one medicinal material, and later gradually developed into a wine prepared with a variety of medicinal materials.

（4）配制原材料品种呈多样化发展：中国配制酒从原有的以草药或动物性原料为主要调制原料，发展到使用各种花卉、果实等原料配制，已成为花色品种最多的酒类。

The varieties of preparation raw materials have diversified development: Chinese has developed from the original herbal or animal raw materials as the main preparation raw materials to the use of various flowers, fruits and other raw materials, and has become the most colorful varieties of wine.

（5）酒精度各有不同：中国配制酒的酒度为20°～45°；一般药酒类酒度较低为20°～

30°；芳香植物类的配制酒度较高，大约为 40°。

The alcohol content is different: the alcohol content of Chinese integrated alcoholic beverage is 20°—45°, general medicine liquor alcohol degree is low, about 20°—30°, the alcoholic content of aromatic plants is higher, about 40°.

（6）含糖量高：我国配制酒多数属甜型酒，含糖量较高。

High sugar content: most of chinese integrated alcoholic beverage wine is sweet wine, with high sugar content.

六、中国配制酒的饮用方式（Drinking Method of Chinese Integrated Alcoholic Beverage）

（一）酒杯与分量（Glass and Portion Size）

中国配制酒一般采用利口酒杯或古典杯盛酒，分量为 45 毫升左右。

Liqueur glasses or classical cups are generally used to integrated alcoholic beverage in China, the weight is about 45 mL.

（二）饮用方法（Drinking Method）

中国配制酒多数宜作为餐后饮用。滋补型配制酒可在进餐、餐后或睡前适量饮用。花类、果实类配制酒可冰镇或加冰块后饮用。

Most Chinese wine is suitable for drinking after meals. Tonic wine can be consumed in moderation during meals, after meals or before going to bed. Flower and fruit wine can be served chilled or with ice.

赛证直通 Competitions and Certificates

一、基础知识部分（Basic Knowledge Part）

（一）专业名词解释（Explanation of Professional Terms）

1. 配制酒（Integrated Alcoholic Beverage）：
2. 勾调配制（Blending）：

二维码 1-4-4：
知识拓展

（二）思考题（Thinking Question）

谈谈中外配制酒的异同点。

Talk about the similarities and differences between Chinese and foreign integrated alcoholic beverage .

二、技能操作部分（Skill Operation Part）

请推荐适合腿脚受寒的老年人饮用的配制酒，并说明原因。

Please recommend the integrated alcoholic beverage suitable for the elderly with cold legs and feet, and explain the reason.

【单元四　反思与评价】Unit 4　Reflection and Evaluation

学会了：_____

Learned to:_____

成功实践了：_____

Successful practice:_____

最大的收获：_____

The biggest gain:_____

遇到的困难：_____

Difficulties encountered:_____

对教师的建议：_____

Suggestions for teachers:_____

单元五　鸡　尾　酒

　　杰弗里·摩根塔勒（Jeffrey Morgentmaler）是现代鸡尾酒领域的杰出人物之一。他是美国俄勒冈州波特兰市著名的调酒师，也是一位博客作者和书籍作者。他所著的名为《酒吧之书：鸡尾酒制作》(*The Bar Book Elements of Cocktail Technique*) 的书籍，成了鸡尾酒制作的标准参考书之一。

单元导入 Unit Introduction

　　鸡尾酒，一种充满创意与魅力的饮品，是酒水文化中的一颗璀璨明星。它是由多种酒和各种辅料按照一定比例混合而成，每一款鸡尾酒都如同一个艺术品，独特而迷人。在本单元中，我们将一起揭开鸡尾酒的神秘面纱，了解它的起源、发展，并学习如何制作和品鉴。我们将探索不同风味的鸡尾酒，感受它们带来的惊喜与愉悦。此外，我们还会讨论鸡尾酒背后的文化和礼仪，如何在享受美酒的同时展现自己的风度和品位。

　　Cocktail, a drink full of creativity and charm, is a bright star in the beverage culture. It is composed of a variety of wines and beverages in accordance with a certain proportion of the mixture, each cocktail is like a work of art, unique and charming. In this unit, we will unravel the mystery of the cocktail, learn about its origin, development, and learn how to make and taste it. We will explore the different flavors of cocktails and feel their surprise and delight. In addition, we will discuss the culture and etiquette behind cocktails, how to enjoy the wine while showing your style and taste.

学习目标 Learning Objectives

➤ 知识目标（Knowledge Objectives）

（1）能描述鸡尾酒的定义、特点以及种类等。

Be able to describe the definition, characteristics and types of cocktails.

（2）能描述鸡尾酒调制的基本原则及配色原则。

To describe the basic principles of cocktail preparation and color matching.

➤ 能力目标（Ability Objectives）

（1）会使用调酒用具及制作装饰物。

Be able to use bartending utensils and make decorations.

（2）能根据所学知识制作创意鸡尾酒。

Can make creative cocktails according to the knowledge.

➤ 素质目标（Quality Objectives）

（1）培养学生具备良好的职业精神、专业精神和工匠精神。

To cultivate students with good professional spirit, professionalism and craftsman spirit.

（2）在制作装饰物的过程中培养学生的劳动精神，在调制鸡尾酒的过程中培养学生的钻研精神并进行挫折教育及双创教育。

Cultivate students' spirit of labor in the process of making decorations, cultivate students' spirit of specialization in the process of making cocktails, and conduct frustration education and entrepreneurship education.

专题一
琳琅满目：鸡尾酒认知

【微课】
琳琅满目：
鸡尾酒

你知道图 1-5-1 中展示的是什么酒水吗？

Do you know what drinks are shown in the picture 1-5-1?

鸡尾酒，作为一种混合酒精饮料，以其独特的魅力与多样性，成了酒水世界中的一颗璀璨明珠。它不仅仅是一种饮品，更是一种文化的象征，一种艺术的表达。鸡尾酒的历史源远流长，其起源可以追溯到 18 世纪的美国，自诞生以来便以其丰富的口感和创新的调制方式，吸引了无数人的喜爱与追捧。

图 1-5-1　酒水（drinks）

Cocktail, as a kind of mixed alcoholic beverage, with its unique charm and diversity, has become a shining pearl in the wine world. It is not only a drink, but also a symbol of culture and an expression of art. The cocktail has a long history, its origin can be traced back to the 18th century in the United States, since its birth with its rich taste and innovative ways of making, attracting countless people's love and pursuit.

一、鸡尾酒的定义（Definition of Cocktail）

鸡尾酒是一种精心调制的混合饮料，其基础成分通常包括一种或数种酒，基酒一般为蒸馏酒，其他酒类还包括酿造酒、配制酒等。这些基酒与汽水、果汁、乳品等丰富的配料相结合，通过特定的调制手法，最终呈现出一种集色泽、香气、口感和形态于一体的独特饮品。

A cocktail is an elaborate mixed drink whose base ingredients usually include one or more spirits, especially distilled, brewed, and prepared spirits. These base wines are combined with rich ingredients such as soft drinks, juices and dairy products, and through specific modulation techniques, they finally present a unique drink that integrates color, aroma, taste and form.

二、鸡尾酒的特点（Characteristics of Cocktail）

（一）种类繁多，调法各异（Many Tricks and Different Mixing Methods）

鸡尾酒作为一种混合酒精饮料，其多样性源于其原料品种的丰富性及调制方法的多样性。这些因素共同作用，使得鸡尾酒的种类繁多。每一种鸡尾酒都展现了其独特的口感、风味和文化背景，成为酒水世界中一道亮丽的风景线。

As a kind of mixed alcoholic beverage, the diversity of cocktail is due to the richness of its raw materials and the diversity of preparation methods. The combination of these factors has resulted in a huge variety of cocktails. Each cocktail shows its unique taste, flavor and cultural background, becoming a beautiful scenery in the wine world.

（二）能增进食欲（Enhancing Appetite）

鸡尾酒独特的口感和风味不仅能够刺激味蕾，提升食欲，还能在一定程度上激发饮者的精神状态，带来一种愉悦的感受。这种特性使鸡尾酒在餐饮场合受到广泛的欢迎，成为许多人喜爱的饮品之一。

Cocktail with its unique pungent taste, and flavor can not only stimulate the taste buds, enhance the appetite, but also stimulate the drinker's mental state to a certain extent, bringing a feeling of pleasure. This characteristic makes the cocktail widely popular in the dining occasions, becoming one of the favorite drinks of many people.

（三）口味丰富（Rich Taste）

鸡尾酒在口感和风味上通常被认为具有优于单一酒类的特点。通过精心调配多种酒类和其他成分，鸡尾酒能够创造出更加丰富、复杂且均衡的口感体验，使整体口味更加引人入胜。

相较于单一酒类，鸡尾酒在层次感和多样性上更具优势，能够满足不同人群对于口感和风味的追求。

Cocktails are generally considered to be superior to single liquors in taste and flavor. By carefully mixing a variety of wines and other ingredients, cocktails can create a richer, more complex and balanced experience, making the overall taste more appealing. Compared with a single wine, cocktails have more advantages in hierarchy and diversity, which can meet the pursuit of taste and flavor of different people.

（四）冷饮性质（Cold Nature）

在调制鸡尾酒的过程中，保持其冷饮性质是确保鸡尾酒口感和风味得以完美呈现的关键步骤。由于鸡尾酒的风味和口感在很大程度上受到温度的影响，因此，调制时保持鸡尾酒处于低温状态至关重要。这有助于保持其清爽的口感和独特的风味，使鸡尾酒呈现出最佳的饮用效果。当然，在现代创意鸡尾酒的制作过程中，也可以根据客人的要求，做成热的鸡尾酒。

In the process of making a cocktail, maintaining its cold drink nature is a key step to ensure that the taste and flavor of the cocktail are perfectly presented. Since the flavor and taste of a cocktail are largely affected by temperature, it is essential to keep the cocktail cold when preparing it. This helps to maintain its refreshing taste and sharp flavor, making the cocktail present the best drinking results. Of course, in the production process of modern creative cocktails, you can also make hot cocktails according to the requirements of guests.

（五）装饰美观，盛装考究（Beautiful Color and Decoration）

鸡尾酒以其色泽优美和盛装考究为特点，展现了其独特的魅力。其色泽通常经过精心调配，呈现出鲜艳、明亮或深沉等不同的色彩，给人以视觉上的享受。同时，鸡尾酒的盛装也较为讲究，通常使用透明或色彩协调的载杯，并配以精美的装饰物，如水果片、薄荷叶等，使整体呈现更加考究和精致。这种对色泽和装饰的重视，进一步提升了鸡尾酒的观赏性和品鉴体验。

The cocktail is characterized by its beautiful color and containing, showing its unique charm. Its color is usually carefully adjusted, presenting different colors such as bright-colored, bright or deep, giving people visual enjoyment. At the same time, the containing of cocktail is also exquisite, usually using transparent or color-coordinated glasses, and with exquisite decoration and ornaments, such as fruit slices, mint leaves, etc., making the overall presentation more sophisticated and delicate. This emphasis on color and decoration further enhances the viewing and tasting experience of cocktails.

三、鸡尾酒的分类（Classification of Cocktail）

（一）餐前鸡尾酒（Per-dinner Cocktail）

餐前鸡尾酒作为餐前开胃的饮品，经过精心调配，旨在激发食欲并为接下来的用餐体验做好铺垫。这些鸡尾酒通常分为甜与不甜两大类，以满足不同用餐者的口味偏好。以马天尼和曼哈顿为代表的餐前鸡尾酒，凭借其独特的风味和口感，成了这一品类的经典之作。马天尼以其清爽的口感和独特的调制工艺，为用餐者带来一种清新的开胃体验；曼哈顿则以其丰富的层次感和辛辣的口感，为用餐增添一份期待。

As a pre-meal appetizer drink, is carefully prepared to stimulate the appetite and prepare for the next dining experience. These cocktails are usually divided into two categories, sweet and not sweet, to meet the taste preferences of different diners. Pre-dinner cocktails, represented by Martini and Manhattan, have become classics of this category with their unique flavor and texture. Martini, with its refreshing taste and unique concoction process, brings diners a fresh and appetizing experience; Manhattan, on the other hand, is fascinating with its rich layers and spicy taste, adding an expectation to the dining.

（二）俱乐部鸡尾酒（Club Cocktail）

俱乐部鸡尾酒在正式餐饮场合中尤为常见，通常与正餐一起提供，或者在某些情况下作为头盆汤菜的替代品。俱乐部鸡尾酒以其鲜艳的色泽、丰富的营养价值和适度的刺激性为特点，为用餐者提供了别样的味觉体验。在这些鸡尾酒中，三叶草俱乐部以其独特的配方和卓越的口感脱颖而出，成了俱乐部鸡尾酒中的佼佼者。

These cocktails are particularly common at formal dining occasions and are usually served with the main meal, or in some cases, they can also be used as an alternative to the first soup dish. Club cocktails are characterized by their bright color, rich nutritional value and moderate irritability, providing diners with a different taste experience. Among these cocktails, Clover Club stands out as a standout among club cocktails with its unique recipe and excellent taste.

（三）餐后鸡尾酒（After Dinner Cocktail）

餐后鸡尾酒多呈现甜美的风味，主要用以满足餐后对甜味的追求和味蕾的享受。在这些酒品中，亚历山大堪称典范，其甜美的口感和细致的调制工艺，使其成为餐后鸡尾酒中备受欢迎的佳选。

This kind of cocktail presents a sweet flavor, mainly to meet the pursuit of sweetness after dinner and the enjoyment of taste buds. Among these wines, Alexander is a good example, and its sweet taste and meticulous preparation make it a popular choice for after-dinner cocktails.

（四）晚餐鸡尾酒（Dinner Cocktail）

晚餐鸡尾酒特别适宜在晚餐时享用，其特点在于口味较为辣烈，能够为晚餐增添一抹独特的刺激感。其中，法国的鸭臣鸡尾酒便是这类酒品中的一款知名之作，以其独特的辣味，为晚餐增添了一分别样的风味体验。

This type of cocktail is particularly suitable for dinner to enjoy, which is characterized by a more spicy taste, can add a unique sense of excitement for dinner. Among them, the French Absinthe cocktail is a well-known wine in this category, with its unique spicy taste and flavor, adding a different flavor experience to the dinner.

（五）香槟鸡尾酒（Champagne Cocktail）

在盛大的庆祝宴会上，香槟鸡尾酒往往作为不可或缺的饮品之一，其独特的口感和制作方式深受人们喜爱。制作香槟鸡尾酒时，首先将各种调制材料预先放入杯中，经过精心调配，以达到理想的口感和风味，随后缓缓倒入适量的香槟酒，使其与其他材料充分融合。这种制作方式既简单又独特，使得香槟鸡尾酒在庆祝场合中备受欢迎。

In the grand celebration banquet, champagne cocktail is often as one of the indispensable drinks, its unique taste and production method is deeply loved by people. When making a champagne cocktail, various ingredients are first put into the glass in advance and carefully prepared to achieve the desired taste and flavor. Then, slowly pour in the appropriate amount of champagne, so that it is fully integrated with the other ingredients. This method of preparation is both simple and unique, making champagne cocktails popular for celebratory occasions.

四、调制鸡尾酒的原料、用具 (Ingredients and Utensils For Mixing Cocktail)

（一）原料 (Raw Materials)

调制鸡尾酒所用的原料包括基酒、辅料和装饰物。

Ingredients used to make a cocktail include base wine, auxiliary materials and ornaments.

1. 基酒 (Base Wine)

鸡尾酒的核心部分是烈酒，这些烈酒被称为"基酒"，常见的基酒包括白兰地、威士忌、金酒、朗姆酒、伏特加和特基拉酒等。然而，也有一些鸡尾酒选择使用中国白酒、葡萄酒或香槟等作为基酒。不过，值得一提的是，极少数鸡尾酒并不包含酒精成分，而是完全由软饮料调制而成。

The core of the cocktail is mainly spirits, which are called "base spirits". Which commonly include brandy, whiskey, gin, rum, vodka, and tequila. However, there are some cocktail options that use chinese spirits, wine or champagne as a base wine. However, it is worth mentioning that there are very few cocktails that do not contain alcohol, but are made entirely from soft drinks.

2. 辅料 (Auxiliary Materials)

（1）加色加味：包括开胃酒类、葡萄酒类、香槟及利口酒类。

Add color and flavors: including aperitif wine, grape wine, champagne and liqueur.

（2）调缓溶液：鸡尾酒中，碳酸饮料和果汁扮演了重要角色，它们的主要作用是降低酒的度数，同时确保酒体的风味不被改变。

Mixing solution: in cocktails, carbonated drinks and fruit juices play an important role, their main role is to reduce the alcohol, while ensuring that the body flavor is not changed.

（3）香料：为增加酒体风味的芳香性植物。

Perfume: aromatic plant to add flavor to the liquor body.

（4）其他辅料：如牛奶、奶油、鸡蛋、砂糖、精盐、咖啡、茶等。

Other auxiliary materials: including milk, cream, eggs, sugar, refined salt, coffee and tea, etc.

（5）冰是鸡尾酒的生命。按冰的大小分为块冰、碎冰、冰坨等。

Ice is the life of cocktail. According to the size of ice, it is divided into block ice, broken ice, and ice lump, etc.

3. 装饰物 (Ornaments)

鸡尾酒不仅仅是一种饮品，更是一种极具装饰性的艺术品。精心挑选的装饰物，能够极大地提升鸡尾酒的观赏性，使其成为一种视觉与味觉的双重享受。樱桃、草莓、柠檬、橙子、菠萝等水果以及芹菜、黄瓜等蔬菜，各种果签、调酒棒，糖及盐霜等都常被用作装饰物。

The cocktail is not just a drink, it is a highly decorative art wine. A carefully selected decoration can greatly enhance the viewing and tasting of the cocktail, making it a dual enjoyment of vision and taste. Fruits such as cherry, strawberry, lemon, orange, pineapple and vegetable such as celery, cucumber, and various fruit sticks, swizzle stick, sugar and salt frosting tend to be used as decorations.

(二) 调酒用具 (Cups for Bartending)

鸡尾酒杯（图1-5-2）作为高脚杯家族的一员，其独特之处在于其三角形的杯身设计。杯底既有尖形也有圆形，其修长的或圆粗的杯脚，均展现出光洁透明的质感。在容量上，它可容纳2至6oz（60至180毫升）的液体，其中41/2oz的容量最为常见。除此之外，鸡尾酒杯还有众多不同款式，但它们都有一个共同点：不带任何花纹、色彩和不可用塑料制作。

Cocktail glass(Figure 1-5-2) as a member of the tall glass family, its unique feature is its triangular body design. The bottom of the cup can be pointed or round, and its slender or round feet show a smooth and transparent texture. In terms of capacity, it can hold 2 to 6oz (approximately 60 to 180 ml) of liquid, with 41/2oz capacity being the most common. There are many different shapes of cocktail glasses, but they all have one thing in common: don't have any patterns or colors, do not use plastic cups.

图1-5-2 不同的鸡尾酒杯（different cocktail glasses）

通过这次专题的学习，我们深入了解了鸡尾酒的丰富多样性并感受到其独特魅力。从餐前到餐后，从庆祝到日常，鸡尾酒以其多变的口味、丰富的层次和独特的文化内涵，成了全球范围内广受欢迎的饮品。随着鸡尾酒文化的不断发展，相信未来还将有更多的创新和精彩等待我们去发现和探索。

Through the study of this topic, we have gained an in-depth understanding of the rich diversity and unique charm of cocktails. From before to after dinner, from celebration to daily life, cocktails have become a popular drink around the world with their varied tastes, rich levels and unique cultural connotations. With the continuous development of cocktail culture, we believe that there will be more innovations and wonderful waiting for us to discover and explore in the future.

赛证直通 Competitions and Certificates

一、基础知识部分 (Basic Knowledge Part)

(一) 专业名词解释 (Explanation of Professional Terms)

基酒（Basic Liquor）：

二维码1-5-1：
知识拓展

1

（二）思考题（Thinking Question）

谈谈对鸡尾酒文化的理解。

Talk about your understanding of cocktail culture.

二、技能操作部分（Skill Operation Part）

在实训室制作鸡尾酒的装饰物。

Making cocktail decorations in the training room.

【微课】
五彩斑斓：
调制原则与
配色原理

专题二
五彩斑斓：调制原则与配色原理

你知道图中鸡尾酒这绚丽多彩的颜色是如何调制出来的吗（图1-5-3）？

_____

Do you know how the colorful cocktail in the picture is mixed (Figure 1-5-3)?

_____

图 1-5-3 鸡尾酒（cocktail）

鸡尾酒，它就是调酒师手中的艺术品。每一款鸡尾酒都有它独有的配方和色彩，这些元素共同构成了鸡尾酒独特的魅力。

A cocktail. It's a work of art in the hands of a bartender. Each cocktail has its own unique formula and color, these elements together constitute the unique charm of the cocktail.

一、调酒的基本原则（Basic Principles of Bartending）

（一）灵活性原则（Flexibility principle）

在调制鸡尾酒时，应遵循配方中规定的材料、质量、种类、分量和步骤，确保饮品的一致性和品质。然而，也存在一种特殊情形，即当客人根据个人喜好要求对传统配方进行更改时，应充分尊重客人的意愿，避免过于坚持原有的配方而与客人产生争论。这种灵活的处理方式体现了对客人个性化需求的尊重和满足，是服务行业中重要的职业素养之一。

While making cocktails, always follow the ingredients, qualities, types, portions and steps specified in the recipe to ensure consistency and quality. However, there is a special case where guests request to change the traditional recipe according to their personal preferences, we should fully respect the wishes of the guests and avoid overly adhering to the original recipe and causing disputes with the guests. This flexible approach reflects the respect and satisfaction of the individual needs of guests, and is one of the important professional qualities in the service industry.

（二）准确性原则（Accuracy principle）

在调制鸡尾酒时，必须确保使用正确的调酒工具、调酒器，以及各类载杯。这些专业工具和设备不仅关系到鸡尾酒调制的准确性和效率，还直接影响到其品质和口感。因此，必须严格遵守规范，不得混用或代用调酒工具和设备，确保调制过程的准确性和饮品的品质。

While making cocktails, you must ensure that you use the correct cocktail tools, shakers, and carrier cups. These professional tools and equipment are not only related to the accuracy and efficiency of beverage preparation, but also directly affect the quality and taste of beverages. Therefore, it is necessary to strictly comply with the regulations and do not mix or substitute bartenders and equipment to ensure the accuracy of the mixing process and the quality of the drink.

（三）标准化原则（Principle of standardization）

在按照既定配方进行鸡尾酒或其他饮品的调制时，所得到的酒品口味应当符合该配方的标准口味要求。这是为了确保饮品的一致性和品质，同时也为了满足消费者对特定口味的需求。因此，在调制过程中，必须严格遵守配方要求，确保各种原料的比例、混合顺序和调制方法均准确无误。

When making cocktails or other drinks according to a given recipe, the resulting wine taste should meet the standard taste requirements of the recipe. This is to ensure the consistency and quality of the drink, but also to meet the needs of consumers for specific tastes. Therefore, in the modulation process, it is necessary to strictly comply with the formula requirements to ensure that the proportion of various raw materials, mixing sequence and modulation methods are accurate.

（四）新鲜度原则（Freshness principle）

在调制鸡尾酒时，对于基酒的选择应着重考虑性价比，即选择价格合理且品质上乘的基酒。对于辅料，尤其是奶、蛋、果汁等原料，应确保其新鲜度和质地优良，以保证最终饮品的风味和品质。新鲜的配料不仅能够为饮品增添丰富的口感和风味，还能够确保饮品的安全性和健康性。因此，在选购和准备调酒原料时，应严格把关，确保所有原料都符合高品质和新鲜度的要求。

While making cocktails, the choice of base wine should focus on its cost performance, that is, choose a reasonable price and high quality base wine. For ingredients, especially raw materials such as milk, eggs, juice, etc., the freshness and quality should be ensured to ensure the flavor and quality of the final drink. Fresh ingredients can not only add rich taste and flavor to the drink, but also ensure the safety and health of the drink. Therefore, when purchasing and preparing raw materials for mixing, strict control should be carried out to ensure that all raw materials meet the requirements of high quality and freshness.

（五）保持清洁（Keep clean principle）

在进行鸡尾酒调制前，应准备足够数量的器皿和工具，并确保这些器皿和工具始终保持清洁状态。清洁的器皿和工具是确保饮品调制过程卫生、安全的基础，同时也有助于提高调制效率，保证连续操作的顺畅进行。因此，在调制过程中，应随时注意检查器皿和工具的清洁状况，在必要时及时更换或清洗。

Before we start making cocktails, prepare a sufficient number of utensils and tools and ensure that these utensils and tools are kept clean at all times. Clean utensils and tools are the basis for ensuring the health and safety of the beverage preparation process, and also help to improve the efficiency of the preparation and ensure the smooth operation of the continuous operation. Therefore, during the modulation process, you should always pay attention to check the cleaning condition of utensils and tools, and replace or clean them in time if necessary.

（六）做好准备工作（Make good preparations）

在进行鸡尾酒调制之前，为了确保调制过程的顺利进行，应当预先准备好所需的酒杯和各种材料。这是鸡尾酒调制过程中不可或缺的一环，旨在确保在调制过程中能够迅速、准确地获取所需物品，从而提高调制效率并减少出错的可能性。

Before cocktail making, in order to ensure the smooth progress of the mixing process, the required glasses and various materials should be prepared in advance. This is an integral part of the cocktail making process and is designed to ensure that the desired items can be obtained quickly and accurately during the mixing process, thereby increasing the efficiency of the mixing and reducing the possibility of errors.

（七）选用合适冰块（Choose appropriate ice cubes）

在调制鸡尾酒时，应选用质地坚硬且不易融化的新鲜冰块。新鲜的冰块不仅能有效冷却饮品，还能确保饮品口感的纯净和清爽。为避免冰块融化过快，不可重复使用冰块，以确保饮品的品质和口感。

When making cocktails, use fresh ice that is firm and does not melt easily. Fresh ice not only effectively cools the drink, but also ensures that the taste of the drink is pure and refreshing. In order to prevent the ice from melting too quickly, it is forbidden to reuse the used ice to ensure the quality and taste of the drink.

（八）起泡材料的正确使用方法（Proper use of Sparkling materials）

在调制鸡尾酒或其他饮品时，若遇到会起泡的辅料，应避免将其直接放入摇酒器、电动搅拌器或榨汁器中。这是因为容易起泡的配料在机械搅拌或摇动过程中可能产生过度的气泡，影响饮品的质量和口感且会溢出。若配方中确实含有会起泡的原料，且需要使用摇和法或搅和法进行调制，应先加入其他材料进行摇晃或搅拌，待其充分混合后最后加入起泡材料。这样可以在确保饮品口感的同时，有效避免过度泡沫的产生。

When making cocktails or other drinks, if you encounter ingredients that can fizz, avoid putting them directly into the shaker, electric mixer or juicer. This is because the foaming ingredients may produce excessive foam during mechanical stirring or shaking, affecting the quality and taste of the drink and will overflow. If the formula does contain foaming raw materials and needs to be modulated by shaking or stirring method, other materials should be shaken or stirred first, and then the foaming material is added after it is fully mixed. This can ensure the taste of the drink at the same time, effectively avoid excessive foam.

（九）载杯的选用（Selection of Drinkware）

鸡尾酒的呈现应当选用合适的载杯。所选的杯型及其容量应与具体配方的要求相符合，以确保调制出的饮品能够呈现出最佳的口感和视觉效果。合适的杯型不仅有助于保持饮品的温度，还能够提升整体的饮用体验。在选择载杯时，需要综合考虑配方特点、饮品风格以及顾客需求等因素，确保所选杯型与饮品相得益彰（图1-5-4）。

The cocktail should be presented in a suitable carrier. The selected cup shape and its capacity should be in line with the requirements of the specific formula to ensure that the concocted drink can present the best taste and visual effect. The right cup shape not only helps to maintain the temperature of the drink, but also enhances the overall drinking experience. When choosing carrier, it is necessary to consider factors such as formula characteristics, drink style and customer needs to ensure that the selected cup type and drink complement each other (Figure 1-5-4).

图 1-5-4　选择适当的载杯（choose suitable cup）

（十）良好的操作习惯（Good operating habits）

在鸡尾酒或其他饮品调配制作完成后，务必养成良好的操作习惯，即确保将瓶子的瓶盖旋紧并复位。在开瓶时，动作应迅速而稳定，要防止饮品洒出。在倒取饮品后，应旋紧瓶盖，确保瓶口密封完好，避免饮品变质或受到污染。

When making cocktails, it is important to develop good operating habits, that is, to ensure that the bottle cap is tightly closed and reset. When opening the bottle, the action should be swift and steady, to prevent drinks from spilling. After pouring the drink, tighten the bottle cap to ensure that the bottle is sealed properly to avoid deterioration or contamination of the drink.

二、配色原理与设计（Color Matching Principle and design）

鸡尾酒的色彩表现丰富且变化多端。三原色，一般是指红色、黄色和蓝色。这三种颜色被普遍认为是构成鸡尾酒色彩体系的"三大基石"或"基础色调"（图1-5-5）。三原色在鸡尾酒的色彩呈现中起着核心且基础的作用，通过它们的巧妙搭配和组合，可以调制出色彩绚丽的鸡尾酒，为宾客带来丰富的视觉享受。

The color of the cocktail is extremely rich and varied. The three primary colors generally refer to red, yellow and blue. These three colors are generally considered to be the "three building blocks" or "foundation tones" that make up the cocktail color

图 1-5-5　鸡尾酒色彩体系的"三大基石"
（"Three cornerstones" of the cocktail color system）

system (Figure 1-5-5). These three colors play a core and fundamental role in cocktail making, and through their clever collocation and combination, you can make a colorful cocktail, which brings rich visual enjoyment to the guests.

例如，红色混合黄色会得出一种新的颜色——橙色（图1-5-6）这种颜色鲜艳且充满活力，常用于表示温暖、欢乐等情感；蓝色混合黄色会得出一种新的颜色——绿色（图1-5-7）绿色通常代表着自然、和平和生长，给人一种清新、宁静的感觉；红色混合蓝色会得出一种新的颜色——紫色（图1-5-8）紫色是一种神秘且高贵的颜色，常用于表达浪漫、优雅等情感。

For example, red mixed with yellow produces a new color, orange (Figure 1-5-6). This color is bright and energetic, and is often used to indicate warmth, joy, and other emotions. Blue mixed with yellow will produce a new color-green (figure. 1-5-7) Green usually represents nature, peace and growth, giving people a fresh, peaceful feeling; Red mixed with blue will create a new color – purple (Figure 1-5-8) Purple is a mysterious and noble color, often used to express romance, elegance and other emotions.

图 1-5-6　黄色＋红色＝橙色
（Yellow + Red = Orange）

图 1-5-7　蓝色＋黄色＝绿色
（Blue + Yellow = Green）

图 1-5-8　红色＋蓝色＝紫色
（Red + Blue = Purple）

此外，通过调整三原色之间的比例，还可以得到更多的颜色。例如，在红色和橙色之间，可以得到不同深浅的橙红色；在黄色和绿色之间，可以得到黄绿色；在蓝色和紫色之间，可以得到蓝紫色等。除了上述的二次色（橙、绿、紫）之外，还可以继续探索更多的色彩变化。比如，当逐渐增加红色在橙色中的比例时，会得到从黄色到橙红色的一系列渐变色彩；调整蓝色在紫色中的比例，会得到从红色到蓝紫色的变化。当然，还可以通过混合不同的二次色或加入其他颜色来得到更为丰富的色彩。例如，将橙色和绿色混合，可以得到一种称为"橄榄绿"的颜色，它带有暖色调的绿色；将紫色和黄色混合，会得到一种带有红色调的棕色等。

在鸡尾酒的配色设计中，对色彩的理解和掌握是非常重要的。通过巧妙地运用三原色的混合和搭配，可以创造出各种不同的氛围和情绪表达。总体来说，暖色调的色彩组合可以带来温暖和欢乐的感觉，而冷色调的组合则会给人一种清凉和静谧的感受。

In addition, by adjusting the ratio between the three primary colors, more colors can be obtained. For example, between red and orange, you can get different shades of orange-red; Between yellow and green, you get yellow-green; Between blue and purple, you get blue and purple and so on. In addition to the above secondary colors (orange, green, purple), you can continue to explore more color variations. For example, when you gradually increase the ratio of red to orange, you get a series of gradual colors from yellow to orange-red; Adjusting the ratio of blue to purple will result in a change from red to blue-purple. Of course, you can also get a richer color by mixing different secondary colors or adding other colors. Mixing orange and green, for example, yields a color called "olive green," which has a warm tinge of green; Mixing purple and yellow gives you a brown color with a reddish tinge.

In the color design of cocktail, it is very important to understand and master the color. Through the clever use of the mixing and matching of the three primary colors, a variety of different atmospheres and emotional expressions can be created. In general, the combination of warm colors can bring a feeling of warmth and joy, while the combination of cool colors can give a feeling of cool and quiet.

赛证直通 Competitions and Certificates

一、基础知识部分（Basic Knowledge Part）

（一）专业名词解释（Explanation of Professional Terms）

1. 雪克壶（Shaker）：
2. 三原色（Three-Primary Colors）：

（二）思考题（Thinking Questions）

如何调制出五颜六色的鸡尾酒？

How to make colorful cocktails?

二维码 1-5-2：
知识拓展

二、技能操作部分（Skill Operation Part）

在实训室调出绿色、紫色、橙色等不同颜色的鸡尾酒。

Mix green, purple, orange and other colored cocktails in the training room.

专题三
匠心匠人：调制器具与调酒技法

【微课】
匠心匠人：
调制器具与
调酒技法

你知道图中的这杯彩虹鸡尾酒采用了什么调制方法吗（图 1-5-9）？

_____

Do you know what is used to make this rainbow cocktail in the picture (Figure 1-5-9)?

_____

要想调好一杯鸡尾酒，就要掌握好调酒器具的使用方法。

It's of paramount importance to master the use of mixdogy tools

图 1-5-9 鸡尾酒（cocktail）

to make a good cocktail.

一、调酒器具（Mixology Tools）

（一）调酒壶（Shaker）

调酒壶（图1-5-10）是一种专业工具，主要由壶盖、滤冰器和壶体三部分组成。其设计初衷在于通过摇动的方式，有效地混合壶内的调酒材料与冰块，从而实现酒液的快速冷却。常见的调酒壶根据使用需求的不同，分为大、中、小三类尺寸，以适应不同场合和调酒配方的需求。

Shaker (Figure 1-5-10) is a professional tool, mainly composed of three parts: lid, ice filter and pot body. The original intention of the design is to effectively mix the mixing ingredients and ice cubes in the pot by artificial shaking, so as to achieve rapid cooling of the liquor. The common shakers are divided into large, medium and small sizes according to the different needs of use, in order to adapt to the needs of different occasions and mixing recipes.

图1-5-10　调酒壶（shaker）

（二）盎司杯（Ounce measuring cup）

盎司杯（图1-5-11），是鸡尾酒调配中不可或缺的专业量酒工具。其设计多样化，以满足不同鸡尾酒调配的需求。常见的量酒器设计包括上部容量为30毫升与下部容量为45毫升的组合，以及30毫升与60毫升、15毫升与30毫升的组合。这些设计旨在确保鸡尾酒调配过程中能够精确计量各种成分，从而制作出口感与风味俱佳的鸡尾酒。

The ounce measuring cup (Figure 1-5-11), is an indispensable professional measuring tool in cocktail mixing. The design is diversified to meet the needs of different cocktail preparations. Common Jigger include a combination of an upper capacity of 30 mL and a lower capacity of 45 mL, as well as a combination of 30 mL and 60 mL, 15 mL and 30 mL. These designs are designed to ensure that the ingredients are precisely measured during the cocktail preparation process, resulting in a cocktail that flavor and tastes great.

图1-5-11　盎司杯
（Ounce measuring cup）

（三）吧匙（Bar Spoon）

吧匙（图1-5-12）是一种专为调制鸡尾酒而设计的工具，特别适用于高身杯的调制过程。该工具具有显著的特点，即其柄部较长，中部呈螺旋形状。这一设计使得调酒师能够方便地在杯中旋转液体和材料，确保鸡尾酒中的各种成分充分混合，从而达到理想的口感和视觉效果。

Bar spoon (Figure 1-5-12) is a professional tool designed for mixing cocktails, especially suitable for the preparation process of tall glasses. The tool has a remarkable feature, that is, it has a long handle and a spiral shape in the middle. This design allows bartenders to

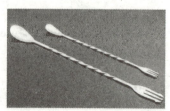

图1-5-12　吧匙（bar spoon）

easily rotate liquids and ingredients in the glass, ensuring that the ingredients in the cocktail are well blended to achieve the desired taste and visual effect.

（四）滤冰器（Strainer）

图 1-5-13　滤冰器（cocktail strainer）

在调制鸡尾酒时，为了确保酒液的纯净度和口感的细腻度，需要使用滤冰器（图 1-5-13）对鸡尾酒进行过滤。经过滤冰操作，能够有效去除鸡尾酒中的冰粒，避免冰粒对口感造成不良影响。滤冰器通常由不锈钢材质制成，这种材质不仅坚固耐用，而且易于清洗和保养，能够确保多次使用后的卫生和安全。在过滤完成后，将得到的纯净酒液倒入预先准备好的载杯中，以便进一步品鉴。

While making cocktails, in order to ensure the purity of the liquor and the delicacy of the taste, it is necessary to use an strainer (Figure 1-5-13) to filter the cocktail. Through the operation of the strainer, the ice particles in the cocktail can be effectively removed to avoid the adverse effect of ice particles on the taste. Strainer are usually made of stainless steel, which is not only strong and durable, but also easy to clean and maintain, ensuring hygiene and safety after multiple uses. After the filtration is complete, the resulting pure liquor should be poured into a pre-prepared carrier cup for further tasting.

（五）冰夹（Ice Tong）

图 1-5-14　冰夹（ice tong）

冰夹（图 1-5-14）作为一种专业的调酒工具，其材质为不锈钢，具有坚固耐用和易于清洁的特点。冰夹的设计旨在方便调酒师夹取冰块，并将其精准地放入酒中或调酒壶内，以调节鸡尾酒的温度和口感。这种工具在调制含有冰块的鸡尾酒时尤为重要，能够帮助调酒师实现更精准的温控和口感调配。

As a professional bartending tool, the ice tong (Figure 1-5-14) is made of stainless steel, which is durable and easy to clean. The ice clamps are designed to make it easy for bartenders to take the ice and place it precisely in the drink or shaker to adjust the temperature and taste of the cocktail. This tool is especially important when making cocktails containing ice, which can help bartenders achieve more precise temperature control and texture preparation.

（六）杯垫（Coaster）

杯垫（图 1-5-15）是一种圆形垫状物，放置在杯子底部以防止杯子直接与桌面接触。其直径通常约为 10 厘米，以适应常见的杯型。杯垫的材质具有多样性，包括纸制、塑料制、皮制和金属制等，这些不同的材质为杯垫提供了不同的特性和用途。在选择杯垫时，考虑到其实用性和功能性，吸水性能好的材质通常被视为最佳选择。这种材质的杯垫能够有效吸收杯子底部的水分，保持桌面的干燥和清洁，同时起到防滑效果并确保杯子能够稳定放置。

A coaster (Figure 1-5-15) is a round cushion that is placed on the bottom of a cup to prevent the cup from coming into direct contact with the table top. Its diameter is usually about 10 cm to accommodate the common cup shape. Coasters have a variety of materials, including paper, plastic, leather and metal, and these different materials provide different characteristics and uses for coasters. When choosing coasters, taking into account their practicality and functionality, thick paper materials with good water absorption properties are usually regarded as the best choice. This material can effectively absorb the water at the bottom of the cup, keeping the table dry and clean, At the same time, it plays a non-slip role and ensures that the cup can be stably plead.

纸制（Paper）　塑料制（Plastic）　皮制（Leather）　金属制（Metal）

图 1-5-15　杯垫（coaster）

二、调酒技法（Mixology Techniques）

鸡尾酒调制是一门富有创新性的技艺，经过调酒师的精湛操作和巧妙构思，能够在短时间内将各种原料和成分融合成色彩鲜艳、香气四溢、味道独特的饮品。调制鸡尾酒的方法主要有以下四种。

Cocktail making is an innovative art. Through the exquisite operation and ingenious conception of the bartender, various ingredients and ingredients can be integrated into colorful, fragrant and unique drinks in a short time. There are four main ways to make a cocktail.

（一）兑和法（Blending Method）

兑和法要求将配方中的酒水按照精确的分量依次直接倒入杯中，无需额外搅拌。这种方法广泛应用于彩虹鸡尾酒的调制中。配制多色酒的基本原理是不同材料的密度不同，密度较大的材料会沉在下方，密度较小的材料会浮在上面，从而形成多层次的视觉效果。因此，在配制多色酒时，应选用密度不同、色泽各异的酒水。

The blending method requires that the wine from the recipe be poured directly into the glass in exact servings without the need for additional stirring. This method is widely used in the preparation of rainbow cocktails. The basic principle of preparing multi-color wine is that the density of different materials is different, the denser material will sink below, and the less dense material will float on top, thus forming a multi-level visual effect . Therefore, in the preparation of multi-color wine, should choose different density, different colors of wine.

在操作时，不能将酒水直接倒入杯中，应使用一把吧匙插入杯内，吧匙背面朝上，然后将酒水倒在吧匙背面上，使其顺着杯内壁缓缓流下。通过这种方式，可以确保各种酒水在杯中形成层次分明、互不混合的效果。要注意的是，配制好的多色酒不宜久放，因为长时间的放置可能导致各酒色互相融合，失去原有的层次感。

During operation, the wine should not be poured directly into the glass. Instead, insert a bar spoon into the glass, back up, and pour the wine over the back of the bar spoon, allowing it to flow slowly down the inside of the glass. In this way, it is possible to ensure that the various liquors form a well-defined and non-mixing effect in the cup. However, it should be noted that the prepared multi-color wine should not be put for a long time, because the long-term placement may cause the wine colors to dissolve with each other and lose the original sense of layer.

（二）调和法（Stirring Method）

调和法具体操作步骤如下：首先，在调酒杯中加入适量的冰块，确保冰块分布均匀。接着，根据鸡尾酒配方，将酒水按照预定的比例倒入调酒杯中。最后，使用吧匙沿调酒杯的内侧进行顺时针搅拌。搅拌的次数一般控制在 10～15 次，以确保酒水能够均匀冷却，同时使各种成分充分融合。

若鸡尾酒调制完成后需要滤除冰块，则采用以下步骤：左手稳定持杯，确保调酒杯稳定不晃动；右手则使用滤水器阻挡冰块，同时倾斜调酒杯，使酒水缓慢流出，冰块则被留在调酒杯中。

The specific operation steps of the harmonic method are as follows: first, fill the glass with the right amount of ice to make sure the ice is evenly distributed. Then, according to the cocktail recipe, pour the wine into the cocktail glass in the predetermined proportion. Finally, use a bar spoon to stir clockwise along the inside of the mixing glass. The number of stirring is generally controlled between 10 and 15 revolutions to ensure that the wine can be evenly cooled and the various ingredients are fully integrated.

If you need to remove the ice after making the cocktail, follow the following steps: hold the glass steadily with your left hand to ensure that the mixing glass is stable and does not shake; With his right hand, he uses an ice filter to block the ice while tilting the mixing glass so that the wine flows slowly and the ice remains in the shaker.

（三）摇和法（Shaking Method）

在调制鸡尾酒时，若遇到某些成分（如牛奶、蛋清、果汁等）无法与基酒稳定混合的情况，需采用手摇调酒壶进行调制。其具体操作步骤如下：

When making cocktails, if some ingredients (such as milk, egg white, juice, etc.) cannot be stably mixed with the base wine, it is necessary to use a hand flask for mixing. The specific steps are as follows:

首先，向调酒壶中加入适量冰块，以确保调制过程中酒水的冷却效果。接着，根据配方要求，将必要的材料按照特定顺序投放至调酒壶内。在拿取调酒壶时，右手拇指应紧压住壶盖，以确保在摇晃过程中壶盖不会松动；其余手指则自然握住壶体。同时，左手中指应抵住壶底，以增加稳定性；其余手指则围住壶体，辅助拿取。

First, add an appropriate amount of ice to the hand shaker to ensure the cooling effect of the wine during the mixing process. Then, according to the recipe requirements, the necessary ingredients are placed into shaker in a specific order.When taking the shaker, the right thumb should be firmly pressed against the lid to ensure that the lid will not come loose during shaking. The rest of your fingers grip the body naturally. At the same time, the middle finger of the left hand should be pressed against the bottom of the shaker to increase stability. The rest of the fingers surround the shaker body to assist in reaching.

摇和时，需反复进行，直至壶内材料充分混合。摇和的时间应适中。一般来说，当调酒壶外部起霜时，即表示酒品已充分冷却并混合均匀。

When shaking the shaker, repeat until the material in the shaker is fully mixed. The shaking time should be moderate. Generally speaking, when there is white frost on the outside of the shaker, it means that the wine has been fully cooled and mixed evenly.

滤酒时，打开壶盖，使调制好的酒水通过滤网滤入杯中，冰块则留在壶内。在实际操作中，可根据个人习惯和熟练程度选择单手或双手的摇壶方式。需要注意的是，在酒壶内不得加入含气材料，以防止摇和过程中产生过多泡沫影响口感。同时，在摇和过程中应大力摇匀，以确保材料充分混合。

To filter the wine, open the back cover of the shaker so that the prepared wine is filtered through the strainer into the glass, while the ice remains in the shaker. In practice, you can choose a one-handed or two-handed shaker according to personal habits and proficiency. However, it should be noted that gas-containing materials should not be added to the shaker to prevent excessive foam from affecting the taste during shaking. At the same time, it should be vigorously shaken during the shaking process to ensure that the material is fully mixed.

（四）搅和法（Blending Method）

此方法适用于基酒与某些固体成分混合的鸡尾酒调制。首先，将酒水与冰块按照配方规定的分量放入电动搅拌机中。接着，启动搅拌机，让其运转约 10 秒，以确保酒水与冰块充分混合且均匀冷却。最后，连冰带酒一同倒入预先准备好的载杯中。

This method is suitable for the preparation of drinks where the base wine is mixed with some solid components. First, place the drinks and ice cubes in an electric blender according to the recipe. Next, start the blender and let it run for about 10 seconds to ensure that the drinks are well combined with the ice and cool evenly. Finally, pour the ice and wine into the prepared container.

赛证直通 Competitions and Certificates

一、基础知识部分（Basic Knowledge Part）

（一）专业名词解释（Explanation of Professional Terms）

1. 滤冰器（Cocktail Strainer）：
2. 摇和法（Shaking Method）：

二维码 1-5-3：
知识拓展

（二）思考题（Thinking Questions）

简述不同鸡尾酒调制方法的特点。

Please briefly describe the characteristics of different cocktail preparation methods.

二、技能操作部分（Skill Operation Part）

在实训室练习鸡尾酒的调制技法。

Practice cocktail making in the training room.

专题四
经典配方：著名鸡尾酒配方选例

你知道图中这款鸡尾酒的名称吗？这款鸡尾酒所用的基酒是什么呢？（图 1-5-16）？

_____

Do you know the name of this cocktail in the picture? What is the base liquor for this cocktail? (Figure 1-5-16)?

_____

鸡尾酒是由至少一种酒精饮料（基酒），辅以果汁、汽水等辅料精心调制而成。它具备独特的口感和多样的搭配，如金酒、伏特加、朗姆酒、龙舌兰、威士忌和白兰地等都可以用作鸡尾酒的基酒。

图 1-5-16　鸡尾酒（cocktail）

Cocktails are actually made up of at least one alcoholic beverage (base liquor), supplemented by juice, soda and other ingredients. It has a unique taste and a variety of combinations, such as Gin, Vodka, Rum, Tequila, whiskey and Brandy can be used as the base wine of cocktails.

一、以金酒为基酒的鸡尾酒（Gin based Cocktails）

（一）红粉佳人（Pink Lady）

这款鸡尾酒得名于美国百老汇著名的同名歌剧《红粉佳人》，剧中的女主角手持这款鸡尾酒亮相，自此，"红粉佳人"鸡尾酒（图 1-5-17）便诞生了。这款鸡尾酒以其独特的辛辣、酸甜口感和迷人的粉红色，深受女性的喜爱，成了专为女性打造的经典鸡尾酒之一。

This cocktail is named after the famous Broadway opera of the same name, Pretty in Pink, the lead actress with this cocktail, appeared since then, the Pink Lady cocktaill (Figure 1-5-17) was born. With its unique spicy, sweet and sour taste and charming pink color, this cocktail is loved by women and has become one of the classic cocktails created for women.

图 1-5-17　红粉佳人（pink lady）

材料：金酒 45 毫升、君度酒 8 毫升、鸡蛋清 1 只、柠檬汁 10 毫升、红石榴糖浆 1 吧匙。

制法：首先，取一只鸡蛋，轻轻将其打开，仔细地将蛋清与蛋黄分离，确保只使用蛋清部分。接着，将冰块以及上述材料依次放入调酒壶中。用力摇动调酒壶，确保所有的材料充分混合并冷却。最后，将调制好的鸡尾酒倒入准备好的载杯中。

载杯：鸡尾酒杯，能完美地展现其色泽和风味。

装饰物：红樱桃。

Ingredients: gin 45 mL, cointreau 8 mL, egg white 1, lemon juice 10 mL, red pomegranate molasses 1 bar spoon.

Preparation method: first, take an egg and gently open it, carefully separating the white from the yolk, making sure to use only the white part. Next, place the ice cubes and the ingredients in a shaker. Shake the shaker vigorously to make sure all the ingredients are well mixed and cooled. At last, pour the cocktail into the prepared carrier.

Container: a cocktail glass that perfectly displays its color and flavor.

Garnish: red cherry.

（二）新加坡司令（Singapore Sling）

新加坡的莱佛士酒店以其悠久历史和独特魅力闻名于世。1915 年诞生于此的"新加坡司令"鸡尾酒（图 1-5-18），成为酒店的标志性饮品，吸引着无数游客。这款鸡尾酒的绚丽色泽和热带风情令人陶醉，深受客人喜爱。

图 1-5-18　新加坡司令
（Singapore Sling）

The Raffles Hotel in Singapore is renowned for its long history and unique charm. Born here in 1915, the "Singapore Sling" cocktail (Figure 1-5-16) has become the hotel's signature drink, attracting countless visitors. The gorgeous colors and tropical flavor of this cocktail are intoxicating, and guests love it.

材料：金酒 45 毫升、樱桃白兰地 15 毫升、柠檬汁 15 毫升、红石榴糖浆 10 毫升、苏打水适量。

制法：第一，将冰块和材料依次放入调酒壶中摇匀；第二，将材料倒入含冰块的载杯中，加入适量苏打水；第三，用酒签串好的菠萝片和红樱桃装饰。

载杯：柯林杯。

装饰物：樱桃、菠萝片、酒签。

Ingredients: gin 45 mL, cherry brandy 15 mL, lemon juice 15 mL, white syrup 10 mL, appropriate amount of soda water.

Preparation method: First, place ice cubes and ingredients in a shaker and shake well. Second, fill a glass with ice and fill it with some soda water. Third, garnish with sliced pineapple and red cherry on a wine stick.

Container: flute shaped champagne glass.

Garnish: cherry, lemon slice, wine tag.

二、以白兰地为基酒的鸡尾酒（Brandy Based Cocktails）

（一）亚历山大（Alexander）

在鸡尾酒的世界中，亚历山大（图 1-5-19）以其独特的口感和风味赢得了广泛的赞誉。这款鸡尾酒巧妙地融合了可可利口酒的甜美滋味与奶油的浓郁口感，使得整体呈现出一种入口即化的柔软滑顺感。可可利口酒的甜蜜与奶油的醇厚在口中交织，带来一种既甜美又浓郁的味觉享受，让人陶醉其中。

材料：白兰地45毫升、可可利口酒30毫升、奶油30毫升。

制法：将冰块和材料依次放入调酒壶中，运用摇和法使各种成分充分融合。最后，将调制好的鸡尾酒倒入载杯中。

载杯：鸡尾酒杯。

装饰物：肉豆蔻粉。

图1-5-19　亚历山大（Alexander）

In the world of cocktails, Alexander (Figure 1–5–19) has won widespread acclaim for its unique taste and flavor. This cocktail cleverly combines the sweet taste of cocoa liqueur with the rich taste of cream, resulting in a soft and smooth feeling that melts in the mouth. The sweetness of the cocoa liqueur and the richness of the cream intertwine in the mouth, bringing a sweet and rich taste enjoyment, which makes people intoxicated.

Ingredients: 45 mL brandy, 30 mL cocoa liqueur, 30 mL cream.

Preparation method: Place the ice cubes and ingredients in a shaker, shaking to combine the ingredients. Finally, pour the cocktail into a glass.

Container: cocktail glass.

Garnish: ground nutmeg.

（二）边车（Side Car）

边车（图1-5-20）是一款口感与风味独具一格的鸡尾酒，其独特的魅力使其成为餐后饮用的理想选择。其口感酸甜清爽，余味悠长。这款鸡尾酒不仅满足了味蕾的享受，也为餐后时光增添了一份别样的风情。

图1-5-20　边车（Side Car）

The side car (Figure 1–5–20) is a cocktail with a unique taste and flavor, and its unique charm makes it an ideal choice for after–dinner drinking. Its taste is sweet and sour and refreshing, with a long aftertaste. This cocktail not only satisfies the taste buds, but also adds a special flavor for the after–dinner time.

材料：白兰地40毫升、君度酒10毫升、柠檬汁10毫升。

制法：将冰块和所有材料依次放入调酒壶中，摇和完成后，注入载杯。

载杯：鸡尾酒杯。

装饰物：柠檬条。

Ingredients: Brandy 40 mL, Cointreau 10 mL, lemon juice 10 mL.

Preparation method: Place the ice cubes and all ingredients in the shaker, shake and finish, and fill the carrier glass.

Container: cocktail glass.

Garnish: lemon sticks.

三、以朗姆酒为基酒的鸡尾酒（Rum based Cocktails）

（一）百家地（Bacardi）

这款鸡尾酒的名字来源于一个朗姆酒品牌。百家地牌朗姆酒是历史悠久的朗姆酒品牌之一。其产品系列不仅在调制鸡尾酒时备受欢迎，也常被用于西式糕点的制作中（图 1-5-21）。

图 1-5-21　百家地（Bacardi）

The name of this cocktail comes from a rum brand. Bacardi Rum is one of the oldest rum brands, and its product range is not only popular in cocktails, but also often used in Western pastries (Figure 1-5-21).

材料：白朗姆酒 30 毫升、柠檬汁 15 毫升、红石榴糖浆 10 毫升。

制法：将冰块和所有材料依次放入调酒壶中摇和，壶身起霜时倒入载杯中，最后用红樱桃装饰。

载杯：鸡尾酒杯。

装饰物：红樱桃。

Ingredients: White rum 30 mL, lemon juice 15 mL, red grenadine 10 mL.

Method: Shake the ice cubes and all the ingredients in the shaker. When the shaker is frosted, pour into the cup. Decorate with red cherries.

Carrying glass: cocktail glass.

Garnish: Red cherry.

（二）椰林飘香（Pina Colada）

"椰林飘香"鸡尾酒（图 1-5-22）起源于 20 世纪 50 年代的美国波多黎各自治邦圣胡安的希尔顿酒店。调酒师用椰浆创造了这款鸡尾酒，它在希尔顿酒店的推广下风靡全球。其名称"Pina Colada"意为"菠萝茂盛的山谷"，象征热带岛屿的自然风光。这款鸡尾酒融合椰奶的香滑、菠萝的酸甜与朗姆酒的清爽，是夏日里的绝佳饮品。

图 1-5-22　椰林飘香（pina colada）

The Pina Colada cocktail (Figure 1-5-22) originated in the 1950s at the Hilton Hotel in San Juan, the common wealth of Puerto Rico, USA. Bartender used coconut milk to create this cocktail, which has taken the world by storm with the promotion of Hilton Hotels. The name "Pina Colada" means "valley full of pineapples", symbolizing the natural beauty of tropical islands. A creamy blend of coconut milk, sweet and sour pineapple and the refreshing taste of rum, this cocktail is perfect for a summer.

材料：白朗姆酒 45 毫升、菠萝 5 片、椰浆 45 毫升。

制法：第一，将冰块及菠萝投入果汁机，随后加入白朗姆酒与椰浆，启动果汁机进行搅拌，待其充分融合；第二，取一片新鲜的菠萝、两片菠萝叶以及一颗红樱桃，巧妙地用酒签将它们串起，然后将这串精美的装饰置于杯沿，为饮品增添一抹热带风情。

载杯：飓风杯。

装饰物：菠萝片、菠萝叶、红樱桃。

Ingredients: 45 mL Branam wine, 5 slices of pineapple, 45 mL coconut milk.

Preparation method: first, put ice cubes and pineapple into the juice machine, then add the white rum and coconut milk, start the juice machine and stir until it is fully integrated; take a fresh pineapple, two pineapple leaves, and a red cherry, string them with a clever skewer, then place the delicate decoration on the rim of the glass to add a tropical touch to the drink.

Container: Hurricane cup.

Garnish: pineapple slices, pineapple leaves, red cherry.

四、以伏特加为基酒的鸡尾酒（Vodka Based Cocktails）

（一）螺丝刀（Screwdriver）

这款鸡尾酒之所以被命名为"螺丝刀"（图 1-5-23），是因为早期的工人们为了解渴和提神，常常将伏特加和橙汁简单地混合在一起饮用，但由于没有合适的搅拌工具，便就地取材，用螺丝刀来搅拌伏特加和橙汁。

The reason why this cocktail is named " screwdriver " (Figure 1-5-23) is because early workers in order to quench their thirst and refresh themselves, often drink vodka and orange juice simply mixed together, but because there is no suitable mixing

图 1-5-23　螺丝刀鸡尾酒
（Screwdriver Cocktails）

tool, they use local materials to stir vodka and orange juice with a screwdriver.

材料：伏特加 50 毫升、橙汁 100 毫升。

制法：准备一个平底杯，向其中加入冰块，随后缓缓倒入伏特加和橙汁，轻轻搅拌使两者充分融合。最后，用一片新鲜的橙子点缀在杯口，增添一丝清新的果香。

载杯：平底杯。

装饰物：橙子片。

Ingredients: Vodka 50 mL, orange juice 100 mL.

Preparation method: get a flat glass with ice, then slowly pour in the vodka and orange juice, stirring gently to combine. Finally, use a fresh orange slice to dot the rim of the glass to add a fresh fruity flavor.

Container: flat cup.

Garnish: orange slices.

（二）大都会（Cosmopolitan）

大都会（图 1-5-24）这款鸡尾酒以其丰富的口感和绚丽的色彩而备受欢迎。君度橙酒的酸甜、蔓越莓汁的果香和柠檬汁的清新在口中交织，形成了一种独特的风味。大都会的名字寓意着这款鸡尾酒像大都市一样，充满了活力和魅力，也代表了奋发向上的精神。

图 1-5-24　大都会（Cosmopolitan）

The Cosmopolitan (Figure 1-5-24) cocktail is popular for its rich taste and brilliant colors. The sweet and sour of Cointreau orange wine, the fruit of cranberry juice and the freshness of lemon juice are intertwined in the mouth, forming a unique flavor. The name Metropolis implies that this cocktail, like the metropolis, is full of vitality and charm, and also represents the spirit of striving.

材料：伏特加 40 毫升、君度橙酒（Cointreau）15 毫升、蔓越莓汁 30 毫升、柠檬汁 15 毫升。

制法：将冰块放入调酒壶，随后加入所有材料，用力摇和 10 到 20 秒以确保所有成分充分混合。

载杯：鸡尾酒杯。

装饰物：柠檬皮条。

Ingredients: Vodka 40 mL, Cointreau 15 mL, cranberry juice 30 mL, lemon juice 15 mL.

Preparation method: Place ice cubes in a shaker, then add all ingredients, shaking vigorously for 10 to 20 seconds to make sure everything is well combined.

Container: Cocktail glass.

Garnish: Lemon zest.

五、以威士忌为基酒的鸡尾酒（Whiskey based Cocktails）

（一）曼哈顿（Manhattan）

曼哈顿鸡尾酒以其独特的配方和口感，成了鸡尾酒世界的经典之一。它既有威士忌的醇厚，又有味美思的甜美，再加上安格斯特拉苦酒的苦味，使得这款鸡尾酒层次丰富，令人回味无穷。

With its unique formula and taste, the Manhattan cocktail has become one of the classics of the cocktail world. It combines the mellowness of whiskey with the sweetness of Vermouth, coupled with the bitterness of Angstra bitters, making this cocktail layered and evocative.

材料：威士忌 45 毫升、甜味美思 15 毫升、安格斯特拉苦酒 1 吧匙。

制法：第一，将冰块和所需的材料依次放入调酒壶中，充分摇匀；第二，使用滤冰器扣住调酒壶，将调制好的酒水缓缓注入载杯中；第三，用酒签刺穿红樱桃，将其作为装饰点缀在酒杯中。

载杯：鸡尾酒杯。

装饰物：红樱桃。

Ingredients: Whisky 45 mL, Sweet Max 15 mL, Angstra bitters 1 bar spoon.

Method: First, put the ice cubes and the required materials into the shaker in turn, and shake well; Second, use the ice filter to fasten the shaker and slowly inject the prepared wine into the carrier cup; Third, use a wine tag to Pierce the red cherry and decorate it in the glass.

Carrying glass: cocktail glass.

Garnish: Red cherry.

（二）威士忌酸（Whisky Sour）

"威士忌酸"这款鸡尾酒起源于 19 世纪 60 年代的美国，最初是由威士忌、柠檬汁、糖和蛋黄混合而成。随着时间的推移，这款鸡尾酒逐渐流行开来，并在 20 世纪 60 年代成为经典鸡尾酒之一。其独特的酸甜口感和丰富的层次感深受人们喜爱，成了酒吧中常见的饮品之一。

The "Whiskey Sour" cocktail originated in the United States in the 1860s and was originally made from a mixture of whiskey, lemon juice, sugar and egg yolks. Over time, this cocktail grew in popularity and became one of the classic cocktails in the 1960s. Its unique sweet and sour taste and rich layers are deeply loved by people and have become one of the common drinks in bars.

材料：威士忌 40 毫升、柠檬汁 20 毫升、糖浆 10 毫升。

制法：第一，将冰块和所需的材料依次放入调酒壶中，然后大力摇和使材料充分混合，最后将调制好的鸡尾酒缓缓注入载杯中；第二，为了更好的视觉效果，可以用柠檬皮条作为装饰物，点缀在酒杯的边缘。

载杯：酸酒杯。

装饰物：柠檬皮条。

Ingredients: Whisky 40 mL, lemon juice 20 mL, syrup 10 mL.

Making method: First, put the ice cubes and the required materials into the shaker in turn, then vigorously shake and mix the materials fully, and finally inject the prepared cocktail slowly into the carrier cup; Second, for a better visual effect, you can use lemon zest as a decoration, dotted on the edge of the glass.

Carrying cup: sour wine glass.

Garnish: Lemon zest.

六、以特基拉酒为基酒的鸡尾酒（Tequila Based Cocktails）

（一）特基拉日出（Tequila Sunrise）

这款鸡尾酒源于墨西哥，特基拉酒的故乡。1972年，滚石乐队巡演中邂逅此酒，其独特风味令人赞叹。凭借滚石乐队的影响力，这款鸡尾酒迅速风靡全球，成为经典之作。其传奇故事与滚石乐队的辉煌历程共同谱写酒与音乐的美妙篇章。

This cocktail is from Mexico, the home of Tequila. In 1972, the Rolling Stones tour came across this wine, its unique flavor is amazing. Thanks to the influence of the Rolling Stones, this cocktail quickly became a worldwide classic. The legendary story and the glorious history of the Rolling Stones together write a beautiful chapter of wine and music.

材料：特基拉酒45毫升、橙汁120毫升、红石榴糖浆10毫升。

制法：在郁金香形香槟杯中放入冰块，随后倒入特基拉酒和橙汁，使用吧匙轻轻搅拌，使其充分混合。然后，沿着吧匙缓缓将红石榴糖浆注入杯底。

载杯：郁金香形香槟杯。

装饰物：红樱桃、橙片。

Ingredients: Tequila 45 mL, orange juice 120 mL, red grenadine 10 mL.

Preparation method: Place ice cubes in a tulip-shaped champagne glass, then pour in the Tequila and orange juice, stirring gently with a bar spoon to combine well. Then, slowly pour the grenadine into the bottom of the glass with the spoon.

Carrying glass: tulip-shaped champagne glass.

Garnish: Red cherry, orange slice.

（二）玛格丽特（Margarita）

玛格丽特鸡尾酒的创作灵感来源于调酒师对逝去情感的纪念，其独特的酸甜口感和浪漫的背景故事使其成为经典（图1-5-25）。

The creation of the Margarita is inspired by the bartenders' remembrance of lost emotions, and its unique sweet and sour taste

图1-5-25　玛格丽特（Margarita）

and romantic backstory make it a classic (Figure 1−5−25).

材料：特基拉酒 30 毫升、君度酒 15 毫升、柠檬汁 15 毫升。

制法：将冰块和所需材料依次放入调酒壶中，用力摇和，使其充分混合。然后，将调制好的鸡尾酒倒入已经修饰有盐圈的杯中。

载杯：玛格丽特杯。

装饰物：盐圈、柠檬片。

Ingredients: Tequila 30 mL, Cointreau 15 mL, lemon juice 15 mL.

Preparation method: Place ice cubes and required ingredients in a shaker, shake vigorously, and mix thoroughly. Then pour the cocktail into a glass that has been decorated with salt rings.

Carrying cup: Margarita cup.

Garnish: Salt ring, lemon slice.

经典鸡尾酒的诞生并非偶然，相信是经过深入研究和学习的必然结果。我们要学会创新，去追求鸡尾酒色、香、味、形的多样化，保持鸡尾酒观赏和品鉴的双重价值，从基酒、辅料、装饰等方面发掘新原料、组合新样式、搭配新品种。在不同文化背景下，创造属于自己的经典鸡尾酒。

The birth of the classic cocktail is not accidental, I believe it is the inevitable result of in−depth research and learning. We should learn to innovate, to pursue the diversification of cocktail color, aroma, taste and shape, to maintain the dual value of cocktail viewing and tasting, and to explore new raw materials, combine new styles and match new varieties from the aspects of base wine, accessories and decoration. Create your own classic cocktails in different cultural contexts.

赛证直通 Competitions and Certificates

一、基础知识部分（Basic Knowledge Part）

二维码 1-5-4：
知识拓展

（一）专业名词解释（Explanation of Professional Terms）

1. 新加坡司令（Singapore Sling）：

2. 特基拉日出（Tequila Sunrise）：

（二）思考题（Thinking Question）

谈谈你对如何做好一杯经典鸡尾酒的理解。

Talk about your understanding of how to make a classic cocktail.

二、技能操作部分（Skill Operation Part）

1. 调制以金酒为基酒的鸡尾酒。Making gin−based cocktails.

2. 调制以白兰地为基酒的鸡尾酒。Make cocktails−based on brandy.

3. 调制以朗姆酒为基酒的鸡尾酒。Make rum−based cocktails.

4. 调制以伏特加为基酒的鸡尾酒。Make vodka−based cocktails.

5. 调制以威士忌为基酒的鸡尾酒。Make whisky-based cocktails.

6. 调制以特基拉为基酒的鸡尾酒。Make tequila-based cocktails.

专题五
别具一格：创意鸡尾酒

【微课】
别具一格：
创意鸡尾酒

请问创意鸡尾酒可以用茅台酒作为基酒吗？

Can you make the creative cocktail with Moutai as the base?

调制创意鸡尾酒同样要掌握鸡尾酒的基本结构，因为创意鸡尾酒遵循着相同的结构原则。

Making a creative cocktail is also about mastering the basic structure of a cocktail, because creative cocktails follow the same structural principles.

一、创意鸡尾酒的基本结构（The Basic Structure of a Creative Cocktail）

　　基酒　　+　　辅料　　+ 装饰物 = 鸡尾酒

basic liquor + auxiliary materials + ornament = cocktail

当然，创意鸡尾酒还需要选择一个适当的载杯。不同的载杯也是不同创意的体现，选择可以多样化，它们不仅用于盛装鸡尾酒，还增强了视觉效果并提升了品鉴体验。

Of course, creative cocktails also need to choose an appropriate carrier. Different carrier glasses are also the embodiment of different creativity, the choice can be diversified, they are not only used to hold cocktails, but also enhance the visual effect and enhance the tasting experience.

（一）基酒（Base Wine）

在调制鸡尾酒时，基酒是构成其核心风味的基础。主要的六大基酒包括：白兰地、威士忌、金酒、伏特加、特基拉以及朗姆酒。除此之外，各类蒸馏酒，如中国白酒等都可以作为基酒。

基酒在鸡尾酒中的作用如下：

When making cocktails, the base wine is the basis that forms the core flavor. The six main base spirits are: brandy, whiskey, gin, vodka, Tequila and rum. In addition, all kinds of distilled spirits, such as Chinese baijiu, can be used as a base wine.

The role of base wine in cocktails as below:

（1）提供基本风味：基酒是鸡尾酒风味的基础，不同的基酒具有不同的风味特点，为鸡尾酒提供了丰富的口感和风味变化。

Provide basic flavor: base wine is the basis of cocktail flavor, different base wines have different flavor characteristics, providing a rich taste and flavor changes for cocktails.

（2）提升口感层次：通过添加果汁、糖浆、苦精等其他辅助材料，基酒可以与这些材料相互作用，产生更加丰富的口感层次和风味变化。

Enhance the taste level: by adding other auxiliary materials such as fruit juice, syrup, bitter, the base wine can interact with these materials to produce a richer taste level and flavor changes.

（3）调节酒精度：基酒的酒精度较高，通过与其他低度酒或饮料的混合，可以调节鸡尾酒的酒精度，使其更加适口。

Adjust the alcohol level: the alcohol level of the base wine is higher, and the alcohol level of the cocktail can be adjusted by mixing with other low-degree wine or drinks to make it more palatable.

（二）辅料（Auxiliary Materials）

辅料的主要作用是增加风味，发挥鸡尾酒的特色，增添鸡尾酒的色彩，提升视觉和口感体验。主要的辅料有以下几类：

The main function of the ingredients is to increase the flavor, play the characteristics of the cocktail, add the color of the cocktail, and enhance the visual and taste experience. The main accessories have the following categories.

（1）力娇酒类：如香蕉力娇酒、蓝橙力娇酒等。

liqueurs: such as bols banana liqueur, bols bule curacao liqueur, etc.

（2）果汁类：如橙汁、柠檬汁等。

Juices: such as orange juice, lemon juice, etc.

（3）汽水类：如苏打水、雪碧等。

Soda: such as soda water, sprite, etc.

（4）其他材料：如鸡蛋、牛奶等。

Other materials: such as egg, milk, etc.

（三）装饰物（Ornament）

创意鸡尾酒的装饰物可以极大地提升鸡尾酒的美感和吸引力。常见的装饰物包括各种水果（如柠檬、樱桃、草莓等）切片或整颗水果，用于增添色彩和口感；花卉与植物（如薄荷叶、迷迭香等）用于增添独特的香气和视觉效果；食品类装饰如橄榄、果冻和奶油等，为鸡尾酒增添风味和口感层次。此外，冰块装饰、鸡尾酒签与小伞、节日特色装饰以及彩色糖珠或盐边等都是创意鸡尾酒装饰的不错选择。

在选择装饰物时，需要考虑其与鸡尾酒口感和风格的搭配，并确保装饰物的卫生情况，以保证鸡尾酒的品质。这些创意装饰物能够为鸡尾酒增添独特的魅力和趣味性，使创意鸡尾酒更具有观赏性。

Creative cocktail garnishes can greatly enhance the beauty and appeal of a cocktail. Common garnishes include various fruits (such as lemons, cherries, strawberries, etc.) sliced or whole to add color and texture; flowers and plants (such as mint leaves, rosemary, etc.) to add a unique aroma and visual effect; food garnishes such as olives, jellies and cream add flavor and texture to cocktails. In addition, ice cube decorations, cocktail tags and umbrellas, festive decorations and colorful sugar beads or salt edges are all good choices for creative cocktail decorations.

When choosing the garnish, it is necessary to consider its match with the taste and style of the cocktail, and ensure the hygiene of the garnish to maintain the quality of the cocktail. These creative decorations can add unique charm and interest to the cocktail, making the creative cocktail has more ornamental value.

二、创意鸡尾酒调制的注意事项（Tips for Creative Cocktail Making）

创意鸡尾酒调制的注意事项有如下几项：

Here are the dos and don'ts of creative cocktail making:

（1）含气体的碳酸饮料不能放入调酒壶中摇和，只能直接倒入载杯，再轻微搅匀。

Carbonated drinks containing gas cannot be shaken in a mixing cocktail shaker, but can only be poured directly into a carrier glass and stirred gently.

（2）在调制鸡尾酒的过程中，斟倒酒液的量需精心把控。一般而言，将酒杯斟至八分满的状态是较为理想的。此举既考虑了服务时的便捷性，也兼顾了饮用时的舒适度，同时也确保了鸡尾酒外观的优雅与美观。若酒杯过分满溢，不仅服务时易泼洒，影响饮用体验，从视觉审美角度而言，也不美观。因此，在调制鸡尾酒时，注意控制酒液的斟倒量，是确保鸡尾酒品质与美感的关键步骤。

In the process of making a cocktail, the amount of liquor needs to be carefully controlled. Generally speaking, it is more ideal to pour the glass until it is eight times full. This takes into account both the convenience of service and the comfort of drinking, while also ensuring the elegant and beautiful appearance of the cocktail. If the glass is too full, not only the service is easy to spill, affecting the drinking experience, but also from the perspective of visual aesthetic, it is not satisfactory. Therefore, when making cocktails, paying attention to controlling the amount of liquor poured is a key step to ensure the quality and beauty of cocktails.

（3）在调制鸡尾酒的过程中，面向宾客时应展现出表演性和观赏性。这要求调酒师在整个调制过程中，无论是选材、配比、搅和还是装饰等各个环节，都应展现出良好的精神风貌，以保持操作的连贯自然和姿态的优美。通过精湛的技艺，为宾客带来一场视觉盛宴。这种表演性和观赏性不仅体现了调酒师的专业素养，也提升了鸡尾酒的文化内涵和艺术价值。

In the process of making cocktails, guest facing should show a high degree of performance and appreciation. This requires the concocter in the entire modulation process, whether it is the selection of materials, matching, mixing or decoration and other links, should show a good spiritual style, maintain the coherent operation of the natural and beautiful posture. Through exquisite skills, we bring a visual feast to the guests. This performance and appreciation not only reflects the professional quality of the concocter, but also enhances the cultural connotation and artistic value of the cocktail.

（4）要注意工作环境的卫生情况，全程保持工作台面的整洁干净。

Pay attention to the hygiene of the working environment, and keep the workbench clean and tidy throughout.

三、创意鸡尾酒的设计与展示（Design and Display of Creative Cocktails）

一般创意鸡尾酒的内容包括鸡尾酒名称、配方、载杯、创意鸡尾酒描述等几个方面。以创意鸡尾酒"看今朝"为例进行说明（表1-5-1）。

The general creative cocktail should include the name of the cocktail, the formula, the carrying cup, the creative cocktail description and so on. Take the creative cocktail "Look at the present" for example (Table 1-5-1).

表 1-5-1　创意鸡尾酒设计表（creative cocktail design table）

鸡尾酒名称 Cocktail name	看今朝 Look at the present	日期 Date	
分量 Amount	配方 Ingredients		
1OZ	中国白酒 Chinese Baijiu		
1OZ	蓝橙力娇酒 Blue curacao		
1/4OZ	糖浆 Syrup		
2OZ	菠萝汁 Pineapple juice		
3/4OZ	柠檬汁 Lemon juice		
装饰物 Garnish			
	薄荷叶 Mint leaves		
	柠檬片 Lemon slice		
杯子 Glasses	白兰地杯 Brandy glass		
创意鸡尾酒描述 Description of the cocktail			

创意背景："看今朝"创意鸡尾酒（图 1-5-26）创意理念的来源是致敬中国古代白酒悠久灿烂的历史发展。从最早的新石器时代晚期开始，经过夏商时期、唐宋时期、明清时期到现代发展时期，中国白酒的发展一共历经五个阶段。中国白酒的发展是一幅深厚灿烂的历史画卷，是传承千年的酿造工艺，是源远流长五千年的岁月沉淀，凝聚着中华民族的智慧与情感，每一滴都充满着历史的厚重感。纵然时光流逝，纵然潮流更迭，但匠心所造就的经典，每一滴都仍然值得我们细细品味。

Creative Background: The creative idea of "Look at the present" creative cocktail (Figure 1-5-26) is to pay tribute to the long and splendid historical development of ancient Chinese baijiu. From the earliest Late Neolithic Age, through the Xia and Shang Dynasties, Tang and Song dynasties, Ming and Qing dynasties to the modern development period, the development of Chinese baijiu has gone through five stages. The development of Chinese baijiu is a profound and splendid historical picture, and it is a brewing process inherited for thousands of years. It is the accumulation of five thousand years of history, which embodies the wisdom and emotion of the Chinese nation, and every drop is full of the depth of history. Even if time passes, even if the trend changes, but the classic created by ingenuity, every drop is still worth our careful taste.

立意创新：层次丰富的口感，借喻中国白酒的发展过程。微酸，是从萌芽到起步之初的技艺稚嫩，虽不完美却初显魅力；微苦，是一路走来，酿造技艺不断锻造发展的历程；回甘，是传承和创新的逐步融合，是历经数千年岁月洗礼、工艺锻造后，在世界酒文化阵营中独具魅力的中国白酒发展的丰硕成果，是每一个中华儿女内心深处为此感到骄傲的那一点微甜。

Intention Innovation: The rich taste of the entrance is a metaphor for the development process of Chinese baijiu. Slightly sour, from the bud to the beginning of the beginning of the immature skills, although not perfect but the initial charm. Micro bitter, is along the way, brewing technology constantly forging development process. Return to sweetness, is the gradual integration of inheritance and innovation, is after thousands of years of baptism, craft forging, in the world wine culture camp of the unique charm of the Chinese baijiu development fruitful results, is the heart of every Chinese proud of that little sweet.

色彩与装饰物创新：鸡尾酒整体分为两个颜色，上半部分为黄色，下半部分为蓝色渐变为浅绿色，由黄至蓝至绿，色彩渐变，美观绚烂，其实质寓意是：虽然岁月变迁，但中国白酒依然拥有蓬勃的生命力和不断发展创新的魄力，此为"看今朝"！同时，用柠檬片进行装饰，也体现了这款鸡尾酒的清爽淡雅，正如中国白酒历经千年沉淀后散发的芳香四溢的中国味。

Color and Garnish Innovation: The cocktail as a whole is divided into two colors, the top half is yellow, the bottom half is blue and gradually becomes light green. From yellow to blue to green, the color changes, beautiful, but the essence is meaning: although the years have changed, but Chinese baijiu still has a vigorous vitality and continuous development of innovation, which is "Look at the present"! At the same time, the garnish with lemon slice also reflects the refreshing elegance of this cocktail, just like the aroma of Chinese liquor after thousands of years of precipitation.

图 1-5-26　"看今朝"

赛证直通 Competitions and Certificates

一、基础知识部分 (Basic Knowledge Part)

(一) 专业名词解释 (Explanation of Professional Terms)

力娇酒 (Liqueur):

(二) 思考题 (Thinking Questions)

含气体的碳酸饮料为什么不能放入调酒壶中摇和?

Why can't carbonated drinks with gas be shaken in a cocktail shaker?

二、技能操作部分 (Skill Operation Part)

自创一款鸡尾酒并提交创意鸡尾酒设计表。

Create a cocktail and submit the creative cocktail design form.

专题六
博文约礼:鸡尾酒历史发展与服务礼仪

你知道鸡尾酒的对客服务需要注意哪些问题吗(图 1-5-27)?

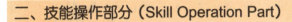

Do you know what to pay attention to in cocktail service (Figure 1-5-27)?

_____

图 1-5-27　鸡尾酒对客服务
(cocktail service)

一、鸡尾酒的历史 (The History of Cocktails)

鸡尾酒的历史源远流长,关于其起源的说法多种多样。

Cocktails have a long history, and theories about their origins vary.

(一) 起源与传说 (Origin and Legend)

"鸡尾酒"一词的起源可以追溯到 17 世纪初期,但具体的发明者和确切时间已无从考证。关于鸡尾酒的起源有多种传说,流传最为广泛的是这样一个小故事。18 世纪末,在纽约州埃尔姆斯福一家用鸡尾羽毛作装饰的酒馆中,一位名叫贝特西·弗拉纳根的女侍者,在酒馆各种酒快卖完时，将剩余的酒混合在一起，并加入一根鸡尾羽毛作为装饰，端给宾客饮用。客人们询问这是什么酒，贝特西随口回答"这是鸡尾酒"。从此，"鸡尾酒"之名便流传开来。

The origins of the term "cocktail" date back to the early 17th century, but the exact inventor and timing are unknown. There are many legends about the origin of the cocktail, but one of the most popular is a short story. Late 18th century, at a tavern in Elmsford, New York, decorated with cocktail-feathers, a waitress named Betsy Flanagan, when the tavern was running out of wine, mixed the remaining wine together, and added a cocktail-feather as a decoration, and served it to the guests. When the guests asked what kind of wine it was, Betsy casually replied, "It's a cocktail." Since then, the name "cocktail" has spread.

（二）早期发展与流行（Early Development and Popularity）

鸡尾酒的发展可以追溯至 18 世纪至 19 世纪。1806 年，美国的一本叫《平衡》的杂志首次详细地解释了鸡尾酒，将其描述为一种由烈酒及其他酒水混合而成的提神饮料。1862 年，法国作家托马斯出版了一本关于鸡尾酒的书籍《如何调配饮料》，对鸡尾酒的发展起到了关键性的作用。19 世纪至 20 世纪初，鸡尾酒在美国逐渐流行起来，成为社交场合中常见的饮品。

The development of cocktails can be traced back to the 18th and 19th centuries. In 1806, an American magazine called Balance explained the cocktail in detail for the first time, describing it as a refreshing drink mixed with spirits and other beverages composed of several spirits. In 1862, French writer Thomas published a book on cocktails, "How to Make Drinks", which played a key role in the development of cocktails. In the 19th century and early 20th century, the cocktail gradually became popular in the United States, becoming an usual drink in social occasions.

（三）现代发展（Modern Development）

进入 20 世纪后，鸡尾酒得到了快速发展和广泛传播。随着鸡尾酒的种类和口味不断丰富和创新，出现了各种新的鸡尾酒配方和调制方法。鸡尾酒不仅仅是一种饮品，更是一种文化和艺术的体现，成了现代社交场合中不可或缺的一部分。随着鸡尾酒的逐渐兴起和发展，调酒师也成为非常受欢迎的职业选择之一。

After entering the 20th century, cocktails have been rapidly developed and widely spread. With the variety and taste of cocktails are constantly enriched and innovative, and a variety of new cocktail recipes and preparation methods have emerged. Cocktail is not only a kind of drink, but also a reflection of culture and art, and has become an indispensable part of modern social occasions. With the gradual rise and development of cocktails, bartending has become one of the most popular career choices.

二、鸡尾酒服务礼仪（The Service Etiquette of Cocktail）

服务礼仪，作为职业礼仪的重要组成部分，是指在服务行业日常工作中逐渐形成的，并得到了广泛认可的礼节与仪式规范。其核心目的在于通过优质的服务和得体的举止，使客人感受到宾至如归的舒适与愉悦，进而有效树立和提升服务人员的个人形象以及所在场所的整体形象。

Service etiquette, as an important part of professional etiquette, refers to the etiquette and ritual norms gradually formed in the daily work of the service industry and have been widely recognized. Its core purpose is to make guests feel comfortable and happy at home through high-quality service and decent manners, and then effectively establish and enhance the personal image of service personnel and the overall image of the place.

(一) 岗位礼仪 (Post Etiquette)

（1）个人卫生：在正式上岗前，员工应确保完成个人卫生的全面检查，确保符合卫生标准，为提供良好的服务奠定基础。

Personal hygiene: before officially taking the job, employees should ensure that they complete a comprehensive inspection of personal hygiene to ensure that they meet health standards and lay the foundation for providing a good service environment.

（2）仪容仪表：员工在上岗前，应自觉进行仪容仪表的自我检查，确保仪容仪表整洁干净，以符合服务业中对于仪容仪表的具体规定和要求。

Self-examination of appearance: employees shall consciously conduct self-examination of appearance before starting work to ensure clean and neat appearance, in order to comply with the specific regulations and requirements of appearance in the service industry.

（3）精神面貌：一旦上岗，员工应坚持站立服务，展现精神饱满的状态，面带微笑，随时准备为每一位宾客提供优质的服务。

mental outlook: once on the job, employees should insist on standing service, showing a full state of spirit, smiling, ready to provide quality service for every guest.

（4）热情迎客：当宾客进门时，员工应微笑并致以亲切的问候。通过友善的语言和灿烂的笑容，使宾客一进门便能感受到温馨与舒适。

Warmly welcome guests: when guests enter the door, the staff should greet them with smiling faces and extend cordial greetings. Through friendly language and bright smile, guests can feel warm and comfortable when they enter the door.

（5）合理引导：同餐厅服务一样，员工应根据宾客的需求和偏好，将他们引领到合适的座位上，确保所有宾客都能得到满意的服务体验。

Guide guests properly: as with restaurant service, staff should guide guests to the appropriate seats according to their needs and preferences to ensure that all guests receive a satisfactory service experience.

(二) 对客礼仪 (Guest Etiquette)

（1）酒单递送与要求记录：在接待宾客时，应礼貌地递上酒水单并耐心等待客人点单。需细心倾听宾客提出的各项具体要求，确保记录完整并向宾客复述点单内容，以避免误解或遗漏。

Wine list delivery and request records: when receiving guests, you should respectfully hand the wine list politely and wait patiently for the order. Listen carefully to the specific requests made by the guests, make sure to record and repeat orders to the guests to avoid any misunderstanding or omission.

（2）尊重宾客个性化需求：对于宾客的具体要求，如"不要兑水"或"多加些冰块"等，应予以高度重视。务必严格按照宾客的要求提供服务，以满足其个性化需求。

Respect for the individual needs of guests: for the small requirements of guests, such as " do not add water " or " add more ice ", should be highly valued and respected. It is important to provide services in strict accordance with the guest's requirements to meet their individual needs.

（3）推荐服务：当宾客对酒水选择犹豫不决时，应热情礼貌地提供推荐，展现周到的服务态度，帮助宾客做出合适的选择。

Recommendation service: when guests hesitate to choose drinks or wine snacks, should warmly and politely provide recommendations, show thoughtful service, and help guests make appropriate choices.

（4）上酒服务的体态要求：在上酒服务时，应避免身体背向宾客。如需转身拿取背后的酒瓶，应侧身操作，以保持对宾客的关注和尊重。

Posture requirements for serving wine: avoid facing away from guests when serving wine. If you need to turn around to take the bottle behind you, you should turn sideways, in order to maintain the attention and respect of the guests.

（5）调制与器皿清洁：在宾客面前调制鸡尾酒时，应举止雅观、态度认真，确保所使用的器皿清洁。不得举止随便、敷衍了事，更不得使用不洁的器皿。

Preparation and utensils cleaning: when making cocktails in front of guests, should be elegant and serious, and ensure that the utensils used are clean. Do not behave casually, perfunctory, and do not use unclean utensils.

（6）用托盘上酒水的正确方法：应使用托盘从宾客的右侧上鸡尾酒，以符合服务礼仪的细节规范。

The correct way to drink on the tray: the tray should be used to serve the cocktail from the right side of the guest, in accordance with the service etiquette.

（7）整瓶酒的确认与开启：若宾客需用整瓶酒，应在斟酒前让客人看清酒标，并经过宾客确认后当面开启瓶塞，让宾客放心饮用。

Confirmation and opening of the whole bottle of wine: if the guests need to use the whole bottle of wine, the guests should see the label on the bottle before pouring wine, and open the bottle cork after confirmation by the guests to ensure that the guests can drink at ease.

（8）结账与送别：当宾客示意结账时，应使用小托盘递上账单，并请客人核查。若宾客无意立即离去，不得催促其提前结账付款。宾客离开时，应热情送别，并表达欢迎他们再次光临的意愿。

Checkout and farewell: when the guest signals to check out, the small tray should be used to pass the bill, and ask the guest to check the payment. If the guest does not intend to leave immediately, he or she shall not be urged to settle the bill in advance. When guests leave, they should send them off warmly and express their willingness to welcome them again.

赛证直通 Competitions and Certificates

一、基础知识部分（Basic Knowledge Part）

思考题（Thinking Question）

谈谈对鸡尾酒服务礼仪的理解。

二维码 1-5-6：
知识拓展

Talk about your understanding of cocktail service etiquette.

二、技能操作部分（Skill Operation Part）

在实训室模拟对客服务过程。

Simulate the customer service process in the training room.

✿【单元五　反思与评价】Unit 5　Reflection and Evaluation

学会了：_____

Learned to:_____

成功实践了：_____

Successful practice:_____

最大的收获：_____

The biggest gain：_____

遇到的困难：_____

Difficulties encountered:_____

对教师的建议：_____

Suggestions for teachers：_____

单元六　葡　萄　酒

葡萄酒是世界上文明的事物之一，也是世界上非常完美的、自然的东西之一。或许它带给我们的享受价值和欣赏价值是无与伦比的，远超其他事物带给我们的纯感官感觉。

——欧内斯特·海明威（美国作家）

 ## 单元导入 Unit Introduction

美国作家艾琳·摩根斯顿曾说"葡萄酒是装在瓶子里的浪漫诗篇"。本单元将带领大家走进美味的葡萄酒世界。

American writer Erin Morgenstern once said, "Wine is a romantic poem in a bottle". This unit will introduce you to the delicious world of wine.

🎯 学习目标 Learning Objectives

➤ 知识目标（Knowledge Objectives）

（1）了解葡萄酒的起源。
Understand the origins of wine.
（2）了解葡萄酒的传播发展。
Understand the communication and development of wine.
（3）认知葡萄酒的产区。
Learn about the wine regions.

➤ 能力目标（Ability Objective）

能根据所学知识掌握葡萄酒的品鉴。
Able to appreciate wine according to the knowledge learned.

➤ 素质目标（Quality Objective）

通过学习和认知葡萄酒的发展及酿造，提升学生的文化素养、专业素养和综合素质。
Enhance students' cultural literacy, professional literacy and comprehensive quality through learning and understanding of wine development and making.

专题一
源远流长：葡萄酒概述

【微课】
源远流长：
葡萄酒概述

一、葡萄酒的起源与传播（The Origin and Spread of Wine）

（一）葡萄酒的起源（The Origin of Wine）

关于葡萄酒的起源，历史上有多种说法。有人认为，葡萄酒起源于格鲁吉亚，因为考古学家在这个地区发掘了约公元前 6000 年世界上最古老的酒窖（图 1-6-1）。

There are many theories about the earliest origin of wine in history. Some people think that it originated in Georgia because archaeologists excavated the world's oldest wine cellar dating back to about 6000 BC in this area (Figure 1-6-1).

也有人认为葡萄酒起源于古波斯（也就是如今的伊朗），原因是 20 世纪 90 年代于伊朗北部扎格罗斯山脉的一个石器时代晚期的村庄里，挖掘出一个公元前 5415 年的罐子，其中有残余的葡萄酒和防止葡萄酒变

图 1-6-1 世界上最古老的酒窖
（the oldest wine cellar in the world）

成醋的树脂。

Others think it originated in ancient Persia (that is Iran today). The reason is that a jar dating from 5415 BC was excavated in a Late Palaeolithic village in the Zagros Mountains of northern Iran in the 1990s, which contained residual wine and the resin that prevented wine from turning into vinegar.

也有不同的说法认为，葡萄酒起源于保加利亚。因为在古代保加利亚人生活过的地区，先后发现了很多公元前 3 000 ~ 6 000 年关于葡萄种植和酿造葡萄酒的痕迹。

There are also different opinions that wine originated in Bulgaria. Because many traces of grape planting and wine making about 3 000-6 000 BC have been found in the area where ancient Bulgarians lived.

还有的人认为葡萄酒起源于古埃及。因为在埃及出土的距今 6000 年前的古墓壁画上，存在大量的描绘古埃及人栽培、采收葡萄和酿造葡萄酒情景的珍贵文物。

Others think that it originated in ancient Egypt. Because there are a large number of precious cultural relics on the murals of ancient tombs unearthed in Egypt 6000 years ago, depict the scenes of ancient Egyptians cultivation, harvesting grapes and brewing wine.

现在，又有一种新的说法，那就是最早酿制葡萄酒可能是在中国。因为在 2004 年，中美考古学家在河南舞阳贾湖遗址（图 1-6-2）挖掘出土的公元前 7000 年的陶器样品中发现了大量的葡萄酒残渣。

图 1-6-2　舞阳贾湖遗址
（Jiahu Site, Wuyang County）

Now, there is a new saying, that is the earliest brewed wine may be in China. In 2004, Chinese and American archaeologists found a large number of wine residues from the 7 000 BC pottery samples, which were excavated at the Jiahu Site, Wuyang County, Henan Province (Figure 1-6-2).

关于葡萄酒的起源，可谓众说纷纭，答案究竟是什么，其实并不重要。因为随着时间的推移，考古技术的发展，可能还会有更多不一样的答案。对于学习者来说，重要的是知道葡萄酒在世界人类历史中扮演着重要的角色，值得好好认知和学习。

There are different opinions about the origin of wine. In fact, it doesn't matter what the answer is. Because with the passage of time and the development of archaeological technology, there may be more different answers. For learners, it is important for us to know that wine plays a very important role in world human history. It is really worth us to recognize and learn it.

（二）葡萄酒的传播（The Spread of Wine）

从目前的记载资料来看，葡萄酒第一次传播是从高加索地区传到了古埃及和古希腊地区，公元 3 000 年前，古希腊的葡萄种植就极为兴盛；第二次传播是通过古罗马帝国的对外扩张，从地中海沿岸地区一直传播到整个欧洲，包括法国、意大利、西班牙、德国等地区；第三次传播是因为宗教以及欧洲开启大航海时代，葡萄酒随着宗教文化以及航海传入南非、澳大利亚、美洲、亚洲等国家和地区；第四次传播，是自第二次世界大战后的六七十年代开始，由于葡萄酒需求量的增加，很多酒厂和酿酒师开始在全世界找寻适合的土壤、相似的气候种植优质的葡萄品种，研发及改进酿造技术，从而促进了全世界葡萄酒产业的兴旺。

According to the records, the first spread of wine was from the Caucasus to ancient Egypt and Greece. Grape planting in Greece was very prosperous before 3000 AD. The second spread was via the external expansion of the ancient Roman Empire. It was from the coastal areas of the Mediterranean to the whole Europe including France, Italy, Spain, Germany and other regions. The third spread was due to religion and the opening of the era of great navigation in Europe. Wine was introduced into South Africa, Australia, America, Asia and other countries and regions with religious culture and the navigation. The fourth spread began in the 1960 s and 1970 s after the World War Ⅱ. Because of the increasing demand for wine, many wineries and winemakers began to find suitable soil and similar climate around the world to grow high-quality grape varieties. And they developed and improved brewing technology. As a result, all these promoted the prosperity of the wine industry all over the world.

二、葡萄酒"新、旧世界"(Wine "New and Old World")

(一)"旧世界"(The Old World)

"旧世界"国家主要包括位于欧洲的传统葡萄酒生产国，如法国、意大利、德国、西班牙和葡萄牙以及匈牙利、捷克斯洛伐克等东欧国家。它们大多位于北纬20°～52°，拥有适合种植酿酒葡萄的自然条件。

The old world countries mainly include the traditional wine producing countries in Europe, such as France, Italy, Germany, Spain and Portugal as well as the Eastern European countries such as Hungary and Czechoslovakia. Most of them lie between 20 and 52 degrees north latitude. They have the natural conditions that are suitable for wine grape planting.

(二)"新世界"(The New World)

工业革命以后，世界经济加速发展，迫使人们开始探索欧洲之外的广大土地。哥伦布发现新大陆之后，欧洲强国开始大肆进行殖民扩张。随着殖民的扩张，欧洲新移民潮带到当地种植的欧洲葡萄品种，传抵至南美洲，进而到达了如今的美国、新西兰等地。葡萄酒产区一直蔓延到以美国、澳大利亚为代表，包括南非、智利、阿根廷和新西兰等欧洲之外的葡萄酒新兴国家，也就是所谓的"新世界国家"。

The accelerated development of world economy force people to explore the vast land outside Europe after the industrial revolution. European strong nations began to carry out colonial expansion after Columbus discovered the new world. With the expansion of colonization, the wave of new immigrants in Europe brought the local European grape varieties to South America, and then to the United States, New Zealand and other places. Wine producing areas have spread to the emerging wine countries outside Europe, such as the United States, Australia, including South Africa, Chile, Argentina and New Zealand, that is, the so-called "new world countries".

所以，世界葡萄酒产区，就像两条红丝带，主要分布在北纬20～52°和南纬15～42°。

目前，最好的葡萄酒都产在这两条"红丝带"上。

Therefore, the world's wine producing areas are like two red ribbons. They are mainly distributed between 20 to 52 degrees north latitude and 15 to 42 degrees south latitude. At present, the best wines are produced on these two "red ribbons".

（三）"新、旧"世界的葡萄酒国家（Wine Countries of the "Old and New" World）

1. "旧世界"葡萄酒国家（"Old World" Wine Countries）

葡萄酒的酿造历史有上千年，据说在公元前1000年，最早的葡萄酒出现在希腊群岛之后遍布整个古希腊大陆。最终，古罗马人将其传遍整个欧洲和中东地区。因此，最早的"旧世界"葡萄酒国家就是欧洲和中东的国家。

The history of winemaking goes back thousands of years, and it is said that in 1 000 BC, the first wine appeared across the whole of ancient Greek mainland after the Greek islands. Eventually, through the Romans, it spread throughout Europe and the Middle East. So the first "old world" wine countries were the countries of Europe and the Middle East.

"旧世界"葡萄酒国家包括法国、西班牙、意大利、德国、葡萄牙、奥地利、希腊、黎巴嫩、克罗地亚、格鲁吉亚、罗马尼亚、匈牙利和瑞士等。这些国家有着丰富的酿酒经验以及悠久的酿酒历史，同时，他们也是最早酿造葡萄酒的国家。

"Old world" wine countries include France, Spain, Italy, Germany, Portugal, Austria, Greece, Lebanon, Croatia, Georgia, Romania, Hungary and Switzerland, among others. These countries have rich wine making experience and a long wine making history, and they are also some of the first countries to make wine grapes.

2. "新世界"葡萄酒国家（"New World" Wine Country）

在18世纪，欧洲国家在世界各地开辟殖民地。与此同时，欧洲的葡萄藤也被带到了各个殖民地，如美洲、澳洲、南美洲等地。自此，这些国家也开始种植葡萄并且酿造葡萄酒，所选用的葡萄藤，也都是从欧洲引进而来。

In the 18th century, European countries established colonies all over the world. At the same time, European vines were also brought to various colonies, such as America, Australia, South America and so on. Since then, these countries have also begun to grow grapes and make wine, using vines imported from Europe.

葡萄酒中的"新世界"是指没有原生葡萄品种的国家，后加入酿酒行列的国家，包括美国、澳大利亚、新西兰、智利、阿根廷、南非以及中国等。

The "new world" of wine is countries without native grape varieties, and later joined the winemaking countries, including the United States, Australia, New Zealand, Chile, Argentina, South Africa and China.

这些"新世界"国家酿酒的历史并不长，来自澳大利亚风靡世界的"奔富"酒庄，成立于1844年，至今也不到200年。中国种植葡萄、酿造葡萄酒也有一定的历史。据司马迁的《史记》记载，中国最早的葡萄酒引进是在公元前138年汉武帝时期，张骞出使西域，带回了葡萄栽培和葡萄酒酿造的技术。虽然中国葡萄酒的引进较早，但由于发展较晚，还是被划分在"新世界"葡萄酒国家的行列中。

The history of winemaking in these "new world" countries is not long, from Australia's popular "Penfolds" winery, founded in 1844, less than 200 years ago. There is also a history of growing grapes and making wine in China. According to Sima Qian's Records of the " Records of the Historian ", the earliest introduction of wine in China was in 138 BC, during the reign of Emperor Wu of Han Dynasty, when Zhang Qian went to the Western Regions and brought back techniques of viticulture and winemaking. Although our Chinese wine was introduced earlier, it is still divided into the ranks of new world wine countries.

三、葡萄酒世界产区（Wine Regions of the World）

世界上的葡萄酒产区主要集中在欧洲西部、地中海沿岸、澳洲南部、南美洲西部、北美洲西海岸和非洲南部。下面介绍世界上最有名的十个葡萄酒产区。

The world's wine regions are mainly concentrated in Western Europe, the Mediterranean coast, southern Australia, western South America, the west coast of North America and southern Africa. Here are ten of the most famous wine regions in the world.

（一）纳帕山谷（Napa Valley）

纳帕山谷位于美国，是当地最大的葡萄种植和葡萄酒产地，虽然整个面积不是很大，但多样的葡萄品种在市场上极具名气，所产出的葡萄酒品质也属上乘。

Napa Valley is located in the United States, is the largest local grape growing and wine producing area, although the entire area is not very large, but the variety of grape varieties are very good in the market reputation, the quality of the wine produced is very good.

（二）卢瓦尔河谷（Loire Valley）

卢瓦尔河谷位于法国，被誉为全球最美的葡萄酒种植区，河流滋润着整片土地，每年产量最多的是白葡萄酒，整个产区也以白葡萄酒而闻名。

Loire Valley is located in France, known as the world's most beautiful wine growing area, the river moistens the whole land, the most annual production is white wine, the entire region is also famous for white wine.

（三）尼亚加拉半岛（The Niagara Peninsula）

尼亚加拉位于加拿大，有全球范围内少见的位于平原地区的葡萄酒产区，产区内充足的水分和优质的土壤成分有利于葡萄产出。

The Niagara Peninsula located in Canada, has one of the rare plains wine regions in the world, where the abundant moisture and high quality soil composition conducive to the production of varietal grapes.

（四）托斯卡纳产区（Tuscany）

托斯卡纳产区位于意大利，以其卓越的葡萄酒品质和独特的地理文化而闻名于世。托斯卡纳产区以其得天独厚的地理环境和气候条件，结合传统的酿造工艺，酿造出了众多品质卓越的葡萄酒。

Tuscany is famous for its excellent wine quality and unique geographical culture.Tuscany region with its unique geographical and climatic conditions, combined with the traditional art of melting-wine, alcohol produced a number of excellent quality wines.

（五）波尔图（Porto）

波尔图位于葡萄牙，是世界上有名的葡萄酒产区，还有着非常悠久的红葡萄酒酿造历史，所产出的单品在风味上很不错。

Porto is located in Portugal, is the world's famous wine region, there is a very long history of red wine production, the output of the single product in the flavor is very good.

（六）安德森谷（Anderson Valley）

安德森谷位于美国加利福尼亚州，在世界葡萄酒十大产区中，安德森谷是比较"低调"的一个产区。近些年来以各种优质的葡萄酒单品让产区被大多数人所了解，酿造出的葡萄酒有着非常浓郁的花香和果味，所以很受欢迎。

Anderson Valley is located in California, USA. Among the top ten wine producing areas in the world, Anderson Valley is a relatively " low-key" producing area. In recent years, a variety of high-quality wine items have made the producing area known by most people, and the wines produced have a very strong floral and fruity flavor, so it is very popular.

（七）门多萨（Mendoza）

门多萨产区是阿根廷最大也是最重要的葡萄酒产区，其葡萄酒产量约占全国总产量的70%。该产区气候干燥，气温较高，灌溉条件优越，土壤丰富多样，以冲积土为主，是世界上少有的种植海拔如此之高的葡萄园。由于高纬度所带来的微气候以及高海拔所特有的地形，这里孕育并生长着不少的葡萄品种，甚至还为一些在其他产区难以种植的葡萄品种提供了适宜的生长环境。

Mendoza is the largest and most important wine region in Argentina, producing about 70% of the country's wine. The producing area has a dry climate, high temperature, superior irrigation conditions, rich and diverse soil, mainly alluvial soil, which is rare in the world to plant vineyards at such a high altitude. Due to the micro-climate brought by the high latitude and the unique terrain of the high altitude, there are many grape varieties bred and grown here, and even provide environment for some grape varieties that are difficult to grow in other regions.

（八）安达卢西亚（Andalucia）

安达卢西亚位于西班牙，是全球种植面积较大的葡萄产区，酿造上的工艺更是可以追溯到十五世纪。其中最有知名度的单品是雪利酒，口味复杂柔和，香气芬芳浓郁。

Andalucia located in Spain, is the world's largest growing grape region, and its winemaking process dates back to the 15 th century. Its most famous wine is sherry, which is clear and transparent; The taste is complex and soft, and the aroma is rich and fragrant.

（九）开普半岛（Cape Peninsula）

开普半岛位于南非，整个面积不大，但却是南半球地区最大的葡萄酒产区，不少单品都被划入名酒之中，有各方面都不错的种植和酿造条件。每年还有不少品牌会选择将酒拿到这里储存。

Cape Peninsula located in South Africa, although the entire area is not very large, but it is the largest wine region in the southern Hemisphere, many single products have been classified into the famous wine, all aspects of good planting and brewing conditions every year there are many brands will choose to get wine here to store.

（十）波尔多（Bordeaux）

波尔多位于法国，是当地甚至全球知名度极高的葡萄酒产区，该区有着优良的种植环境，种植的葡萄品种繁多。

Bordeaux is located in France, is the local and even the world's very famous wine producing areas. The area has an excellent planting environment, planting a wide variety of tree species.

 ## 赛证直通 Competitions and Certificates

一、基础知识部分（Basic Knowledge Part）

（一）专业名词解释（Explanation of Professional Terms）

1. "旧世界"葡萄酒国家（"Old World" wine country）：
2. "新世界"葡萄酒国家（"New World" wine country）：

二维码 1-6-1：
知识拓展

（二）思考题（Thinking Question）

请简述葡萄酒的分类。

Please briefly describe the classification of wine.

二、技能操作部分（Skill Operation Part）

请说出葡萄酒"新世界"著名的葡萄酒产区以及代表酒品。

Name the famous wine regions and representative wines in the "new world" of wine.

专题二

匠心之美：葡萄酒的酿造

【微课】
匠心之美：
葡萄酒的酿造

一、常见酿造葡萄品种（Common Vinification Grape Varieties）

俗话说"七分原料三分工艺，好葡萄酒源于好葡萄"。葡萄品质很大程度决定了葡萄酒的品质。

As the saying goes, " Seven points of raw materials three points of process, good wine comes from good grapes." Grape quality largely determines the quality of wine.

（一）红葡萄酒酿造葡萄品种（Red wine Grape Variety）

1. 赤霞珠（Cabernet Sauvignon）

赤霞珠是法国波尔多地区传统的酿制红葡萄酒的良种，也是世界著名的红色酿酒葡萄品种。酿造特性：口感严密紧涩，香味强烈，酒精度中等，单宁含量高，味道复杂丰富；有黑莓果香或薄荷、青菜、青叶、青豆、青椒、破碎的紫罗兰的香气和烟熏味；未成熟时有明显的青椒味。

It is a fine variety of traditional red wine made in Bordeaux, France, is also the world famous red wine grape variety. Wine making characteristics: tight and astringent taste, strong aroma, medium alcohol, high tannin content, complex and rich taste; Notes of blackberry or peppermint, green vegetables, green leaves, green beans, green peppers, crushed violets and smoky notes. When unripe, it has a distinct green pepper flavor.

2. 品丽珠（Cabernet franc）

品丽珠最早是在法国波尔多和罗亚河区种植，也被称为卡伯纳·佛朗或卡门耐特，是法国古老的酿酒葡萄品种，与赤霞珠、蛇龙珠是姊妹品种。酿造特性：深宝石红色，结构柔和，风味纯正，酒体完美，低酸、低单宁、酒质极优，充满了优雅、和谐的果香和细腻的口感。

It was first cultivated in Bordeaux and the Loire region of France, also known as Cabernet Franc or Camenette, is an ancient wine grape variety in France, with Cabernet Sauvignon and Serpentine are sister variety. Wine making characteristics: deep ruby red, soft structure, pure flavor, perfect body, low acid, low tannin, excellent quality, full of elegant, harmonious fruit and delicate taste.

3. 梅洛（Merlot）

梅洛在法国波尔多种植，是世界著名的红色酿酒葡萄品种。酿造特性：宝石红色，酒体丰满，柔和，果香浓郁，清爽和谐，单宁含量低，酒体柔和顺口，有李果香气，其酒体优劣程度与其土壤品质密切相关。

It was grown in Bordeaux, France and is the world famous red wine grape variety. Wine making characteristics: ruby red, full bodied, soft, fruity, fresh and harmonious, tannin content is low, the wine body is soft and smooth, with plum aroma, and the quality of the wine body is closely related to the quality of the soil.

4. 黑皮诺（Pinot Noir）

黑皮诺在法国勃艮第种植。黑皮诺被誉为"红葡萄皇后"，对生长环境非常挑剔。酿造特性：酒体轻盈，果香持久，并在悠长的余味中散发着果香的气息。此外，随着酒的陈年，黑皮诺还可能展现出玫瑰花、泥土、野味和蘑菇等复杂的香味。

It was grown in Burgundy, France. Known as the " Queen of red grapes ", Pinot Noir is very picky about the growing environment. Brewing characteristics: light bodied, fruity and long-lasting, with fruity notes in a long finish. In addition, as the wine ages, Pinot Noir may also develop complex aromas of rose, earth, game and mushroom.

5. 西拉（Selah）

西拉在法国北罗讷河谷种植，以其独特的香气和口感备受赞誉。酿造特性：复杂口感，酒体丰满，果香浓郁，并带有顺滑的余味和细腻的单宁感。西拉也在澳大利亚广泛种植，这里的西拉葡萄酒通常具有更多的黑巧克力、胡椒和香料风味。

It was grown in the Northern Rhone Valley of France, and highly praised for its unique aroma and taste. Brewing characteristics: complex, full-bodied, fruity, with a smooth finish and fine tannins. Selah is also widely grown in Australia, where syrah wines often have more dark chocolate, pepper and spice flavors.

6. 佳美（Camry）

佳美在法国勃艮第种植，目前博若莱种植面积较大，主要用于酿造博若莱风格的葡萄酒。酿造特性：酒液通常呈紫罗兰色或紫粉色，单宁含量极低，酒体轻盈，果香异常浓郁且清新自然。

It was grown in Burgundy, France and is currently planted in Beaujolais, which it is mainly used to make Beaujolais style wine. Brewing characteristics: the wine is usually violet or purplish-pink in color, with very low tannin content, light body, unusually fruity and fresh and natural.

7. 歌海娜（Grenache）

歌海娜在西班牙里奥哈地区种植。法国南部、澳大利亚和美国加利福尼亚州也有种植。酿造特性：风味独特，含有覆盆子、草莓、红醋栗和蓝莓以及白胡椒的香味。陈年后的歌海娜葡萄酒还含有太妃奶糖和皮革的风味。

It was grown in the Rioja region of Spain. It is also grown in southern France, Australia and California. Brewing characteristics: unique flavor, with raspberry, strawberry, red currant and blueberry and white pepper aromas. Aged Grenache wine also contains notes of toffee and leather.

8. 仙粉黛（Zinfandel）

仙粉黛原产于克罗地亚，现在主要种植在美国加利福尼亚州。酿造特性：酒精含量高，单宁柔和、口感甜润，并带有浆果香味，如草莓、覆盆子、蓝莓等。

It is native to Croatia, and now grown mainly in California. Brewing characteristics: high alcohol content, soft tannins, sweet taste, and a berry flavor, such as strawberry, raspberry, blueberry.

9. 佳丽酿（Carignan）

佳丽酿原产于西班牙，也称法国红。酿造特性：瑰丽的宝石红色，口感醇和且协调，酒体丰满，具有典型的特性。由于其个性不突出，在混酿时，常被用来增强酒的颜色和骨架感。

It is native to Spain, also known as French red. Brewing characteristics: magnificent ruby red color, mellow and harmonious taste, full body, with typical characteristics. Because its personality is not prominent, it is often used to enhance the color and skeleton of the wine when blend.

10. 丹魄（Tempranillo）

丹魄原产于西班牙，种植面积超过20万公顷，在全球的种植面积排名第四，因此，西班牙被誉为"丹魄王国"。酿造特性：有覆盆子、蓝莓、樱桃、松树、甘草、桂皮、丁香花蕾、皮革、巧克力和烟熏等多种风味。

It is native to Spain, the cultivation area of more than 200 000 hectares, in the world's planting area ranks fourth, so Spain is known as the "Tempranillo Kingdom". Brewing characteristics: raspberry, blueberry, cherry, pine, licorice, cinnamon, clove bud, leather, chocolate and smoked flavors.

11. 内比奥罗（Nebbiolo）

内比奥罗原产于意大利皮埃蒙特产区，被誉为"雾葡萄"。酿造特性：香气复杂，高单宁，高酒精，高酸，有柏油、玫瑰、松露、覆盆子、红醋栗、蓝莓、樱桃、烟熏、巧克力和胡椒的风味。

It is native to Piedmont, Italy, known as " fog grape ". Wine making characteristics: complex aroma, high tannin, high alcohol, high acid, with notes of tar, rose, truffle, raspberry, red currant, blueberry, cherry, smoke, chocolate and pepper.

（二）白葡萄酒酿造葡萄品种（White Wine Grape Variety）

1. 霞多丽（Chardonnay）

霞多丽原产于法国勃艮第。酿酒特性：酒色金黄，香气清新优雅，果香柔和悦人，酒体协调强劲、丰满，尤其是在橡木桶内发酵的干酒，酒香特别，干果香十分典型。

It is native to Burgundy, France. Wine making characteristics: golden color, fresh and elegant aroma, soft and pleasant fruit, wine body coordination, strong, full, especially in oak barrels fermented dry wine, wine aroma is special, dry fruit is very typical.

2. 雷司令（Riesling）

雷司令原产德国，是欧亚种。酿酒特性：味醇厚，酒体丰满，柔和爽口，高雅细腻，有持久的浓郁果香。

It is native to Germany, vitis vinifera. Wine making characteristics: mellow taste, full body, soft and refreshing, elegant and delicate, with lasting rich fruit flavor.

3. 长相思（Sauvignon Blanc）

长相思原产法国波尔多和卢瓦尔河谷，是一种芳香型品种，以其独特的绿色草本芳香和早熟特性而受到青睐。酿造特性：酸度高，酒体适中，香气浓郁复杂，包括青草、芦笋、青苹果和接骨木花等风味。

It is native to France, Bordeaux and the Loire Valley. It is an aromatic variety favored for its unique green herbal aroma and early maturing characteristics. Wine making characteristics: high acidity, medium bodied wine with a rich and complex aroma, including grass, asparagus, green apple and elderflower.

4. 白玉霓（Ugni Blanc）

白玉霓原产于法国卢瓦尔河谷地区。酿造特性：酸度较高，质地圆润，有鲜桃、青草和药草的芳香。

It is native to Loire Valley region, France. Brewing characteristics: high acidity, round texture, fresh peach, grass and herb aromas.

5. 灰皮诺（Pinot Gris）

灰皮诺原产于法国阿尔萨斯产区。酿造特性：酒精度较高，酸度低，有桂皮、甘草等多种香料的风味。

It is native to Alsace, France. Brewing characteristics: high alcohol, low acidity, cinnamon, liquorice and other spices flavor.

二、葡萄部位（Grape Part）

成熟的葡萄是葡萄酒酿造的主要原料，其各部分所含的成分不同，在酿造过程中扮演的角色也不同。葡萄在结果后大约需要一百天的时间成熟。随着葡萄的体积变大，糖分增加，酸味降低，红色素和单宁等酚类物质增加使颜色加深。同时，潜在的香味也逐渐形成，经发酵后散发出来。葡萄的大小、形状、颜色等都会因为品种、产地、环境而不同。

Ripe grapes are the most important raw materials for wine making, and each part of them contains different ingredients and will play different roles in the winemaking process. Grapes generally take about one hundred days to ripen after fruiting. As the size of the grape increases, the sugar content increases, the acidity decreases, and the phenols such as red pigment and tannin increase to deepen the color. At the same time, the underlying fragrance is gradually formed, which will be released after fermentation. The size, shape, color, etc. of grapes will vary depending on the variety, place of origin, and environment.

（一）葡萄梗（Grape Stems）

葡萄梗含有丰富的单宁，但其所含单宁收敛性强且较粗糙，常带有刺鼻的草味，通常酿造之前会先去除。如果需要增加酒的单宁含量，有时也会加进葡萄梗一起发酵。葡萄梗还含有不少钾，具有去酸的功能。

Grape stems are rich in tannins, but the tannins they contain are astringent and coarse, often with a pungent grassy taste, which is usually removed before brewing. If the tannin content of the wine needs to be strengthened, grape stems are sometimes added to fermentation. Grape stems also contain a lot of potassium, which has the function of removing acid.

（二）果肉（Flesh）

果肉占葡萄质量的 80% 左右，一般供食用葡萄的肉质较丰厚，而酿酒葡萄较多汁，其主要成分有水分、糖分、有机酸和矿物质。其中糖分是酒精发酵的主要成分，包括葡萄糖和果糖，有机酸则以酒石酸、乳酸和柠檬酸三种为主。

Flesh accounting for about 80% of the weight of grapes, the flesh quality of general food grapes is rich, and wine grapes are more juice, its main components are water, sugar, organic acids and minerals. Sugar is the main component of alcohol fermentation, including glucose and fructose, and organic acids are mainly tartaric acid, lactic acid and citric acid.

（三）葡萄籽（Grape Seeds）

葡萄籽内部含有许多单宁和油脂，单宁收敛性强、不够细腻，而油脂又会破坏酒的品质，所以在葡萄酒酿造的过程中必须避免弄破葡萄籽而影响酒的品质。

Grape seeds contain many tannins and oils, the tannin convergence is strong and not delicate enough, the oil will destroy the quality of the wine, so in the process of winemaking must avoid breaking the grape seeds and affecting the quality of the wine.

（四）葡萄皮（Grape Skin）

葡萄皮仅占葡萄整体的十分之一，但对品质的影响很大。它含有丰富的纤维素、果胶、单宁和香味物质。黑葡萄的皮还含有红色素，是红酒颜色的主要来源。葡萄皮中的单宁较为细腻，是构成葡萄酒结构的主要元素。其香味物质存于皮的下方，分为挥发性香和非挥发性香，后者须待发酵后才会慢慢形成。

Grape skin accounts for only one tenth of the total grape, but has a great impact on quality. It is rich in cellulose, pectin, tannins and flavor substances. The skin of black grapes also contains red pigment, which is the main source of red wine color. The tannins in grape skin are more delicate and are the main elements that constitute the structure of wine. Its aroma substances are stored under the skin and are divided into volatile and non-volatile fragrances, the latter of which must be slowly formed after fermentation.

三、红葡萄酒的生产（Red Wine Production Steps）

（一）采摘（Picking）

采摘分为人工采摘和机器采摘。手工采摘方式比较慢，人力消耗大，但可以对葡萄进行挑选，对葡萄造成的损坏也较小，机器采摘则反之。

It is divided into manual picking and machine picking. Manual picking is relatively slow, manpower consumption is large, but the grapes can be selected, and the damage to the grapes is less, the machine picking is vice versa.

（二）破碎（Crushing）

破碎是将葡萄浆果压破，以利于果汁流出，除梗是将葡萄浆果与果梗分离的过程。这两者往往使用破碎除梗机器进行。这个阶段，需要考虑是否保留部分果梗为葡萄酒增加更多单宁。

Peeling is when the grape berries are crushed to allow the juice to flow out, destemming is the process of separating the grape berry from the stem, both of which are often carried out using a crushing machine. At this stage, it is necessary to consider whether to retain some stems to add more tannins and bones to the wine.

（三）发酵（Fermentation）

葡萄碎除梗后，会被转移到不锈钢桶、水泥槽或橡木桶内，葡萄汁在桶内开始华丽蜕变。葡萄果浆里的糖分在酵母菌的作用下慢慢转化为酒精。果皮上的色素在浸渍过程中得以释放。葡萄酒获得色素、单宁、酸类物质与浸渍的时间需要根据葡萄酒的风格而定，如果想酿造果香、清新感十足的即饮型葡萄酒，则应缩短浸渍时间，降低单宁含量，保持酸度；如果想酿造陈年型优质红葡萄酒，则需要加强浸渍，提高单宁含量。一般情况下，红葡萄酒的发酵温度比白葡萄酒略高，为 26 ～ 32℃，较高的温度有利于单宁和颜色的提取，发酵时间从几天到几周不等。

After the grapes are crushed and destemed, they are transferred to stainless steel barrels, cement tanks or oak barrels, where the grape juice begins to sharpen. The sugar in the fruit pulp is slowly converted into alcohol by the action of yeast. The pigments on the peel are released during the maceration process, and the time for the wine to obtain pigments, tannins and acids depends on the style of the wine. if you want to make fruit with acid and fresh ready drink wine, you should shorten the maceration time, reduce tannins and maintain acidity; if you want to make high-quality aged red wine with acid, it is necessary to strengthen the impregnation and increase the tannin content. Under normal circumstances, the fermentation temperature of red wine is slightly higher than that of white wine, 26–32℃, and the higher temperature is conducive to the extraction of tannins and colors, and the fermentation time varies from a few days to a few weeks.

（四）榨汁（Juicing）

榨汁指将发酵后存于皮渣中的果汁或葡萄酒通过机械压力压榨取汁的过程。发酵结束后葡萄酒的汁液通常分为两种：一种是未经压榨自然流出的汁液，被称为自流汁；另一种是第一次和第二次压榨后所得到的汁液，被称为压榨汁。自流汁分离完毕，待容器内二氧化碳释放完成后就可以将发酵容器中的皮取出。对于红葡萄酒而言，压榨酒约占15%。压榨汁与自流汁相比，果皮挤压，口感发涩，酒体较为粗糙，而自流汁柔和圆润。酒厂通常按照一定比例直接混合或处理后将两者混合调配，以提高葡萄酒利用率，同时，用来打造不同风格与质量等级的葡萄酒。

Juicing is the process of extracting juice or wine from the marc after fermentation by mechanical pressure pressing. The juice of the wine after fermentation is usually divided into two types, one is the juice that flows naturally without pressing, known as free juice, and the other is the juice obtained after the first and second pressing, known as press juice. After the separation of the free juice is completed, the skin in the fermentation container can be extracted after the cartan dioxide release in the container is completed. For red wines, the press accounts for about 15%. Compared with the free flow juice, the fruit skin is squeezed, the taste is astringent, the wine body is rough, and the free flow juice is soft and round. Wineries usually blend the two directly or after treatment in a certain proportion to improve the efficiency of the wine, and at the same time, to create different styles and quality levels of wine.

（五）苹果酸—乳酸发酵（Malic-lactic Fermentation）

大部分红葡萄酒的酿造通常都会采用这种发酵工序，红葡萄酒只有在苹果酸—乳酸发酵结束，并进行恰当的二氧化硫处理后，才具有生物稳定性。酒精发酵后的红葡萄酒保持高浓度的酸度，酸度锋利敏锐，所以酿造红葡萄酒通常会采用苹果酸—乳酸发酵的方式，把生硬尖锐的苹果酸转化为柔和的乳酸，同时为葡萄酒增加烤面包、饼干、奶香等香气。白葡萄酒大部分不进行这一过程，以保留清新的果香以及脆爽的酸度。

Most red wine is usually made using this fermentation process, and red wine is only biostable after malic-lactic fermentation is completed and proper sulfur dioxide treatment is carried out. Red wine after alcohol fermentation to maintain a high concentration of acidity, acidity is sharp and sensitive, so the brewing of red wine is usually malic acid-lactic acid fermentation, the blunt and sharp malic acid into soft lactic acid, while increasing the aroma of toast, biscuits, milk and other wine. White wines mostly do not undergo this process in order to retain fresh fruit and crisp acidity.

（六）调配（Blending）

葡萄酒的调配混合是很多地区的酿酒惯例，品种之间相互调和可以形成特性互补、风味加持，对葡萄酒增加香气、酸度、酒体与色泽也有帮助。

The blending of wines is the practice of wine making in many regions, and the blending of varieties can form complementary characteristics, enhance flavor, and increase the aroma, acidity, body and color of the wine.

（七）陈酿（Aging）

高品质的红葡萄酒都经过橡木桶的培养，通过提供适度的氧气使酒更圆润、和谐，且补充红酒香味。培养时间依据酒质而定，通常不会超过两年。

High-quality red wines are cultivated in oak barrels, which provide moderate oxygen to make the wine more rounded and harmonious, and complement the aroma of red wine. The cultivation time depends on the wine quality, usually not more than two years.

（八）过滤（Filtration）

使用各种试剂去除固体杂质，常用的物质有果胶、黏土、鸡蛋清等。但一些酿酒者认为过度过滤会使葡萄酒流失香气及其他有益物质，所以，不少现代酒厂也会不过滤澄清而是直接装瓶。

Use a variety of reagents to remove solid impurities, commonly used substances are pectin, clay, egg white and so on. However, some winemakers believe that excessive filtration can cause the wine to lose aroma and other beneficial substances, so many modern wineries will not filter the clarification but directly bottled.

（九）装瓶（Bottling）

从橡木桶中提取酒液，装入深色玻璃瓶中保存。深色玻璃瓶能够阻碍 90% 左右的光线，更好地保护酒体，还要注意温度、湿度、振动等其他因素。

The wine is extracted from the oak barrels and stored in dark glass bottles. Dark glass bottles can block about 90% of the light, better protect the body of the wine, but also pay attention to temperature, humidity, vibration and other factors.

（十）瓶内陈年（Maturation）

部分葡萄酒在正式发售之前会进行一段时间的瓶内熟成，熟成结束后，塑帽并贴标发售。

Some wines are aged in the bottle for a period of time before they are officially released, after which they are capped and labeled.

四、白葡萄酒的生产（White Wine Production）

酿造白葡萄酒的品种一般是白葡萄，也可以使用红葡萄。在酿造时，先榨汁获得清澈的葡萄汁，再进行发酵酿造，果皮中的单宁与色素便不会渗入。

White wine varieties are generally white grapes, you can also use red grapes. In the brewing, first juice to obtain clear grape juice, and then fermentation of the original, the tannin and pigment in the peel will not penetrate.

（一）采收（Harvest）

白葡萄的采收时间一般会比红葡萄早，特别是香槟产区，这是为了保留天然的酸度，一般提早采摘。采收方式有人工采摘和机器采收。

White grapes are generally picked earlier than red varieties, especially in Champagne, in order to retain their natural acidity. Harvesting methods include manual harvesting and machine harvesting.

（二）接收与分选（Reception and Selection）

白葡萄与红葡萄接收分选过程一致。

Consistent with the red grape receiving sorting process.

（三）破碎除梗（Crushing and Destemming）

大部分白葡萄的破碎除梗程序与酿造红葡萄酒是一致的，对白葡萄来说，有些酒庄会进行整串压榨。破碎后的葡萄原料现在多进行冷浸工艺处理，以提取果皮中的芳香物质，冷浸温度通常在 5 ～ 10℃进行，浸渍时间需根据原料特性及质量而定，通常为 10 ～ 20 小时。冷浸工艺结束后，分离自流汁。

For most white grapes, the crushing and destemming process is the same as for red wines. For white grapes, some wineries press the whole bunch. The crushed grape raw materials are now mostly processed by cold soaking process to extract the aromatic substances in the peel, the cold soaking temperature is usually carried out at 5−10 ℃, and the impregnation time needs to be determined according to the characteristics and quality of the raw materials, usually 10−20 hours. After the cold soaking process, the free juice is separated.

（四）压榨（Pressing）

与红葡萄的压榨不同，生产白葡萄酒时，压榨是对新鲜葡萄的榨汁过程。这一程序需要尽快处理，尤其是使用红葡萄酿造白葡萄酒时，更需要速战速决，减少果皮与果汁接触的时间。白葡萄酒的压榨汁约占 30%。同样，榨取的汁液根据质量情况通常会与自流汁根据比例调配使用。

Unlike the pressing of red grapes, in the production of white wine, the pressing is the juicing process of fresh grapes. This process needs to be handled as soon as possible, especially when red grapes are used to make white wine, and it needs to be done quickly to reduce the time of contact between the peel and the juice. The pressed juice of white wine accounts for about 30%. Similarly, the extracted juice is usually proportionally blended with the free juice according to its quality.

（五）澄清（Clarification）

通常酿造白葡萄酒需要先澄清，然后进入发酵阶段。发酵汁中如果含有较多的果皮、种子、果梗残留物构成的悬浮物，会影响酒液发酵后的香气。因此，在酒精发酵之前应该将这些物质去掉，但不要过度澄清，以免影响酒精发酵的正常进行。

White wine is usually made by clarification and then fermentation. If the fermented juice contains more suspended matter composed of fruit peel, seed and fruit stem residue, it will affect the aroma after alcohol fermentation. These substances should therefore be removed before the alcohol ferments. However, do not over-clarify, so as not to affect the normal progress of alcohol fermentation.

（六）发酵（Fermentation）

为了保留白葡萄酒中的水果果香，白葡萄酒发酵温度一般在 15 ～ 20℃，比红葡萄酒低，葡萄酒香气更加优雅细致，发酵时间 2 ～ 4 周不等。为了更好地控制温度，多使用不锈钢桶发酵，也有部分白葡萄酒会使用橡木桶发酵。发酵虽然在红葡萄酒酿造中非常普遍，但对白葡萄酒却不一定合适。苹果酸 - 乳酸发酵通常会为葡萄酒的酿造增加香气，但同时也会减少新鲜的果味，这对那些果香型葡萄品种如雷司令、长相思是致命的打击。

In order to retain the fruit aroma of white wine, the fermentation temperature of white wine is generally 15-20℃, which is lower than that of red wine, and the wine aroma is more elegant and detailed, and the fermentation time varies from 2-4 weeks. In order to better temperature control more use of stainless steel barrel fermentation, there are also some white wine will use oak barrel fermentation, fermentation is common in red wine, but not necessarily suitable for white wine. Malolactic fermentation usually gives wine wine non-aroma, but also reduces the fresh fruit flavor, which is for those fruity grape varieties, such as Riesling, Sauvignon Blanc was a fatal blow.

（七）调配（Blending）

与红葡萄酒一样，很多产区都会使用不同的品种进行混酸，但单一品种酿造更为常见。

As with red wine, many regions use different varieties for acid mixing, but single varieties are more common.

（八）熟成（Maturation）

白葡萄酒比红葡萄酒脆弱很多，是否成熟需要根据不同品种、不同质量以及不同风格进行区分。大部分白葡萄酒为了保留其新鲜的酸度与果香会直接装瓶；也有部分白葡萄酒会转移到橡木桶（多使用旧桶）内进行陈年，为葡萄酒增加酒体、香气与质感。

White wine is much more fragile than red wine, so it needs to be matured according to different varieties, different qualities and different styles. Most white wines are bottled directly in order to retain their fresh acidity and fruit. Some white wines are also transferred to oak barrels (mostly old barrels) for aging, adding body, aroma and texture to the wine.

（九）澄清过滤（Clarification）

与红葡萄酒相似，白葡萄酒装瓶之前，会先冷却澄清，再过滤酒石酸，让酒变得比较稳定，否则葡萄酒很容易出现白色结晶状的酒石酸。

Similar to red wine, white wine is cooled and clarified before bottling to filter tartaric acid and stabilize the wine, otherwise the wine is prone to white crystalline tartaric acid.

（十）装瓶（Bottling）

装瓶前要确定使用软木塞还是螺旋盖，"新世界"很多葡萄酒产区螺旋盖的使用频率较高。

Before bottling, decide whether to use cork or screw cap, screw cap is used more frequently in many "new world" wine regions.

（十一）瓶内陈年（Maturation）

部分白葡萄酒在正式发售之前会进行一段时间的瓶内熟成，熟成结束后塑帽并贴标发售。

Some white wines are aged in the bottle for a period of time before they are officially released, after which they are capped and labeled.

五、桃红葡萄酒的生产（Rose Wine Production）

桃红葡萄酒是含有少量红色素的葡萄酒，最常见的颜色有玫瑰红、橙红、黄玫瑰红、紫玫瑰红等色泽，其颜色深浅及风味特征与葡萄使用品种、发酵时间、酿造方法都有很大关系，其口感介于红葡萄酒和白葡萄酒之间。优质的桃红葡萄酒多呈现新鲜的果香、活泼愉悦的酸度以及平衡的质感。桃红葡萄酒不易陈年。

Rose wine is a small amount of red pigment slightly red wine, the most common colors are rose red, orange red, yellow rose red, purple rose red and other colors, its color depth and flavor characteristics and the use of varieties, fermentation time, brewing methods have a great relationship, its taste between red wine and white wine. High quality rose wines often show fresh fruit, lively and pleasant acidity and balanced texture. Rose wine is not easy to age.

虽然可以通过调配红、白葡萄酒进行桃红葡萄酒的酿造，但大部分的桃红葡萄酒还是在红、白葡萄酒酿造方式的基础上酿造而成的。葡萄酒颜色的萃取与发酵的温度、时长是分不开的，因此，在酿造红葡萄酒的基础上降低发酵的温度或压缩发酵的时间便可以生产出桃红葡萄酒。

Although rose wine can be made by blending red and white wine, but most rose wine is still made on the basis of red and white wine brewing. The extraction of wine color is inseparable from the temperature cond duration of fermentation, so it is possible to produce rose wine by reducing the fermentation temperature or compressing the fermentation time on the basis of making red wine.

（一）直接压榨（Direct Press）

直接压榨的方法更适合葡萄原料色素含量高的葡萄品种的酿造。直接采用白葡萄酒的酿造程序即可酿造桃红葡萄酒，用这种方法酸出的桃红葡萄酒，颜色往往过浅，因此，适合高色素含量的葡萄品种，如佳丽酿。该方法的流程为：原料接收→破碎→ SO_2 处理→分离→压榨→澄清→发酵。

This method is more suitable for the limited production of grape varieties with high pigment content of grape raw materials. Directly using the brewing process of white wine can be made rose wine, with this method of acid rose wine, the color is often too light, so suitable for grape varieties with high pigment content, such as carignan. This method process is: raw material reception → crushing → SO_2 treatment → separation → pressing → clarification → fermentation.

(二) 放血法 (Saignee Method)

与红葡萄酒的酿造方法一样，当破皮的葡萄浸渍数小时后，在酒精发酵之前，分离出部分葡萄汁酿造桃红葡萄酒，剩余部分酿造正常的红葡萄酒。用这种方法酿成的桃红葡萄酒，颜色比前者略深，有更多果香。该方法流程为：原料接收→破碎→ SO_2 处理→浸渍 2 ～ 24 小时→分离→压榨→澄清→发酵。

As with red wine, when broken skins of grapes were macerated for several hours, before alcohol fermentation, some of the grape juice is separated to make rose wine, and the rest is restricted to normal red wine. The rose produced in this way is slightly darker and more fruity than the former. This method process is: raw material reception → crushing → SO_2 treatment → impregnation for 2–24 hours → separation → pressing → clarification → fermentation.

(三) 排出法 (Drawing off)

与放血法相似，在红葡萄酒酿造流程中，通过缩短发酵时间实现桃红葡萄酒的酿造。在发酵 6 ～ 48 小时后，将发酵的葡萄酒排出，转移至低温环境中继续发酵。这种方法由于和果皮接触时间长，酿成的桃红葡萄酒颜色更理想。

Similar to Saignee method, in the red wine brewing process, the fermentation time is shortened to achieve rose wine brewing. After 6–48 hours of fermentation, the fermented wine is discharged and transferred to a low temperature environment to continue fermentation. This method, due to the long contact time with the peel, produces more ideal rose wine color.

赛证直通 Competitions and Certificates

一、基础知识部分 (Basic Knowledge Part)

二维码 1-6-2:
知识拓展

(一) 专业名词解释 (Explanation of Professional Terms)

1. 破碎除梗 (Crushing and Destemming):
2. 调配 (Blend):

(二) 思考题 (Thinking Question)

1. 简答葡萄酒的生产程序。
Give a brief answer to the production procedure of wine.
2. 结合葡萄酒酿造的学习，谈谈你对职业精神的理解。
Combination with the study of wine making, talk about your understanding of professionalism.

二、技能操作部分（Skill Operation Part）

请分析长城干红的酿造特点。

Please analyze the brewing characteristics of Great Wall dry red.

【微课】
成人之美：
葡萄酒的侍酒
服务

专题三
成人之美：葡萄酒的侍酒服务

一、葡萄酒侍酒服务的基本知识（Basic Knowledge of Wine Sommelier Service）

（一）葡萄酒的储藏（Wine Storage）

要让餐桌上的葡萄酒表现出最好的一面，葡萄酒的前期储存非常重要。

In order to bring out the best in the wine on the table, the early storage of the wine is very important.

储存葡萄酒的 5 个基本准则：

5 Basic Rules for Storing Wine:

（1）保持凉爽恒定的温度。葡萄酒的最佳储存温度是 10～15℃，温度过高会导致葡萄酒加速老化，无法发展出更丰富、细腻的风味。当环境温度过低时，葡萄酒的成熟速度就会放缓，香气也会变得相对封闭。

Keep a cool and constant temperature. The best storage temperature of wine is 10～15 ℃, and too high a temperature will lead to accelerated aging of wine, unable to develop a richer and more delicate flavor. When the ambient temperature is too low, the maturation of the wine will slow down, and the aroma will become relatively closed.

长期储存葡萄酒时，除了要保证适宜的温度，还要确保温度恒定。在频繁急剧的温度变化下，软木塞容易出现热胀冷缩现象，导致过量氧气从瓶口的缝隙进入酒瓶内部，进而破坏酒质。此外，温度骤变也会打乱葡萄酒的化学反应进程，扰乱其陈年步伐。因此，葡萄酒不适合存放在厨房和阁楼等温度变化大或常有光线直射的地方。

When storing wine for a long time, in addition to ensuring a suitable temperature, it is also necessary to ensure that the temperature is constant. Under frequent and sharp temperature changes, the cork is prone to thermal expansion and cold contraction, resulting in excessive oxygen entering the inside of the bottle from the gap in the bottle mouth, and then destroying the wine quality. In addition, sudden temperature changes can also disrupt the chemical process of wine and disrupt its aging pace. Therefore, wine is not suitable for storage in the kitchen and attic, such as large temperature changes or often direct light places.

（2）维持适宜且稳定的湿度。葡萄酒储存的最佳环境湿度应维持在 70% 左右。湿度太高，软木塞和酒标容易腐烂、发霉；湿度太低又会导致软木塞变干失去弹性，无法密封瓶口，加速葡萄酒的氧化进程。需要强调的是，冰箱在制冷过程中会降低其内部环境的湿度，所以并不适合长期存放葡萄酒。

Maintain Appropriate and Stable Humidity. The optimal ambient humidity for wine storage should be maintained at about 70%. The humidity is too high, corks and wine labels easily rot and mold. Too low humidity will cause the cork to dry out and lose elasticity, unable to seal the bottle, and accelerate the oxidation process of the wine. It should be emphasized that the refrigerator will reduce the humidity of its internal environment during the cooling process, so it is not suitable for long-term storage of wine.

（3）避免强光照射。不论是自然光还是人造光线，任何一种来源的紫外线都会加速葡萄酒老化，人造光线还有可能催生出令人不悦的异味。长期处于光线照射下，葡萄酒的"骨架"会逐渐松散，香气和风味也会日趋寡淡。对于细腻雅致的香槟和酒体较轻的白葡萄酒而言，光线的杀伤力更为致命。大多数葡萄酒都采用深色酒瓶盛装的原因正是在于此。使用无色玻璃瓶盛装的葡萄酒对光线最为敏感，储存时应尤其注意避光。

Avoid bright light. Whether natural or artificial light, ultraviolet light from any source will accelerate the aging of the wine, and artificial light may also cause unpleasant odors. Under long-term light exposure, the "backbone" of the wine will gradually loosen, and the aroma and flavor will gradually fade. For delicate Champagne and lighter white wines, the light is even more deadly. This is why most wines are served in dark bottles. Wine in colorless glass bottles is the most sensitive to light, and special attention should be paid to avoiding light when stored.

（4）避免振动和摇晃。频繁的摇晃或振动会加快葡萄酒中各种化学物质的反应，从而破坏酒质，还会搅起陈年老酒中的沉淀物，使其口感变得粗糙。因此，在日常生活中，应尽量让葡萄酒远离洗衣机、扬声器以及其他"动感"区域。

Avoid vibration and shaking. Frequent shaking or vibration can accelerate the reaction of various chemicals in the wine, which can destroy the quality of the wine. It can also stir up sediment in the old wine, making the taste rough. Therefore, in daily life, we should try to keep wine away from washing machines, speakers and other "dynamic" areas.

（5）水平放置酒瓶。水平放置酒瓶的目的在于保证软木塞能接触到酒液，以防止其干瘪变形，让氧气有机可乘。如果是采用螺旋盖封瓶的葡萄酒，则无需讲究是否水平放置。在存放采用软木塞封瓶的香槟或其他起泡酒时，也不一定要水平放置。一方面，香槟等起泡酒的瓶内气压较大，可以让软木塞保持足够的湿度。另一方面，起泡酒直立存放时，其中比氧气更重的二氧化碳气体会停留在酒液表面，可有效隔绝氧气与酒液的接触。

Horizontal wine bottle. The purpose of placing the bottle horizontally is to ensure that the cork can contact the wine, to prevent it from drying out and deformation, so that oxygen can be used. If it is a screw cap bottle of wine, there is no need to pay attention to whether the horizontal placement. It is also not necessary to store champagne or other sparkling wines in corked bottles horizontally. On the one hand, sparkling wines such as champagne have higher air pressure inside the bottle, which allows the cork to maintain sufficient humidity. On the other hand, when sparkling wine is stored upright, carbon dioxide gas, which is heavier than oxygen, will stay on the surface of the wine, which can effectively isolate the contact between oxygen and wine.

（二）葡萄酒杯的选择（The Choice of Wine Glasses）

葡萄酒杯的选择，也关乎客人饮酒体验。

The choice of wine glass is also related to the guest's drinking experience.

（1）一般优先考虑郁金香杯，其腹大口小的设计利于香气聚集，高脚的设计既便于手持，又能避免手温影响到酒温。

Generally give priority to the tulip cup, the design of the belly and small mouth is conducive to the aroma gathering, and the design of the tall foot is easy to hold, but also to avoid the hand temperature affecting the wine temperature.

（2）其次葡萄酒杯要轻薄，薄壁的酒杯利于香气扩散，也更容易观察酒的颜色。

Secondly, wine glasses should be light and thin, thin-walled wine glasses are conducive to the diffusion of aroma, and it is easier to observe the color of wine.

（3）正式场合宜选大酒杯，减少倒酒次数，并提供足够空间进行摇杯醒酒。

Formal occasions should choose large wine glasses, reduce the number of pouring, and provide enough space for shaking the cup to sober up.

（4）无色透明酒杯为佳，避免有色或装饰性杯子分散注意力，影响品鉴。

Colorless transparent glass is preferred, avoid colored or decorative cups to distract attention and affect tasting.

（三）葡萄酒适饮温度（Good Wine Temperature）

不同类型的葡萄酒有不同的饮用温度，选择最合适的温度可以让葡萄酒更加充分地绽放魅力。

Different types of wine have different drinking temperatures, and choosing the most appropriate temperature can make the wine more fully bloom its charm.

1. 基本原则（Basic Principles）

白酒的适饮温度低于红酒，清淡型酒的适饮温度低于浓郁型的酒，甜白酒的适饮温度低于干白酒，酒精强化型酒的适饮温度略高于浓郁型干红酒。

The suitable drinking temperature of white wine is lower than red wine, light wine is lower than full-bodied wine, sweet white wine is lower than dry white wine, alcohol fortified wine is slightly higher than full-bodied dry red wine.

2. 具体原则（Specific Principles）

白葡萄酒的适饮温度为 7 ～ 10℃；红葡萄酒为 12 ～ 18℃。

The suitable drinking temperature of white wine is 7-10℃, red wine is 12-18℃。

甜白葡萄酒的适饮温度为 4 ～ 5℃；气泡酒为 5 ～ 9℃；清淡型白葡萄酒为 6 ～ 9℃；浓郁型白葡萄酒为 8 ～ 12℃；清淡型红葡萄酒为 10 ～ 12℃；浓郁型新红葡萄酒为 14 ～ 16℃；浓郁型陈红葡萄酒为 16 ～ 18℃。

4-5℃ for sweet white wine, sparkling wine is 5-9℃, light white wine is 6-9℃, full-bodied white wine for 8-12℃, light red wine is 10-12℃, rich new red wine is 14-16℃, rich old red wine is 16-18℃.

（四）醒酒（Decanting）

对于红葡萄酒而言，要使其充分绽放魅力，还需要进行唤醒服务。通常红葡萄酒在饮用前30分钟就要开瓶醒酒。醒酒的方式有：

For red wine, in order to fully bloom its charm, it also needs a wake-up service. Usually red wine should be decanted 30 minutes before drinking. The ways to sober up are:

（1）唤醒，让葡萄酒"呼吸"新鲜空气。通过氧化，柔化单宁丰富、又尚未完全成熟的葡萄酒，使酒的口感变得更为顺畅。

Wake up and let the wine "breathe" fresh air. By oxidation, the tannin-rich, but not fully ripe wine is softened, so that the wine taste becomes smoother.

（2）滗酒，通过换瓶的方式分离沉淀物，保证口感，尤其是陈年葡萄酒。

Decanting, separating sediment by decanting to ensure taste, especially for aged wines.

（3）过滤，如果开瓶前沉淀物没有沉入瓶底，可利用专业过滤网过滤掉沉淀物。

Filter, if the sediment does not sink to the bottom of the bottle before opening, the sediment can be filtered out by a professional filter.

二、侍酒服务基本流程（Basic Process of Serving Wine）

（一）新年份红葡萄酒（New Vintage Red Wine）

（1）点单：为客人呈递酒单，服务主人点酒。进行酒品介绍和推荐，根据客人需要做好点酒记录，如果客人所点酒品需要醒酒，向客人提出醒酒建议，同时复述所点酒的年份、酒名等重要信息。

Order: present the wine list for guests, serve the host to order wine. Make wine introduction and recommendation, make wine records according to the needs of the guests, if the guests need to decanting, make suggestions to the guests, and repeat the important information such as the year and name of the wine ordered.

（2）准备所需用具：准备酒杯，以及醒酒服务所需要的醒酒器、小餐碟、开瓶器等，并将这些醒酒服务器皿放于酒水车或移动工作台上，推向客人餐桌旁，为后续服务做准备。

Prepare the necessary utensils: prepare wine glasses, decanters, small dishes, bottle openers, etc. required for the decanter service, and place these decanter servers in the wine cart or mobile work table and push them to the guests' table to prepare for the subsequent service.

（3）取酒：于酒柜中轻取酒品，检查确认酒品无误，擦拭干净酒瓶。

Take the wine: gently take the wine from the wine cabinet, check the wine and clean the bottle

（4）示酒：左手持餐布，右手持酒，端于主人面前，保持正标朝上，向客人重复酒名、年份及出产地信息。待客人确认无误后，在示意之下进行下一步服务。

Show the wine: hold the cloth in left hand, and the wine in right hand, put it in front of the host, keep the positive label facing up, and repeat the wine name, year and origin information to the guests. After the guest confirms that nothing is wrong, the next service is carried out under the signal.

（5）开瓶：以海马刀开瓶后，轻轻取出酒塞，轻闻瓶塞，确认葡萄酒酒质。擦拭瓶口，保持瓶口清洁卫生。同时把盛放软木塞的餐碟端于主人右侧。

Open the bottle after opening the bottle with a waiter's Friend, gently remove the cork and smell the cork to confirm the quality of the wine. Wipe the mouth of the bottle to keep it clean and hygienic. At the same time, the plate with the cork is placed on the right side of the host.

（6）鉴酒：在征得客人同意的情况下，向服务酒杯内倒入 30 mL 葡萄酒，协助品鉴服务；确认酒品无误后，再倒少量葡萄酒请主人品鉴。

Wine tasting: with the consent of the guests, pour 30 mL wine into the service glass to assist the tasting service. After confirming the wine is correct, pour a small amount of wine and invite the host to taste.

（7）醒酒：左手握醒酒器，右手握瓶，确保酒标朝向客人，将葡萄酒缓缓倒入醒酒器。注意在瓶底处稍做余留，避免将酒渣倒入醒酒器内。

Decanting: hold the decanter in your left hand, hold the bottle in your right hand, make sure the wine label is facing the guest, and slowly pour the wine into the decanter. Be careful to leave a little bit at the bottom of the bottle to avoid pouring the residue into the decanter.

（8）斟酒：左手拿餐布，右手持醒酒器，遵循女士优先原则，为主人倒酒。

Pour wine: hold the cloth in the left hand, hold the decanter in the right hand, follow the principle of ladies first, and pour wine for the master.

（9）祝酒：祝客人用餐愉快，通常除醒酒器及空酒瓶外，撤走所有器皿。

Toast: wish guests a happy meal, usually remove all utensils except decanters and empty bottles.

（二）老年份红葡萄酒（Aged Red Wine）

（1）点单：为客人呈递酒单，服务主人点酒。进行酒品介绍和推荐，根据客人需要做好点酒记录，如果客人所点酒品需要醒酒，向客人提出醒酒建议，同时复述所点酒的年份、酒名等重要信息。

Order: present the wine list for guests, serve the host to order wine. Make wine introduction and recommendation, make wine records according to the needs of the guests, if the guests need to decanting, make suggestions to the guests, and repeat the important information such as the year and name of the wine ordered.

（2）准备所需用具：准备酒杯以及醒酒服务所需要的醒酒器、小餐碟、开瓶器、蜡烛、火柴等，并将醒酒服务器皿放于酒水车或移动工作台上，推向客人餐桌旁，为后续服务做准备。

Prepare the necessary utensils: prepare wine glasses, decanters, small dishes, bottle openers, candles, matches, etc. required for decanter service, and place the decanter server on the wine cart or mobile work table and push it to the guests' table to prepare for the subsequent service.

（3）取酒：于酒柜中轻取酒品，检查确认酒品无误，擦拭干净酒瓶后轻放至垫有餐布的酒篮。

Take the wine. gently take the wine in the wine cabinet, check the wine is correct, wipe the bottle clean, gently put the wine in a wine basket lined with a table cloth.

（4）示酒：左手托住酒篮底部，右手持酒篮把手，端于主人面前，保持正标朝上，向客人重复酒名、年份及出产地信息。待客人确认无误后，在示意之下进行下一步服务。

Show the wine: hold the bottom of the wine basket with left hand, and hold the handle of the wine basket with right hand before the host. Keep the label up and repeat the wine name, vintage and country of origin to guests. After the guest confirms that nothing is wrong, the next service is carried out under the signal.

（5）开瓶：用海马刀直接为酒篮里的葡萄酒开瓶，注意保持酒篮不要晃动，开瓶后，轻轻取出酒塞，轻闻瓶塞确认葡萄酒酒质。擦拭瓶口，保持瓶口清洁卫生。同时把盛放软木塞的餐碟端于主人右侧。

Open the bottle: use the waiter's Friend to directly open the bottle of wine in the wine basket, pay attention to keep the wine basket doesn't shake, after opening the bottle, gently remove the wine stopper, smell the bottle stopper to confirm the wine quality. Wipe the mouth of the bottle to keep it clean and hygienic. At the same time, the plate with the cork is placed on the right side of the host.

（6）鉴酒：在征得客人同意的情况下，向服务酒杯内倒入 30 mL 葡萄酒，协助品鉴服务，确认酒品无误后，再倒少量葡萄酒请主人品鉴。

Wine tasting: with the consent of the guests, pour 30 mL wine into the service glass to assist the tasting service. After confirming the wine is correct, pour a small amount of wine and invite the host to taste.

（7）点燃蜡烛，并把使用过的火柴棒放于火柴盒一端或餐碟内，用来后续熄灭蜡烛。

Light the candle and place the used matchstick in the end of the matchbox or in a serving dish to extinguish the candle later.

（8）醒酒：从酒篮中轻取出葡萄酒，左手握醒酒器，右手握瓶，确保酒标朝向客人；将酒在蜡烛的照亮下缓缓倒入醒酒器中，注意观察沉淀，及时停止倾倒，避免将酒渣倒入醒酒器内。

Decanting: gently remove the wine from the wine basket, hold the decanter in left hand and the bottle in right hand, ensure that the wine label is facing the guests. Slowly pour the wine into the decanter under the light of the candle, pay attention to the precipitation, stop the pouring in time, and avoid pouring the wine residue into the decanter.

（9）斟酒：左手拿餐布，右手手持醒酒器，遵循女士优先原则，最后为主人倒酒。

Pour wine: hold the cloth in the left hand, hold the decanter in the right hand, follow the principle of ladies first, and finally pour wine for the master.

（10）祝酒：祝客人用餐愉快，通常除醒酒器及空酒瓶外，撤走所有器皿。

Toast: wish guests a happy meal, usually remove all utensils except decanters and empty bottles.

（三）起泡酒服务（Sparkling Wine）

（1）点单：为客人呈递酒单，服务主人点酒。进行酒品介绍和推荐，根据客人需要做好点酒记录，如果客人所点酒品需要醒酒，向客人提出醒酒建议，同时复述所点酒的年份、酒名等重要信息。

Order: present the wine list for guests, serve the host to order wine. Make wine introduction and recommendation, make wine records according to the needs of the guests, if the guests need to decanting, make suggestions to the guests, and repeat the important information such as the year and name of the wine ordered.

（2）准备所需用具：准备冰桶、餐布、酒杯、小餐碟、开瓶器等，并将这些服务器皿放于酒水车或移动工作台上，推向客人餐桌旁，为后续服务做准备。

Prepare the necessary utensils: prepare ice bucket, table cloth, wine glass, small plate, bottle opener, etc., and put these servers on the wine cart or mobile workbench, and push them to the guests' table to prepare for the subsequent service.

（3）选择酒杯：为客人选择合适的起泡酒杯，对照光线，注意检查酒杯的清洁程度。递送酒杯时，通常使用托盘或手持呈上，抓握过程切不可直接触摸杯口与杯身。

Choose the glass: choose the right sparkling glass for the guests, compare the light, pay attention to check the cleanliness of the glass. When the glass is delivered, it is usually presented on a tray or hand−held, and the mouth and body of the glass must not be directly touched during the grasping process.

（4）冰桶服务：将冰桶（含冰水混合物）放在餐布上方，端于客人餐桌一旁，一般靠近主人位，放于右侧。

Ice bucket service: place the ice bucket (containing ice water mixture) over the table cloth, and put it beside the guest table, usually near the host seat, on the right side.

（5）取酒：在酒柜内为客人取酒，查看酒标信息，保证葡萄酒与客人所点一致。左手持餐布，右手将葡萄酒倾斜托于手上，如果瓶身温度过低有水雾，瓶底垫上餐布托送。使用右侧胳膊做一定支撑，平稳地走向主人位右侧。

Pick up wine: pick up wine for guests in the wine cabinet, check the wine label information to ensure that the wine is consistent with the customer's order. Hold the cloth in left hand, and tilt the wine on right hand. If the temperature of the bottle is too low and there is water mist, put the cloth on the bottom of the bottle. Use the right arm to do some support, and smoothly walk to the right side of the host.

（6）示酒：保持正标朝上，向客人重复酒名、年份及出产地信息，待客人确认无误后，在主人示意之下进行下一步服务。

Show the wine: keep the positive label facing up, repeat the wine name, year and origin information to the guests. After the guest confirms that nothing is wrong, the next step of service is carried out under the host's signal.

（7）开瓶：按照起泡酒开瓶方式，通常在事先准备好的酒水车或可移动餐台上进行开瓶；注意开瓶的过程中，瓶口不准朝向任何客人，同时开瓶声不宜过大；轻闻瓶塞，确认酒质，把软木塞与铁丝圈分开，同时放于事先准备好的餐碟内，并端向主人位右侧餐桌上。

Open the bottle: according to the way sparkling wine is opened, usually in the prepared wine car or mobile table to open the bottle, pay attention to the process of opening the bottle, the bottle mouth is not allowed to face any guests, and the opening sound should not be too large. Sniff the cork to confirm the quality of the wine, separate the cork from the wire ring, place it in the prepared plate, and bring it to the right table of the host.

（8）擦拭瓶口，保持瓶口清洁卫生。

Wipe the mouth of the bottle to keep it clean and hygienic.

（9）主人鉴酒：左手持餐布，为主人位斟倒少量葡萄酒（约为1盎司），请主人鉴酒。

Host tasting wine: holding the table cloth in left hand, pouring a small amount of wine (about 1 thief) for the host.

（10）斟酒：待主人鉴酒无误，且示意后正式为客人倒酒，遵循女士优先的原则，最后为主人斟酒。

Pour wine: wait for the host to taste wine without error, and signal the formal pour for the guests, follow the principle of ladies first, and finally pour wine for the host.

（11）祝酒：祝客人用餐愉快，把酒瓶放于冰桶中，倾斜瓶身，酒瓶上放上白色餐布，带走盛放软木塞的餐碟及酒车。

Toast: wish guests a happy meal, place the wine bottle in the ice bucket, tilt the bottle, place a white dinner cloth on the wine bottle, take the cork plate and wine cart.

赛证直通 Competitions and Certificates

一、基础知识部分（Basic Knowledge Part）

（一）专业名词解释（Explanation of Professional Terms）

1. 鉴酒（Wine tasting）：
2. 醒酒（Decanting）：

（二）思考题（Thinking Question）

请写出陈年葡萄酒的服务流程。

Please write the service process of aged wine.

二维码 1-6-3：
知识拓展

二、技能操作部分（Skill Operation Part）

请完整练习起泡酒的服务流程。

Please practice the sparkling wine service process completely.

专题四
耐人寻味：葡萄酒品鉴与佐餐礼仪

【微课】
耐人寻味：
葡萄酒品鉴与
佐餐礼仪

一、品鉴基础知识（Tasting Basics）

（一）品鉴温度（Wine Tasting Temperature）

不同类型的葡萄酒需要不同的饮用温度，即使是同样类型的葡萄酒由于产地不同、风格不同、年份不同，适饮温度也会随之而异。如果葡萄酒的饮用温度过低会使酒的内涵不能得到充分体现；温度过高，则会让原本品质细腻的葡萄酒变得粗糙。

Different types of wine need different drinking temperatures, even the same type of wine due to different origin, different styles, different years, suitable drinking temperatures will also vary. If the wine drinking temperature is too low, the connotation of the wine can not be fully reflected. The temperature is too high, it will make the original quality of delicate wine become rough.

白葡萄酒的酸度比较高，以清爽的口感和果香为主要特色。温度过高会使酸味变重，所以白葡萄酒的饮用温度要比红葡萄酒低。

The acidity of white wine is relatively high, with refreshing taste and fruit as the main characteristics. The temperature is too high will make the sour taste heavier, so the drinking temperature of white wine is lower than that of red wine。

红葡萄酒则以复杂的香气和丰富、厚实的口感为主要特色。温度过低会使香气被封闭起来；温度过高则酒精味会变得很重，从而影响香味体现。

Red wine is mainly characterized by complex aroma and rich, thick taste. The temperature is too low will make the aroma is closed. The temperature is too high, the alcohol flavor will become very heavy, thus affecting the flavor.

各类型葡萄酒的适饮温度如下：加强酒：17 ～ 20 ℃，干红葡萄酒：16 ～ 18℃，干白葡萄酒：12 ～ 16℃，桃红酒：10 ～ 12℃，香槟、起泡酒、甜酒：5 ～ 9℃。

The optimum temperature for each type of wine is as follows：fortified wine: 17–20 ℃, dry red wine: 16–18 ℃, dry white wine 12–16 ℃, rose wine 10–12 ℃, Champagne, sparkling wine, dessert wine 5–9 ℃.

（二）味觉信息（Taste Information）

品尝葡萄酒讲究一定的方法。人的味蕾分布在口腔之中，而非胃肠内。人的舌头上分布着四大味蕾源，舌尖感知甜味，舌两侧对酸味极其敏感，舌后部为咸味感知区，舌根是苦味敏感区。所以，品尝菜肴需要细细咀嚼品味；品酒则是让葡萄酒进入口腔后，吸入一点空气，通过合理转动，使葡萄酒酒液充满整个口腔，仔细感受分辨葡萄酒的酸度、甜度、香气、酒精等的强度。另外，通过口腔对葡萄酒的加温，可以更好地释放香气分子，当香气充满口腔，链接嗅觉器官，就可进一步确认香气的类型与浓郁度。

Wine tasting requires a certain approach. Our taste buds are located in the mouth, not the stomach. There are four taste bud sources distributed on our tongue: the tip of the tongue senses sweetness, the sides of the tongue are extremely sensitive to sour taste, the back of the tongue is a salty perception area, and the base of the tongue is a bitter sensitive area. Therefore, tasting dishes need to chew and taste, wine tasting is to let the wine into the mouth, inhale a little air, through reasonable rotation, so that the wine liquor filled with the whole mouth, carefully feel the acidity, sweetness, aroma, alcohol and so on. In addition, by heating the wine in the mouth, it can better release aroma molecules, and when the aroma fills the mouth, linking the olfactory organs, we can further confirm the type and intensity of the aroma.

1. 甜度（Sweetness）

甜度通常可以指代糖分残留。残留糖分多的葡萄酒，根据糖分含量可以分为干型、半干、半甜、甜型葡萄酒。来自炎热产区的葡萄酒，葡萄成熟度高，舌尖的糖分感知明显高于冷凉产区的葡萄酒。

Sweetness can usually refer to the residual sugar. Residual sugar wine, according to the sugar content can be divided into dry, semi-dry, semi-sweet, sweet wine. Wines from hot regions, with high grape ripeness, have a significantly higher sugar perception on the tongue than wines from cool regions.

2. 酸度（Acidity）

葡萄富含酸性物质，给葡萄酒带来新鲜度，白葡萄酒比红葡萄酒、桃红葡萄酒含酸度更高。品尝高酸的葡萄酒时，舌头两侧很容易产生唾液，因此，可以通过唾液流出的速度与数量来感知一款葡萄酒酸度的高低。一般情况下，来自冷凉气候产区的葡萄酒比温暖炎热气候产区的葡萄酒酸度更高，口感清爽，酒体清瘦；相反，来自炎热气候产区的葡萄，酸度往往不足，葡萄酒显得较为肥美、顺滑。昼夜温差大的产区也有利于酸度的维持。在酿酒厂里，可以通过人工加酸，解决酸度不足的问题。酸度可以描述为低酸、中酸与高酸。

Grapes are rich in acidity, which brings freshness to wine. White wine contains more acidity than red wine or rose wine. When tasteing high-acid wine, it is easy to produce saliva on both sides of the tongue, therefore we can sense the acidity of a wine by the speed and amount of saliva. In general, wines from cold and cool regions have higher acidity, refreshing taste and leaner body than those from warm and hot regions. On the contrary, the acidity is often insufficient, and the wine is more round and smooth. Producing areas with large temperature differences between day and night are also conducive to maintaining acidity. In the brewery, the problem of insufficient acidity can be solved by artificially adding acid. Acidity can be described as low acid, medium acid and high acid.

3. 单宁（Tannin）

单宁是一种让口腔收敛、发干、褶皱、粗糙的物质。单宁存在于很多物质里，如树皮、果皮、茶叶等。葡萄的单宁来自果皮，如果葡萄酒经过橡木桶陈年，葡萄酒也会吸收橡木桶里的单宁。因为红葡萄酒带皮发酵，浸渍颜色的同时单宁一并被浸泡出来，所以红葡萄酒比白葡萄酒富含更多单宁。白葡萄酒先榨汁再发酵，除了经过橡木桶陈酿的类型，白葡萄酒单宁一般很少，品酒时也很少提及。单宁含量的多少与品种直接挂钩，单宁还需要区分成熟单宁与粗糙单宁。如果口感明显苦涩，这可能是因为葡萄酒比较年轻、葡萄成熟度不高或过度榨汁而导致的，成熟的单宁则口感相对顺滑、柔和。一般而言，温暖炎热的产区更有利于单宁的成熟。葡萄酒的单宁含量一般描述为低、中等、高单宁。

Tannin is a substance that makes the mouth astringent, dry, wrinkled and rough. Tannin is found in many substances, such as bark, fruit peel and tea. The tannins of grapes come from the skins, and if the wine is aged in oak barrels, the wine will also absorb the tannins in oak barrels. Because red wine ferments with the skin, the tannins are soaked out along with the color, so red wine has more tannins than white wine. White wine is first juiced and then fermented, except for the type of wine that has been aged in oak barrels, white wine generally has very little tannin and is rarely mentioned when tasting wine. The amount of tannin is directly related to the variety, and tannins also need to distinguish between ripe tannins and rough tannins. If the taste is obviously bitter, it may be because the wine is young, the grapes are not ripe or over-juiced. Ripe tannins are smooth and soft on the palate. Generally speaking, warm and hot producing areas are more conducive to the maturation of tannins. The tannins of wines are generally described as low, medium and high tannins.

4. 酒精（Alcohol）

葡萄的果糖与酵母发生化学反应，即可生成酒精，也就是葡萄酒。不同的气候环境、不同的品种，糖分含量都不同，加上酿酒方法的多样性，酒精的多少也大不相同。一般来说，葡萄酒的最佳酒精度范围通常在 12.5% ～ 14% 范围内。当品尝酒精度在 13%vol（中高酒精）以上的葡萄酒时，喉咙能感觉到明显的灼热感；品尝酒精度在 15%vol（高酒精）以上的葡萄酒时，感知更加明显，胃肠温度甚至会快速升高，面部血管流动加速。相反，品尝酒精度在 12%vol（低酒精）以下的葡萄酒，感知会明显减弱，饮用一般较为顺畅，口腔内没有压力。酒精的感知与香气、酸度的平衡与否也有一定的关系，所以不能完全靠灼热感判断酒精含量的高低。通常，温暖产区的葡萄酒的酒精含量明显高于冷凉产区的葡萄酒。酒精的含量可以用低、中、高描述。

The fructose in grapes reacts with yeast to produce alcohol, also known as wine. Different climates, different varieties, sugar content are different, coupled with the diversity of brewing methods, the amount of alcohol is also very different. The optimal alcohol range for wine is usually between 12.5% and 14%. When tasting wine with alcohol content above 13%vol(medium and high alcohol), the throat can feel obvious burning sensation. When tasting wine with alcohol content above 15%vol(high alcohol), the perception is more obvious, and even the gastrointestinal temperature increases rapidly, and the facial blood vessel flow accelerates. On the contrary, tasting wines with an alcohol content below 12%vol(low alcohol), this perception will be weakened a lot, and basically drinking is very smooth and there is no pressure in the mouth. The perception of alcohol also has a certain relationship with the balance of aroma and acidity, so the level of alcohol content can not be judged entirely by the burning sensation. In general, wines from warm regions have significantly higher alcohol content than wines from cool regions. The alcohol content can be described as low, medium or high.

5. 酒体（Body）

酒体是酒在舌头上的重量的感觉，指葡萄酒被含入口中的饱和度、浓郁度与压迫感。酒精含量高的葡萄酒，压迫与饱和感更加强大；酒精含量低的葡萄酒，则显得比较轻盈。果香丰富的葡萄酒，一般酒体感觉相对饱满、浓郁，相反则会比较清脆，酒体寡淡。单宁的高低也会影响对酒体的感知，成熟的单宁，浓郁感较强。酒体的描述，可以用酒体轻盈、中等、浓郁表示。

Body is the feeling of the weight of the wine on the tongue, which means that the wine is contained in the entrance of saturation, intensity and pressure. High alcohol content of wine, pressure and saturation sense is more powerful, and low alcohol content of wine, it is relatively light. Fruity rich wine, general wine body feeling relatively full, rich, on the contrary will be more crisp, wine body light. The level of tannin will also affect the perception of the wine body, mature tannins, strong flavor. The description of the wine body can be expressed by light, medium and full-bodied.

6. 回味（Finish）

回味指葡萄酒的风味在口腔内持续的时间，果香很快消失说明葡萄酒酒质相对较差。优质的葡萄酒，其果香风味会持续数秒，甚至数十秒以上，让人回味无穷。

Finish refers to how long the flavor of the wine lasts in the mouth, and the quick disappearance of the fruit notes indicates that the wine is relatively poor in quality. Quality wine, its fruit flavor will last for a few seconds, or even more than tens of seconds, let people feel endless aftertaste.

二、品鉴方法（Appreciation Approach）

（一）观色（Observing Color）

将装有 1/3 酒样的杯子倾斜 30 ～ 45°，在明亮的白色背景下观察酒液的色调（暗淡还是鲜亮）、澄清度（是否有混浊）和黏度（酒液对抗流动的程度）。

First tilt the wine glass containing 1/3 of the wine sample by 30–45 degrees, and observe its color tone (dark or bright), clarity (whether there is turbidity) and viscosity (the degree of resistance to flow) against a bright white background.

1. 红葡萄酒（Red Wine）

红葡萄酒主要以紫红色、宝石红、棕红色三种色系为主（图 1-6-3）。红酒随着酒龄的增长，酒液中的紫色会慢慢变淡。因此，红葡萄酒年份越老颜色越浅，年份越年轻则颜色越深。

Red wine is mainly red violet, ruby red and red brown (Figure 1-6-3). As the wine age increases, the purple tone will gradually fade. Therefore, the older the red wine, the lighter the color, and the younger the red wine, the darker the color.

2. 白葡萄酒（White Wine）

白葡萄酒的颜色主要以柠檬黄、金黄色、琥珀色、棕色四种色系为主（图 1-6-4）。随着酒龄的增长，白葡萄酒的颜色会越深。

The color of white wine is mainly lemon yellow, golden yellow, amber and brown (Figure 1-6-4). White wines tend to get darker as they age.

图 1-6-3　红葡萄酒（Red wine）

图 1-6-4　白葡萄酒（white wine）

（二）闻香（Sniffing Aroma）

摇转酒杯，让酒中的香气物质完全释放出来。先在杯口嗅闻摇杯后的香气，接着再往深处闻杯里的香气。第一层香气取决于葡萄酒品种，第二层香气主要来源于发酵质量，第三层香气来源于橡木桶陈酿和瓶内熟成。酒的香气越丰富、醇厚，通常酒的品质就越好。

Shake the wine glass to fully release the aroma in it. First sniff the aroma from the rim of the cup, and then stay close to smell the aroma in the cup. The top note of the wine aroma lies in the wine variety, the middle note mainly comes from the quality of fermentation, and the base note comes from oak barrel aging and bottle maturation. The richer and mellower the wine aroma, the better the wine quality.

（三）品味（Tasting）

饮一口样酒（6 ～ 10 毫升），让葡萄酒在口腔内打转，使其接触到舌头、上腭以及口腔内所有的表面。葡萄酒一般有四种基本味感：甜、酸、咸、苦。舌尖对甜味最敏感；接近舌尖的两侧对咸味最敏感；舌头的两侧对酸味最敏感；舌根则对苦味最敏感。

Take a sip of the wine (6–10 mL) and allow the wine to swirl in our mouth, so that it touches the tongue, the palate and all the surfaces in our mouth cavity. It has four basic wine tastes: sweet ness, sour, saltiness, and bitter. Sweetness is the most sensitive taste on the tip of tongue; both sides near the tip of tongue are most sensitive to salty taste; both sides of the tongue are most sensitive to sour taste; the base of the tongue is most sensitive to bitter taste.

(四) 评价 (Appraisal)

1. 平衡感 (Sense of Balance)

不管酒中的味觉成分是什么，最关键的是各种元素要能形成良好的平衡，例如，葡萄酒会因为酸味低而口感平淡，也会因为单宁高显得味苦。

No matter what the taste ingredients in the wine are, it's critical to make all elements balanced, for example, one may feel the wine taste insipid due to the low acidity, or bitter due to the high tannins.

2. 回味 (Aftertaste)

如果回味短，通常品质一般；回味长，一般品质较高。回味包括收敛感，灼热感，刺痒或刺麻感、金属感。酒越年轻，单宁越多，收敛感越强。酒精度数越高、灼热感相对明显。二氧化碳过量会产生刺麻感。

If the aftertaste is short, the quality is usually average; the aftertaste is long, the quality is generally higher. Aftertaste includes astringent sensation, burning sensation, itching or tingling, metallic sensation. The younger the wine, the more tannins and the stronger the astringency. The higher the alcohol degree, the burning sensation is relatively obvious. Excess carbon dioxide produces a tingling sensation.

3. 酒体 (Wine Body)

酒体分为轻盈、中等、醇厚等，可理解为葡萄酒在口中的"重量"和"质地"。它主要由舌头来感觉，而非葡萄酒本身的重量，与酒的实际品质及黏度无关，而是与酒的酒精度、甘油及酒中的酸有关。丰满的葡萄酒通常酒精度或干浸出物含量较高，黏稠的酒液会在杯壁上留下明显的"酒泪"给人一种"醇厚"的感觉；相反，轻盈的葡萄酒通常酒精度或干浸出物含量较低，不易在杯壁留下黏稠的痕迹，会给人一种"清淡"的感觉。

The wine body is divided into light, medium, mellow, etc., which can be understood as the "weight" and "texture" of the wine in the mouth. It is mainly felt by the tongue, not the weight of the wine itself, and has nothing to do with the actual quality and viscosity of the wine, but with the alcohol, glycerin and acid in the wine. Full wine usually has a high alcohol or dry extract content, and the viscous liquor will leave obvious "wine tears" on the wall of the cup to give people a "mellow" feeling; in contrast, light wines usually have a lower alcohol or dry extract content, which is not easy to leave a sticky mark on the wall of the glass, and will give people a "light" feeling.

三、餐酒搭配 (Wine Pairing Dining Etiquette)

(一) 通用法则 (Generic Rule)

在葡萄酒的餐酒搭配中，通用法则是白酒配白肉、红酒配红肉。即肌肉纤维细腻、脂肪含量较低、脂肪中不饱和脂肪酸含量较高、肉色较浅的白肉类选用白葡萄酒，因为白葡萄酒酒体轻柔，又能利用果酸去腥；肌肉纤维粗硬、脂肪含量较高、饱和脂肪酸含量较高、味道醇厚、颜色偏深的红肉选用红葡萄酒，因为红葡萄酒酒体醇厚且酒中的单宁可去油。

In the wine paired with food, the general rule is white wine with white meat, red wine with red meat. That is, the muscle fiber is delicate, the fat content is low, the unsaturated fatty acid content in the fat is high, the meat color is lighter white meat choose white wine, because the white wine is soft, and the fruit acid can be used to remove the fishy. Red wine is chosen for red meat with coarse muscle fibers, high fat content, high saturated fatty acid content, mellow taste and dark color, because red wine is mellow and the tannin in the wine is better to remove oil.

但是，这只是通用方法，想要餐酒搭配出绝佳风味，在选择合适的葡萄酒搭配食物时，还需要充分考虑食物中添加的配料以及香料。因为不同的配料与香料，会直接影响食物的味道和口感，一些口味较重的调料也会盖过葡萄酒本身的风味。

However, this is only a general method, in order to achieve excellent flavor with table wine, you need to fully consider the ingredients and spices added to the food when choosing the right wine to match the food. Because different ingredients and spices will directly affect the taste and texture of the food, some flavors of heavier spices will also overpower the flavor of the wine itself.

（二）具体法则（Specific Rule）

（1）注重味觉协调，不过分突出某一元素。一般来说，酸、甜、辣、咸或油腻的食物，可以按照下面的法则适配。

Focus on taste coordination, but not overly highlight a certain element. Generally speaking, the sweet, sour, spicy, salty or greasy foods can be paired as per the following rules.

较酸的食物搭配酸度较高的葡萄酒，较辣的食物搭配甜型葡萄酒，较咸的食物搭配甜酒或起泡酒，较甜的食物搭配甜度更高的葡萄酒，较油腻的食物搭配单宁充沛的葡萄酒。

Sour foods with wines which higher acidity, spicy food with sweet wine, salty foods with sweet wine or sparkling wine, sweet food with wine containing higher sweetness, oily food with wine containing abundant tannins.

（2）匹配浓郁度相当的食物和葡萄酒。

Match foods and wines with comparable intensity.

将浓郁度相当的食物和葡萄酒进行搭配，是一个简单直接的方法，它能在一定程度上保证餐酒搭配的平衡和谐。

It is also a simple and direct way to match the food and wine with comparable intensity. It can somewhat ensure the balance and harmony of food and wine pairing.

沙拉蔬菜，鱼肉和水煮蛋，清淡的食物，匹配清新轻盈的葡萄酒，比如酒体较轻、果香活泼的红葡萄酒。例如博若莱的佳美和各类鱼肉搭配起来也很出色。

For salads, vegetables, fish, boiled eggs and light foods, match fresh and light wines, such as the wine which with light body and lively fruity aroma. The match of Beaujolais's Gamay with diverse fish looks excellent.

用酒体和风味适中的葡萄酒佐以烤鸭、奶油意面等食物。

Pair wine with moderate body and flavor to roast duck, cream, pasta and other foods.

将酒体强劲、醇香浓郁的葡萄酒搭配口感丰满、浓郁的食物，如烩牛肉、香辣咖喱等。

Wines with strong wine body and mellow taste to matched full-bodied and rich foods like stewed Beef, spicy curry, etc.

（3）以食物的主要风味决定搭配用酒，如果所要搭配的食物分类比较复杂，在挑选搭配葡萄酒时，建议以食物最突出的风味比较权衡，从而决定最终的搭配用酒。

To determine the main flavor of the food with wine, if the classification of food to be matched is more complex, in the selection of matching When it comes to wine, it is recommended to compare and weigh the most prominent flavor of the food, so as to decide the final wine to use.

赛证直通 Competitions and Certificates

一、基础知识部分（Basic Knowledge Part）

（一）专业名词解释（Explanation of Professional Terms）

1. 酒体（Body）：

2. 平衡感（Balance）：

（二）思考题（Thinking Question）

请写出葡萄酒的品鉴流程。

Please write the tasting procedure of the wine.

二维码 1-6-4：
知识拓展

二、技能操作部分（Skill Operation Part）

请任意品鉴一款葡萄酒，并写出品鉴总结。

Please taste any wine and write a summary.

【单元六　反思与评价】Unit 6　Reflection and Evaluation

学会了：＿＿＿＿＿＿＿＿＿＿＿＿＿＿＿＿＿＿＿＿＿＿＿＿＿

Learned to: ＿＿＿＿＿＿＿＿＿＿＿＿＿＿＿＿＿＿＿＿＿＿＿

成功实践了：＿＿＿＿＿＿＿＿＿＿＿＿＿＿＿＿＿＿＿＿＿＿

Successful practice: ＿＿＿＿＿＿＿＿＿＿＿＿＿＿＿＿＿＿＿

最大的收获：＿＿＿＿＿＿＿＿＿＿＿＿＿＿＿＿＿＿＿＿＿＿

The biggest gain: ＿＿＿＿＿＿＿＿＿＿＿＿＿＿＿＿＿＿＿＿

遇到的困难：＿＿＿＿＿＿＿＿＿＿＿＿＿＿＿＿＿＿＿＿＿＿

Difficulties encountered: ＿＿＿＿＿＿＿＿＿＿＿＿＿＿＿＿＿

对教师的建议：＿＿＿＿＿＿＿＿＿＿＿＿＿＿＿＿＿＿＿＿＿

Suggestions for teachers: ＿＿＿＿＿＿＿＿＿＿＿＿＿＿＿＿＿

单元七　其他酿造酒

葡萄美酒夜光杯，欲饮琵琶马上催。
醉卧沙场君莫笑，古来征战几人回？

——唐·王翰

单元导入 Unit Introduction

除了前面单元中学习过的各类酒，还有其他种类丰富的酿造酒类，如黄酒、啤酒、清酒等。这些酿造酒的制作材料是什么？生产的过程和工艺是怎样的？有何代表名品或知名品牌？这个单元将一一解密。

Translation: In addition to the various types of alcoholic beverages that we have learned in the previous unit, there are also a wide variety of other brewed wines, such as huangjiu, beer, sake, etc. What are the materials used for making these fermented alcoholic beverages? What is the production process and craftsmanship like? What are the representative famous products or well-known brands? This unit will unveil the secrets for you one by one.

学习目标 Learning Objectives

➤ 知识目标（Knowledge Objectives）

（1）了解黄酒的分类、制作工艺、饮用方法等。

Understand the classification, production process, and drinking methods of huangjiu.

（2）了解啤酒的分类、制作工艺、饮用方法等。

Understand the classification, production process, and drinking methods of beer.

（3）了解米酒的分类、制作工艺、饮用方法等。

Understand the classification, production process, and drinking methods of rice wine.

（4）了解清酒的分类、制作工艺、饮用方法等。

Understand the classification, production process, and drinking methods of sake.

➤ 能力目标（Ability Objectives）

（1）能判断黄酒品质。

Able to judge the quality of huangjiu.

（2）能判断啤酒品质。

Able to judge the quality of beer.

（3）能介绍米酒制作过程。

Able to introduce the process of making rice wine.

（4）能区分日本清酒和中国黄酒。

Able to distinguish between Japanese sake and Chinese huangjiu.

➤ 素质目标（Quality Objectives）

（1）树立文化自信，培养学生对中华传统文化的自豪感。

Establish cultural confidence and cultivate students' sense of pride in Chinese traditional culture.

（2）培养创新精神。

Foster innovative spirit.

【微课】
中华国酿：
黄酒

专题一
中华国酿：黄酒

你听说过绍兴酒吗？绍兴酒属于什么酒（图1-7-1）？

_____

Have you heard of Shaoxing rice wine? What type of liquor is Shaoxing rice wine (Figure1-7-1)?

_____

图 1-7-1 绍兴酒
（**Shaoxing rice wine**）

一、黄酒简介 (Introduction to Huangjiu)

黄酒是世界上古老的酒类之一，源于中国且唯中国有之，与啤酒、葡萄酒并称世界三大发酵古酒。黄酒是以稻米、黍米、小米、玉米、小麦、水等为主要原料，经加曲或部分酶制剂、酵母等糖化发酵剂酿制而成的发酵酒。由于色泽多呈黄色，故称"黄酒"。它保留了发酵过程中产生的各种营养成分和活性物质，具有低耗粮、低酒度、高营养、性格温和的特点，是最具健康价值、内外兼修的酒种。黄酒酒精度不高，一般为8°～20°，多数在15°左右。

Huangjiu is one of the oldest alcohol beverages in the world, originating in China and only in China. It is known as one of the three ancient fermented beverages along with beer and wine. Huangjiu is a fermented drink made from main ingredients such as rice, sorghum, millet, corn, wheat, and water, through the fermentation process using distiller's yeast or partial enzyme preparations, yeast, and other saccharifying agents. Due to its yellowish color, it is called "huangjiu." It retains various nutrients and active substances produced during the fermentation process, characterized by its low consumption of grains, low alcohol content, high nutrition, and mild personality, making it one of the most healthful and well-balanced types of alcohol. The alcohol content of huangjiu is not high, generally ranging from 8° to 20°, with most around 15°.

二、黄酒的分类 (Classification of Huangjiu)

黄酒种类繁多，根据产地、原料和酿造工艺的不同，可以分为以下几类。

Yellow wine comes in a wide variety, and can be categorized into the following types based on origin, ingredients, and brewing techniques.

（一）按甜度分类 (By Sweetness Level)

干型黄酒：含糖量低，口感清爽。
半干型黄酒：含糖量适中，口感柔和。
半甜型黄酒：含糖量较高，口感甘甜。
甜型黄酒：含糖量高，口感浓郁。

Dry Yellow Wine: Low sugar content with a refreshing taste.

Semi-Dry Yellow Wine: Moderate sugar content with a smooth and mellow taste.

Semi-Sweet Yellow Wine: Higher sugar content with a sweet taste.

Sweet Yellow Wine: High sugar content with a rich and intense taste.

（二）按产地分类 (By Origin)

绍兴黄酒：中国最著名的黄酒产地，以花雕酒、女儿红为代表。
福建黄酒：以红曲米为原料，酒体呈红色，风味独特。
山东黄酒：以即墨老酒为代表，口感醇厚。

Shaoxing Yellow Wine: The most famous yellow wine in China, represented by Huadiao wine and Nü'erhong.

Fujian Yellow Wine: Made with red yeast rice, giving it a red hue and a unique flavor.

Shandong Yellow Wine: Represented by Jimo Laojiu, known for its mellow and robust taste.

（三）按用途分类 (By Usage)

饮用型黄酒：直接饮用，适合搭配菜肴。
烹饪型黄酒：用于烹饪，去腥增香。

Drinking Yellow Wine: Consumed directly, often paired with dishes.

Cooking Yellow Wine: Used in cooking to remove fishy odors and enhance flavors.

三、黄酒的酿造工艺 (Huangjiu Brewing Process)

黄酒的传统酿造工艺是一门综合性技术，根据现代学科分类，它涉及食品学、营养学、化学、和微生物学等多种学科知识。我们的祖先在几千年漫长的实践中逐步积累经验，不断完善，使之形成极为纯熟的工艺技术。今天的黄酒生产技术有了很大的提高，新原料、新菌种、新技术和新设备的融入为传统工艺的改革、新产品的开发创造了机遇，产品不断创新，酒质不断提高。

The traditional brewing process of Huangjiu (yellow rice wine) is a comprehensive technology that, according to modern academic classification, involves knowledge from various disciplines such as food science, nutrition, chemistry, and microbiology. Over thousands of years of practice, our ancestors gradually accumulated experience and continuously refined the process, developing it into an extremely sophisticated craft. Today, Huangjiu production technology has seen significant advancements. The integration of new raw materials, microbial strains, techniques, and equipment has created opportunities for the reform of traditional methods and the development of new products. As a result, the products are continuously innovated, and the quality of the wine is constantly improving.

黄酒生产流程主要包括原料选择、浸米与蒸饭、拌曲与落缸发酵、开耙与坛发酵、压榨与澄清、煎酒与灌坛等关键步骤。

The production process of Huangjiu (yellow rice wine) mainly includes key steps such as raw material selection, soaking and steaming of rice, mixing with yeast and fermentation in vats, stirring and fermentation in jars, pressing and clarification, pasteurization, and bottling.

黄酒的主要原料是糯米，也可选用粳米、籼米、黍米和玉米等。糯米因其富含淀粉，易于发酵，是黄酒甜润口感的主要来源。同时，制酒的水源也非常关键。将选好的糯米浸泡后上锅蒸熟，蒸饭时要掌握好火候和时间，确保米饭熟而不糊、内无白心、透而不烂、均匀一致，为后续的发酵打下良好的基础。蒸熟的糯米凉至适宜温度后，拌入酒曲后放入发酵缸中，让酒液在适宜的温度和湿度下慢慢发酵。

The primary raw material for Huangjiu is glutinous rice, although japonica rice, indica rice, millet, and corn can also be used. Glutinous rice, rich in starch and easy to ferment, is the main source of the sweet and mellow taste of Huangjiu. Additionally, the water source used in the brewing process is also crucial. After soaking the selected glutinous rice, steam it until fully cooked. It is crucial to control the heat and timing during the steaming process to ensure that the rice is thoroughly cooked without becoming mushy, has no white core inside, is translucent but not overly soft, and is uniformly consistent, laying a solid foundation for the subsequent fermentation. Once the steamed glutinous rice has cooled to an appropriate temperature, mix it with yeast and transfer it to a fermentation vat. Allow the mixture to ferment slowly under suitable temperature and humidity conditions.

在发酵过程中，需要定期开耙，即搅拌酒液。用木耙在缸中搅拌，使其均匀接触空气和水分，以促进微生物的均匀分布和充分发酵。之后，酒液会被转移到坛中进行进一步的发酵和陈酿。发酵完成后，通过压榨将酒液与酒糟分离，再将酒液加热，达到杀菌、稳定的作用。最后装入消毒过的土陶坛中，储藏在阴凉干燥通风的地方，使黄酒进一步熟化和提升。

During the fermentation process, it is necessary to regularly open and stir the wine liquid, known as " KaiBa ". A wooden paddle is used to stir the liquid in the vat, ensuring it is evenly exposed to air and moisture, which promotes the uniform distribution of microorganisms and thorough fermentation. After this stage, the wine liquid is transferred to jars for further fermentation and aging. Once fermentation is complete, the wine liquid is separated from the lees through pressing, and then heated to achieve sterilization and stabilization. Finally, the wine is poured into sterilized earthenware jars and stored in a cool, dry, and well-ventilated place, allowing the yellow wine to further mature and enhance its quality.

四、黄酒代表（Huangjiu Representative）

中国过去有南酒北酒一说。但所谓的北酒，并不是指白酒，而是指产于北方的各类黄酒为主的酒，山西、山东都是黄酒的重要产区。南酒则是指南方以江浙为核心出产的黄酒。目前，在北方只有山东、山西还有少量黄酒出产，而绍兴黄酒就产业来说是保存得最为完整的一脉。

In the past, China had a saying about southern and northern wines. However, the so-called northern wine does not refer to baijiu, but to the various types of huangjiu produced in the north, Shanxi and Shandong were both important production areas. While southern wine is the huangjiu produced in the Jiangsu and Zhejiang areas of the south. At present, in the north, only Shandong and Shanxi still produce a small amount of rice wine, and Shaoxing rice wine is the most well-preserved industry.

（一）绍兴黄酒（Shaoxing Rice Wine）

绍兴黄酒历史源远流长，从春秋时期的《吕氏春秋》记载起，历史文献中绍兴酒的名字屡有出现。尤其是清代饮食名著《调鼎集》对绍兴酒的历史演变、品种和优良品质进行了较全面的阐述，在当时绍兴酒已风靡全国，在酒类中独树一帜。酒液黄亮有光，香气芬芳馥郁，滋味鲜甜醇厚，越陈越香，久藏不坏。绍兴黄酒由于酒精浓度较低，酒味醇和，营养丰富，被公认为"东方名酒之冠"。绍兴酒品牌中的古越龙山、塔牌、会稽山，是地理标志产品绍兴酒（绍兴黄酒）国家标准 GB/T 17946—2008 的主要制订单位，足以体现他们工艺的权威性。

The history of Shaoxing rice wine is long-standing, starting from the " Lü shi Chun qiu " of the Spring and Autumn period, the name of Shaoxing wine has appeared repeatedly in historical documents. In particular, the Qing Dynasty's famous food book " Diaodingji " provides a comprehensive account of the historical evolution, varieties, and excellent quality of Shaoxing wine. At that time, Shaoxing wine was popular throughout the country and unique in the category of wine. The wine is bright yellow with a luster, fragrant and rich in aroma, fresh and mellow in taste, the older the better, and does not spoil for a long time. Shaoxing rice wine is considered the " Crown of Eastern Famous Wines " because of its low alcohol content, mellow taste, and rich nutrition. In the Shaoxing wine brand, Gu Yue Longshan, Ta Pai, and Kuijishan are the main drafting units of the national standard GB/T 17946-2008 for Shaoxing wine (Shaoxing rice wine), which fully reflects the authority of their craftsmanship.

（二）北派黄酒（Northern Style Huangjiu）

北派黄酒以厚实绵密而著称，其入口润涩并存、回味则甜苦交织。其代表是山东即墨老酒，这也是黄酒国标的主要制订单位之一。即墨老酒是我国最古老的黄酒品种，有着四千年的悠久历史，被尊称为"黄酒北宗"。与南方以大米、糯米入酒不同，即墨老酒使用的主要原料是黍米，酒度为 12°，含糖较高，是一种甜型黄酒，陈酿贮存一年以上的风味更加醇厚甘美。即墨老酒产于山东青岛即墨区，酒液清澈透明，酒香浓郁，口味醇厚，微苦而香，回味悠长。即墨老酒还具有显著的医疗功效，能祛风散寒、活血化瘀。

Northern style huangjiu is known for its thick and dense texture, with a smooth and astringent taste, and a sweet and bitter aftertaste. Its representative is Jimo millet wine from Shandong, which is also one of the main drafting units of the national standard for huangjiu. Jimo millet wine is the oldest variety of rice wine in China, with a long history of four thousand years, and is honored as the "Northern Ancestor of Huangjiu". Unlike the south that uses rice and glutinous rice in wine, the main raw material used for Jimo millet wine is millet, with an alcohol content of 12 degrees and a higher sugar content, making it a sweet type of rice wine. The flavor of Jimo millet wine that has been stored for more than one year becomes more mellow and delicious. Jimo millet wine is produced in Jimo District, Qingdao, Shandong Province, with a clear and transparent wine, rich aroma, mellow taste, slightly bitter and fragrant, and a long aftertaste. Jimo millet wine also has significant medical effects, can expel wind and cold, and activate blood and resolve stasis.

五、黄酒的饮用方法（The Method of Drinking Huangjiu）

黄酒是一种低酒度的酿造酒，适饮人群广泛，营养丰富，有益于健康。黄酒最传统的饮法是温饮。温饮的显著特点是酒香浓郁，酒味柔和。温酒的方法一般有两种：一种是将盛酒器放入热水中烫热，另一种是隔火加温。但黄酒加热时间不宜过久，否则酒精挥发酒淡而无味。最佳品评温度是在38℃左右。在黄酒烫热的过程中，黄酒中含有的极微量对人体健康无益的甲醇、醛、醚类等有机化合物，会随着温度升高而挥发，同时，脂类芳香物则随着温度的升高而蒸腾。

Huangjiu is a low-alcohol fermentation wine, suitable for a wide range of drinkers, rich in nutrition, and beneficial to health. The most traditional way to drink huangjiu is warm. The significant characteristic of warm drinking is the rich aroma, the mellow taste. There are generally two methods of warming wine: one is to put the wine container into hot water, and the other is to warm it by the fire. However, huangjiu should not be heated for too long, otherwise, the alcohol will evaporate and the wine will be bland and tasteless. The best tasting temperature is around 38℃. In the process of heating huangjiu, the extremely small amounts of methanol, aldehydes, ethers and other organic compounds in the huangjiu that are not beneficial to human health will evaporate with the increase of temperature, at the same time, the aroma of esters will rise with the increase of temperature.

此外，冰镇和佐餐也是非常受欢迎的黄酒饮用方法。年轻群体偏好冷饮，喜欢将黄酒冷藏或加冰后饮用，这不仅降低了酒精度数，还带来冰爽口感。也可根据个人口味，在酒中放入话梅、柠檬等，或兑雪碧、可乐、果汁，既增加风味，又有消暑、促进食欲的功效。吃蟹时搭配黄酒，是黄酒很经典的饮用方法。在《红楼梦》中每出现吃蟹场景，必是要温一壶黄酒的。因为蟹性寒凉，多食易加重体内湿寒之气，而黄酒性暖，搭配起来吃可以中和蟹寒，同时黄酒的酒香更能激发蟹本身的鲜美。

Additionally, serving chilled and with meals are also very popular methods of drinking huangjiu. Younger groups prefer cold drinks, enjoying chilling huangjiu or drinking it with ice, which not only reduces the alcohol content but also brings a refreshing taste. You can also add plums, lemon, and other ingredients to the wine according to personal taste, or mix it with sprite, cola, and fruit juice to increase the flavor while having the effects of cooling, promoting appetite, and other health benefits. Pairing huangjiu with crab is a very classic way of drinking it. In "Dream of the Red Chamber," whenever there is a scene of eating crab, a pot of warm huangjiu is necessary. Crabs are cold in nature, and eating too much can increase the dampness and cold in the body, while huangjiu is warm in nature. Drinking it with crab can neutralize the coldness of the crab, and the aroma of the wine can better stimulate the delicacy of the crab itself.

赛证直通 Competitions and Certificates

一、基础知识部分（Basic Knowledge Part）

（一）名词解释（Explanation of Professional Terms）

开耙（stir the wine liquid）

（二）思考题（Thinking Question）

说说黄酒的南北派系及典型代表酒。

Discuss the northern and southern systems of Huangjiu (Chinese rice wine) and its typical representative wines.

二维码 1-7-1：知识拓展

二、技能操作部分（Skill Operation Part）

品鉴不同黄酒，区分干型黄酒、半干型黄酒、半甜型黄酒、甜型黄酒。

Taste different types of yellow wine to distinguish between dry yellow wine, semi-dry yellow wine, semi-sweet yellow wine, and sweet yellow wine.

【微课】"液体面包"：啤酒

专题二
"液体面包"：啤酒

你知道啤酒是以什么为原料制成的吗（图 1-7-2）？

_____

Do you know what beer is made from (Figure 1-7-2)?

_____

图 1-7-2　啤酒（beer）

一、啤酒的历史（The History of Beer）

　　啤酒是一种古老的酒精饮料，啤酒的历史可以追溯到公元前 5 000 ～ 6 000 年，当时的人们开始使用麦芽和水来制作啤酒。苏美尔人被认为是早期酿造啤酒的文明之一，他们在楔形文字泥板上记录了啤酒的配方和酿造过程。啤酒在古代中东地区得到广泛的普及和发展，成为当时社交和文化活动的重要组成部分。随着时间的推移，啤酒的酿造技术和文化逐渐传播到欧洲、美洲和亚洲等地，成为一种全球性的饮品。它不仅是全球消费量较大的酒精饮品之一，而且因其多样的风味和类型而广受欢迎。

Beer is an ancient alcoholic beverage with a history that can be traced back to around 5 000-6 000 BC, when people began to use barley and water to make beer. The Sumerians are considered one of the early civilizations to brew beer, and they recorded the recipes and brewing process on cuneiform clay tablets. Beer was widely popularized and developed in the ancient Middle East, becoming an important part of social and cultural activities at the time. Over time, the brewing technology and culture of beer gradually spread to Europe, the Americas, and Asia, becoming a global beverage. It is not only one of the better consumed alcoholic beverages in the world but is also widely welcomed for its diverse flavors and types.

中文里的"啤酒"这个词是外来语的音译，来源于英语"Beer"或德语"Bier"的音译。在中国，啤酒作为一种舶来品，在19世纪末20世纪初随着西方文化的传入开始流行起来。啤酒被称为"液体面包"，其酒精含量较低，含有二氧化碳、多种氨基酸、维生素、低分子糖、无机盐和各种酶。1升12°的啤酒，可产生400～600千卡的热量，相当于4～8个鸡蛋所产生的热量，一个轻体力劳动者，如果一天能饮用1 L啤酒，即可获得所需热量的三分之一。

The Chinese word for "beer" is a transliteration of a foreign language, derived from the English word "Beer" or the German word "Bier". In China, beer, as an imported product, began to gain popularity in the late 19th and early 20th centuries with the introduction of Western culture. Known as "liquid bread," beer has a low alcohol content and contains carbon dioxide, various amino acids, vitamins, low molecular sugars, inorganic salts, and various enzymes. One liter of 12-degree beer can produce 400-600 kilocalories of heat, equivalent to the heat produced by 4-8 eggs. A light manual laborer who drinks one liter of beer a day can obtain one-third of the required calories.

二、啤酒的分类（Beer Classification）

啤酒种类繁多，可以根据不同的标准进行分类。

There are a wide variety of beers, which can be classified according to different criteria.

（一）按发酵方式分类（According by the Fermentation Method）

按发酵方式，可分为艾尔啤酒和拉格啤酒。

According to the fermentation method, beer can be divided into Ale and Lager.

1. 艾尔啤酒（Ale）

艾尔啤酒使用艾尔酵母在较温暖条件下进行发酵，比较古老，采用上层发酵法，储藏时间短，适合小批量、小作坊式生产。其入口先是酒花、酵母及其他配料的香气，其次才是酒体的麦芽香气，通常口感更加复杂，风味多样。它大约占啤酒种类的80%。大多数的精酿啤酒属于艾尔啤酒。

Ale uses Ale yeast for fermentation at a warmer temperature, which is relatively old and uses the upper fermentation method. The storage time is short, suitable for small-batch, small-scale production. The taste starts with the aroma of hops, yeast, and other ingredients, followed by the malt aroma of the body. It usually has a more complex taste and a variety of flavors. It accounts for more than 80% of the types of beer, and most craft beers belong to Ale.

2. 拉格啤酒（Lager）

拉格啤酒使用拉格酵母在较冷温度下进行发酵，比较年轻，采用底层发酵法，储藏时间长，适合大批量、工业化生产。其入口是麦芽香气，酒花、酵母及配料香气残存较少，口感通常更加平滑。为了配合工业化生产需求，保证品质的稳定，风味通常是中规中矩。国内所见到的大绿瓶子，以及部分瓶装、罐装进口啤酒属于拉格。

Lager uses Lager yeast for fermentation at a colder temperature, which is relatively young and uses the bottom fermentation method. The storage time is long, suitable for large-scale, industrial production. The taste starts with the malt aroma, with less residual aroma of hops, yeast, and ingredients, and the taste is usually smoother. To meet the needs of industrial production and ensure the stability of quality, the taste is usually conservative. All the big green bottles you can see domestically, as well as some bottled and canned imported beers, belong to Lager.

（二）按色泽分类（According to the Color）

按照色泽，可分为淡色啤酒、浓色啤酒和黑色啤酒。

According to the color, it can be divided into light beer, dark beer, and black beer.

1. 淡色（Light）

淡色色度在 5 ~ 14EBC（European Brewery Convention，一种用于描述啤酒颜色的度量单位），淡色啤酒又分为浅黄色啤酒、金黄色啤酒。淡色啤酒为啤酒产量最大的一种。

Light color is between 5-14EBC (European Brewery Convention, a unit of measure used to describe the color of beer), which is further divided into pale yellow beer and golden yellow beer. Light beer is the largest production type of beer.

（1）浅黄色啤酒：大多采用色泽极浅、溶解度不高的麦芽为原料，糖化周期短，因此，啤酒色泽浅。其口味多属淡爽型，酒花香味浓郁。

Pale yellow beer: most of them use malt with very light color and low solubility as raw materials. The saccharification period is short, so the beer color is light. The taste is mostly light and refreshing, with a strong hop aroma.

（2）金黄色啤酒：采用的麦芽溶解度相较浅黄色啤酒略高，因此，色泽呈金黄色，其产品商标上通常标注"Gold"一词，以便消费者辨认。口味清爽而醇和，酒花香味突出。

Golden yellow beer: the malt used has a slightly higher solubility than pale yellow beer, so the color is golden. Its product trademark usually indicates the word Gold, so that consumers can recognize it. The taste is refreshing and mellow, with a prominent hop aroma.

（3）棕黄色啤酒：采用溶解度高的麦芽，烘烙麦芽温度较高，因此，麦芽色泽深，酒液黄中带棕色，实际上已接近浓色啤酒，其口味也较粗重、浓稠。

Brown yellow beer: it uses malt with high solubility, and the malt is roasted at a higher temperature, so the malt color is deep. The beer is yellow with a brown tint, and it is actually close to dark beer. Its taste is also heavier and thicker.

2. 浓色（Dark）

浓色色度在 14 ~ 40EBC，色泽呈红棕色或红褐色，含固形物较多，酒体透明度较低，产量与淡色啤酒相比较少。根据色泽的深浅，又可分为棕色、红棕色和红褐色。浓色啤酒麦芽汁浓度大，麦芽香味突出、口味醇厚，酒花口味较轻。

Dark color is between 14–40EBC, with a reddish–brown or reddish–brown color, containing more solids, and the beer body is relatively opaque. The production is less than light beer. According to the depth of the color, it can be further divided into brown, reddish brown, and bronzing. Dark beer has a high malt juice concentration, prominent malt fragrance, mellow taste, and lighter hop taste.

3. 黑色（Black）

黑色色泽呈深红褐色乃至黑褐色，酒体透明度很低或不透明，一般原麦汁浓度高，酒精含量在 5.5% 左右，产量较低。黑色啤酒麦芽香味突出、口味浓醇、泡沫细腻，苦味根据产品类型而有较大差异。

Black color is dark reddish–brown or black brown, and the beer body is very low in transparency or opaque. The original malt juice concentration is generally high, and the alcohol content is about 5.5%. The production is low. Black beer has a prominent malt fragrance, mellow taste, delicate foam, and the bitterness varies greatly depending on the product type.

（三）按酒精度分类（According to the Alcohol）

按照酒精度，可分为低酒精度啤酒、中等酒精度啤酒、高酒精度啤酒。低酒精度啤酒酒精度通常在 2% ~ 4%，中等酒精度啤酒酒精度为 4% ~ 6%，高酒精度啤酒酒精度超过 6%。

According to the alcohol content, it can be divided into low–alcohol beer, medium–alcohol beer, and high–alcohol beer. Low alcohol is usually between 2%–4%, medium–alcohol beer has an alcohol content between 4%–6%, and high–alcohol beer has an alcohol content above 6%.

此外，啤酒还可以按照风格，分为比利时风格、英式风格、美式风格；按照口味和添加物，分为果啤、香料啤、烟熏啤；按照包装，分为瓶装啤酒和罐装啤酒；按照杀菌方式，分为未经高温杀菌的生啤和经过高温杀菌的熟啤。

In addition, it can also be classified according to style, such as Belgian style, English style, American style; according to taste and additives, such as fruit beer, spice beer, smoked beer; according to packaging, such as bottled beer and canned beer; according to sterilization method, such as raw beer that has not been sterilized at high temperatures and pasteurized beer that has been sterilized at high temperatures.

上述这些分类并不是互斥的，一款啤酒可能同时属于多个分类。例如，一款啤酒既可以是艾尔，又是淡啤，同时还是美式风格。不同的分类方式有助于消费者根据自己的口味偏好选择适合的啤酒。

These classifications are not mutually exclusive, and a beer may belong to multiple categories at the same time. For example, a beer can be both an Ale and a light beer, and also an American style. Different classification methods help consumers choose the right beer according to their taste preferences.

三、啤酒品鉴（Beer Tasting）

啤酒品鉴主要从以下方面进行：外观、风味、口感。

Beer tasting mainly from the following aspects: appearance, flavor, mouthfeel.

（一）外观（Appearance）

1. 颜色光泽（Color and Luster）

啤酒的颜色可以从浅黄色到琥珀色再到深黑色。不同类别的啤酒呈现不同的颜色，通过颜色能基本判断啤酒的风格。此外，还需要观察啤酒的透明度。

The color of beer can range from light yellow to amber to dark black. Different types of beer have different colors, and the color can essentially determine the style of the beer. In addition, the transparency of the beer should also be observed.

2. 酒头（Wine Head）

啤酒的酒头即指泡沫，是由麦芽中的蛋白质和啤酒花、酵母反应形成的。不同类型的啤酒会形成不同形态的泡沫，有些啤酒的泡沫滋滋作响并不断地破裂消失，有些泡沫细腻洁白如一层厚厚的奶油，有些泡沫有大有小且酒头表面凹凸不平。啤酒倒入透明玻璃杯中后，应观察啤酒泡沫的形态和持久性，能否挂杯，能否在杯壁上形成漂亮的花边等。

The head of beer refers to the foam, which is formed by the reaction of proteins from malt with hops and yeast. Different types of beer form different shapes of foam. Some beers have foam that crackles and continuously breaks and disappears, while others have foam that is delicate, pure white, and thick like cream. Some foams come in different sizes and have an uneven surface. After pouring beer into a transparent glass, observe the shape and persistence of the beer foam, whether it can stick to the glass, and whether it can form Brussels on the glass wall, etc.

（二）风味（Flavor）

1. 香气（Aroma）

啤酒的香气一般来自三个方面：麦芽、啤酒花和酵母。麦芽带来的香气有甜味、饼干、巧克力、黑色水果（如黑李子、葡萄干、无花果等）、坚果、焦糖、烟熏、烘烤等香味。啤酒花带来的香气包括苦味、水果（西柚、柠檬、橙子、橘子等）、花朵、青草、泥土、草本植物、松树、香料、树脂等香味。酵母带来的香气包括酸味、脂类物质（如花香、果香或辛香料香）等香味。闻香气的时候还要考虑香气的浓郁程度，香气是清淡、中等还是浓郁。

The aroma of beer generally comes from three aspects: malt, hops, and yeast. Malt brings a sweet, biscuit, chocolate, black fruit (such as plums, raisins, figs, etc.), nutty, caramel, smoky, and baked aroma.Hops bring a bitter, citrus fruit (grapefruit, lemon, orange, tangerine, etc.), floral, grassy, earthy, herbal, pine, spicy, and resinous aroma.Yeast brings an acidic, ester (such as floral, fruity, or spicy) aroma. When smelling the aroma, it is also important to consider the intensity of the aroma, whether it is light, medium, or strong.

2. 味道（Taste）

通常人能闻到的香气也能在口中尝出来，将啤酒含进口中并吞下肚时，品味其是否有麦芽、啤酒花和酵母的香味，这些香味组合是否协调平衡。

The flavors we can smell can also be tasted in the mouth. When you hold the beer in your mouth and swallow it, taste whether there is a flavor of malt, hops, and yeast, and whether the combination of these flavors is harmonious and balanced.

（三）口感（Mouthfeel）

1. 杀口感（Carbonation）

"杀口感"是啤酒行业的常用术语，主要指啤酒中含有的二氧化碳在口腔中释放时带来的刺激感，这种感觉类似于一种轻微的刺痛感，它能够让人感觉到啤酒的爽口和清凉。啤酒中的二氧化碳含量、啤酒的泡沫质量以及啤酒的温度都会影响"杀口感"的强烈程度。"杀口感"是评价啤酒好坏的一个重要指标，如果啤酒中的二氧化碳含量不足，则会缺乏"杀口感"；反之，则"杀口感"十足，能带来非常清爽、刺激的感觉。

" Carbonation " is a common term in the beer industry, mainly referring to the tingling sensation brought by the release of carbon dioxide contained in the beer in the mouth, similar to a slight tingling sensation. This feeling can make people feel the refreshing and coolness of the beer. The content of carbon dioxide in the beer, the quality of the beer foam, and the temperature of the beer will all affect the intensity of the " carbonation ". " Carbonation " is an important indicator for evaluating the quality of beer. If the carbon dioxide content in the beer is insufficient, the carbonation will be lacking; on the contrary, if the " carbonation " is strong, it can bring a very refreshing and stimulating feeling.

2. 质地（Texture）

啤酒风味物质和"杀口感"一起出现时，感受其在口中带来的是奶油的绵密感，还是丝绸般的顺滑感，又或是油润的黏腻感。

When the flavor substances of beer and the " carbonation " appear together, feel whether it brings a creamy thick feeling, a silk−like smoothness, or an oily greasy feeling.

3. 酒体（Body）

啤酒的酒体指的是舌头能感受到含在口中酒的重量，其酒体可以从轻盈、中等到厚重。

The body of beer refers to the weight that the tongue can feel when holding the wine in the mouth. The body of the wine can range from light to medium to heavy.

4. 余味（Aftertaste）

余味是指酒吞下肚后残留在口中的味道和感觉，具体包括：干涩感，不一定是苦的，但一定不甜；湿润感，口中残留有甜味或其他浓郁的味道，需要用水或其他东西清洁口腔才能让味蕾轻松起来；温暖感，因高酒精度而带来的热感。

Aftertaste refers to the taste and feeling that remains in the mouth after swallowing the wine, including dryness, not necessarily bitter, but definitely not sweet; moistness, with a residual sweetness or other rich tastes in the mouth, which requires cleaning the mouth with water or other substances to make the taste buds relaxed; warmth, the hot feeling brought by high alcohol content.

上述三个方面是品鉴啤酒时最直观的感受，也是衡量啤酒品质的重要指标。一款优秀的啤酒，泡沫应细腻挂杯，泡持性达到相应标准，酒花香气明显，酒体醇厚柔和，具有一定的"杀口感"，有回甘。但是品鉴也有很强的主观感受，不同的品鉴者可能会有不同的感受和评价。

The above three aspects are the most intuitive feelings when tasting beer, and they arc also important indicators for measuring the quality of beer. An excellent beer should have delicate foam that adheres to the glass with good persistence, obvious hop aroma, rich and soft body, a certain degree of carbonation, and a sweet aftertaste. However, tasting also has a strong subjective feeling, and different tasters may have different feelings and evaluations.

赛证直通 Competitions and Certificates

一、基础知识部分（Basic Knowledge Part）

（一）专业名词解释（Explanation of Professional Terms）

1.酒花（Hops）：

2.艾尔发酵（Ale fermentation）：

3.拉格发酵（Lager fermentation）：

（二）思考题（Thinking Question）

如何品鉴啤酒？

How to appreciate beer?

二维码 1-7-2：
知识拓展

二、技能操作部分（Skill Operation Part）

鉴别一款啤酒的品质，写下其外观、风味和口感的特征。

Identify the quality of a beer and document its characteristics in terms of appearance, flavor, and taste.

专题三
"玉液琼浆"：米酒

【微课】
"玉液琼浆"：
米酒

你的家乡有米酒吗？你知道米酒是怎么制成的吗？

_____

Do you have rice wine in your hometown? Do you know how it's made?

_____

一、米酒概述（Introduction to Rice Wine）

这里所述的米酒，特指糯米酒，又称为江米酒、甜酒、酒酿，是一种使用糯米为原料，通过发酵制成的酒类，以其独特的甜味和香气而被喜爱。我国用优质糙糯米酿酒，已有千年以上的悠久历史。不同地区和文化中的糯米酒有不同的制作方法和风味，名称也各不相同。

The rice wine referred to in this specifically refers to glutinous rice wine, also known as Jiangmi wine, sweet wine, or fermented glutinous rice. It is a type of alcoholic beverage made from glutinous rice through fermentation, and is beloved for its unique sweet taste and aroma. Our country has a long history of over a thousand years in brewing high-quality glutinous rice wine. In different regions and cultures, glutinous rice wine has different production methods and flavors, and the names also vary.

糯米酒的酒精含量比较低，为 1% ～ 15%，口味清甜。糯米酒含有丰富的氨基酸、维生素和一些矿物质，适量饮用被认为有益健康。

The alcohol content of glutinous rice wine is relatively low, ranging from 1%–15%, with a sweet and light taste. Glutinous rice wine is rich in amino acids, vitamins, and some minerals, and moderate consumption is considered beneficial to health.

糯米酒含有较高的碳水化合物，可以提供能量。其在发酵过程中会产生多种氨基酸，有助于人体的营养吸收。糯米酒中还含有 B₁、B₂ 等多种维生素和钙、磷、铁等对人体有益的矿物质。适量饮用糯米酒有助于血液循环，具有一定的活血化瘀作用。

The glutinous rice wine contains a high level of carbohydrates, which can provide energy. During the fermentation process, it produces various amino acids that help with the absorption of nutrients in the human body. Glutinous rice wine also contains various vitamins such as B_1 and B_2, as well as beneficial minerals for the human body, such as calcium, phosphorus, and iron. Drinking glutinous rice wine in moderation can help with blood circulation and has a certain effect on promoting blood circulation and removing blood stasis.

二、米酒制作工序 (Rice Wine Production Process)

1. 备料 (Material Preparation)
选择优质糯米，颗粒饱满，无杂质，无霉变。

Select high-quality glutinous rice, which is plump, free from impurities, and not moldy.

2. 清洗 (Cleaning)
将糯米用清水冲洗干净，去除表面的灰尘和杂质。

Rinse the glutinous rice with clean water to remove surface dust and impurities.

3. 浸泡 (Soaking)
将清洗好的糯米浸泡在水中，根据不同的工艺要求，浸泡时间可能有所不同。

Soak the cleaned glutinous rice in water. Depending on the process requirements, the soaking time may vary.

4. 蒸煮 (Steaming)
使用工业化的蒸煮设备，对糯米进行蒸煮，确保糯米熟透但不糊。

Use industrial steaming equipment to steam the glutinous rice, ensuring it is cooked thoroughly but not mushy.

5. 冷却 (Cooling)
蒸煮后的糯米通过冷却装置迅速冷却至适宜的温度，为下一步发酵做准备。

Cool the steamed glutinous rice quickly to a suitable temperature through a cooling device, preparing for the next step of fermentation.

6. 加曲 (Adding Distiller's Yeast)
将冷却后的糯米与适量的酒曲混合均匀。酒曲中含有酵母菌和酒药曲霉等微生物，是发酵的关键。

Mix the cooled glutinous rice evenly with an appropriate amount of distiller's yeast. Distiller's yeast contains microorganisms such as yeast and aspergillus oryzae, which are key to fermentation.

7. 发酵（Fermentation）

将加曲后的糯米转移到发酵罐中，控制好发酵条件，如温度、时间和 pH 值，进行发酵。

Transfer the added distiller's yeast glutinous rice to a fermentation tank and control the fermentation conditions, such as temperature, time, and pH value, for fermentation.

8. 糖化（Saccharification）

在发酵过程中，微生物会将糯米中的淀粉转化为糖，然后进一步转化为酒精，形成糯米酒特有的甜味和香味。

During the fermentation process, microorganisms convert the starch in the glutinous rice into sugar, which is then further converted into alcohol, creating the unique sweetness and aroma of rice wine.

9. 停止发酵（Terminate Fermentation）

当发酵达到预定的酒精度数和风味时，通过降温、过滤等方法终止发酵。

When the fermentation reaches the predetermined alcohol and flavor, terminate the fermentation by methods such as cooling and filtering.

10. 过滤（Filtering）

将发酵好的糯米酒过滤，去除酒糟和杂质，得到清澈的糯米酒。

Filter the fermented rice wine to remove the lees and impurities, resulting in a clear rice wine.

11. 杀菌罐装（Sterilization and Bottling）

为了防止后续微生物的污染，确保产品在常温下的稳定性，通常要对糯米酒进行杀菌处理，将杀菌后的糯米酒灌装到瓶中或其他包装容器中。

To prevent contamination by subsequent microorganisms and ensure the product's stability at room temperature, it is usually necessary to sterilize the rice wine and then bottle it into bottles or other packaging containers.

三、各地米酒（Regional Rice Wine）

中国米酒传统文化源远流长，各地区米酒的名称虽然不同，风味各有千秋，但做法大体一致。然而，物产和气候的不同导致所酿制的米酒在味道上也有些微差别。例如，北方人多用圆糯米酿制米酒，而南方人则喜欢用长糯米酿制米酒。再加上饮用水的不同，所以米酒在中国各地区之间的差异不小。

Chinese rice wine culture has a long and rich history. Although the names and flavors of rice wine in different regions is different, but the general method of production remains largely consistent. However, differences in local products and climate lead to subtle variations in taste. For example, people in the north often use round glutinous rice for brewing, while those in the south prefer long-grain glutinous rice. The type of water used also differs, resulting in significant variations in rice wine across different regions in China.

（一）湖北（Hubei）

湖北孝感的米酒可谓一绝，在清朝光绪年间《孝感县志》中就有关于米酒的记载。孝感米酒选用的是当地糯米，采用民间配方调制而成的蜂窝酒曲中含有 10 多味中药材。酿制米酒用的水液颇有讲究，要选用孝感城关西门外城隍潭的地下水。酿制好的米酒，白如玉液，清香宜人，甜润爽口。糯米仍保留了完整的颗粒状，饮后生津暖胃，回味悠长。

The rice wine from Xiaogan, Hubei province is considered a local delicacy, with records of its production in the Annals of Xiaogan County during the reign of guangxu in Qing Dynasty. The Xiaogan rice wine is made from local glutinous rice, using honeycomb distiller's yeast containing over ten types of traditional Chinese medicinal herbs, prepared according to a folk recipe. The water used for brewing the rice wine is also quite particular; it is sourced from the underground water at the Chenghuang Pond outside the west gate of Xiaogan city. The resulting rice wine is as white as jade, with a pleasant aroma, sweet and refreshing taste. The glutinous rice retains its intact grains, and after drinking, it stimulates saliva production, warms the stomach, and leaves a lasting aftertaste.

（二）湖南（Hunan）

湖南桃源坪出产的西山米酒，被称为"土茅台"，据记载已有 500 多年的酿酒历史。西山环境优良，没有工业的污染，水质清凉甘洌，呈天然的弱碱性，且含有对人体非常有益的多种微量元素，这是酿制西山米酒的最佳水质。正宗的西山米酒，自然酒香纯正，香气迷人。

Xishan rice wine from Taoyuanping, Hunan province is known as the "local Moutai" and has a brewing history of over 500 years. The environment in Xishan is excellent, free from industrial pollution. The water quality is cool, sweet, and naturally weakly alkaline, containing various trace elements that are highly beneficial to human health, making it the optimal water for brewing Xishan rice wine. The authentic Xishan rice wine has a pure and natural fragrance with a captivating aroma.

（三）广西（Guangxi）

桂林三花酒被称为"米酒之王"。据说正宗的三花酒入坛堆花，入瓶堆花，入杯也要堆花，故名"三花酒"。其最大的特点就是，蜜香清雅、入口柔绵、落口爽利、回味怡畅。广西桂林东兰县的糯米酒，采用当地富含硒元素的墨米为原料，经过壮族特有的酿酒工艺，精制而成，最大程度上保留了墨米的原色和醇香。东兰墨米酒是中国地理标志产品。

Guilin Sanhua Jiu is known as the "King of Rice Wine." It is said that the authentic Sanhua Jiu is named for its characteristic of "flowering" in the jar, in the bottle, and in the cup, hence the name "Sanhua Jiu." Its biggest feature is its honey-like fragrance and elegant taste, with a soft and mellow mouthfeel, smooth and refreshing taste, and a pleasant aftertaste. Glutinous rice wine in Donglan county, Guilin city, Guangxi province is made from locally sourced black rice rich in selenium, through a unique Zhuang ethnic brewing process, to preserve the original color and rich aroma of the black rice to the greatest extent. Donglan black rice wine is a Chinese geographical indication product.

（四）贵州（Guizhou）

贵州镇远金堡乡是远近闻名的米酒之乡，这里的米酒最为出名。镇远古城依靠独特的地形优势和气候优势，酿造了贵州大山原生态、高品质的米酒。糯米酒配方纯净天然，无任何添加，主要采用来自无污染的喀斯特岩溶裂隙水和优质的云贵高原特产糯米以及生长在大自然青山绿水深处众多珍贵的本草植物作为特制酒曲配方，工艺上传承千年苗族酿造工艺，采用传统手法匠心酿制，是不可多得的优质糯米酒。

Jinbao Township in zhenyuan county, Guizhou province, is a well-known rice wine region, where the most famous rice wine is produced. Relying on its unique terrain and climate advantages, the ancient town of Zhenyuan has brewed a high-quality, original ecological rice wine in the mountains of Guizhou. The glutinous rice wine recipe is pure and natural, with no additives. It mainly uses uncontaminated karst fissure water and high-quality Yunnan-Guizhou Plateau specialty glutinous rice, as well as precious herbal plants grown in nature, which are deep in the green mountains and clear waters as a special wine yeast formula. The process inherits the thousand-year-old Miao brewing process and is crafted with traditional craftsmanship, making it a rare high-quality glutinous rice wine.

赛证直通 Competitions and Certificates

一、基础知识部分（Basic Knowledge Part）

二维码 1-7-3：知识拓展

（一）名词解释（Explanation of Professional Terms）

三花酒（Sanhua Jiu）：

（二）思考题（Thinking Question）

为什么米酒在民间备受喜爱？

Why is rice wine so popular among the people?

二、技能操作部分（Skill Operation Part）

搜集你家乡的米酒资料并介绍。

Collect information about the rice wine from your hometown and introduce it.

专题四
源起中国：清酒

【微课】源起中国：清酒

你听说过清酒吗？了解哪些清酒相关的知识？

___

Have you heard of sake? What do you know about sake?

___

一、清酒的来历 (The Origin of Sake)

今天提起清酒，首先联想到的是日本清酒。日本清酒是借鉴中国黄酒的酿造方法而发展起来的日本国酒。在大型宴会上，结婚典礼中，在酒吧间或寻常百姓的餐桌上，人们都可以看到清酒的身影。

When mention sake today, the first thing that comes to mind is Japanese sake. Japanese sake, the national drink of Japan, developed based on the brewing method of Chinese huangjiu. For over a thousand years, sake has been the most commonly consumed beverage in Japan. You can see it at large banquets, wedding ceremonies, bars, or on the tables of ordinary people.

清酒起源于3世纪到4世纪，在中国的水稻传入日本之后，中国将用米酿酒的技术随着大米种植技术一起传到了日本，从此日本才有了用米酿酒的技术，这是日本清酒的开端。中国西晋时的《魏志·倭人传》，全书不到两千字，但提到了日本清酒。这是史书上第一次对日本清酒的记载，时间在公元280年左右。最初，日本人仿照了中国人酿造黄酒的技术，酿造出了浊酒，浊酒经沉淀过滤，上面清澈的液体就是清酒了。"清酒"这一名称得以成立，也是相对于"浊酒"之谓。

Sake originated between the 3rd and 4th centuries, after rice from China was introduced to Japan. China brought the technology of rice wine brewing along with rice cultivation to Japan, and since then, Japan has had the technology to brew rice wine, which is the beginning of Japanese sake. The " Wei Zhi Woren Zhuan " of the Western Jin Dynasty in China, which has less than two thousand words, mentioned Japanese sake. This is the first record of Japanese sake in history, around 280 AD. Initially, the Japanese imitated the Chinese technology for brewing rice wine and produced turbid wine. After sedimentation and filtration, the clear liquid on top is sake. The name " sake " was established in contrast to " turbid wine. "

除了学习借鉴中国黄酒的酿造方法，清酒的命名也和中国有着千丝万缕的联系。清酒名有着"中国古典迷你百科"的说法，许多清酒的名字更是引经据典，为酒赋予了深刻含义。例如，"上善如水"是一款纯米吟酿高级清酒，有着如丝般的口感。"上善如水"出自老子《道德经》"上善若水，水善利万物而不争"一节，这款酒名是对中国古贤老子的致敬。类似的清酒名，还有出自《庄子》的"明镜止水"，出自《史记》的"国士无双"。

In addition to learning and drawing on the Chinese huangjiu brewing method, the naming of sake is also closely related to China. The names of sake have the reputation of being a " mini encyclopedic classic of Chinese culture, " and many sake names are based on classical references, giving the wine profound meanings. For example, " Shang Shan Ru Shui " is a high-grade sake made from pure rice with a silky texture. " Shang Shan Ru Shui " comes from Laozi's " The Scripture of Ethics " " Highest good is like water. Because water excels in benefiting the myriad creatures without contending with them ". This sake name is a tribute to the ancient Chinese sage Laozi. Other similar sake names include " Ming Jing Zhi Shui " from " Zhuangzi " and " Guo Shi Wu Shuang " from " Records of the Historian ".

中国古代诗人也是日本酒文化中的重要组成部分。例如，岐阜县生产的纯米酒"梅乃寒山"，便得名于中国唐代白话诗人寒山。寒山在中国籍籍无名，甚至连生卒都未留下，但在日本却作为大诗人备受推崇。岛根县产的纯米吟酿酒"李白"，相对来说就耳熟能详得多。该酒厂甚至直接取名"李白酒造有限公司"。

Ancient Chinese poets are also an important part of Japanese alcohol culture. For example, the pure rice wine "Meinahan Mountain" produced in Gifu Prefecture is named after the Tang Dynasty Chinese vernacular poet Han Shan. Han Shan is unknown in China, and even his birth and death dates are not recorded, but he is highly regarded as a great poet in Japan. The pure rice wine "Li Bai" produced in Shimane Prefecture is much more widely known. The brewery even directly named itself Li Bai Brewery Co., Ltd.

二、清酒的类别 (Categories of Sake)

日本把优质清酒的等级分类统称为"特定名称酒"，一般指纯米酒、本酿造酒和吟酿酒。除了这三种具有特定名称的清酒外，其他的都是普通酒。

Japan classifies high-quality sake into a general term "specific name sake," which usually refers to Junmai, Honjozo, and Ginjo. In addition to these three types of sake with specific names, the rest are considered ordinary sake.

纯米酒指只使用大米和酒曲酿造出来的酒。由于没有加入食用酒精，这种酒的口感大多比较醇厚。

Junmai: refers to a type of sake brewed solely from rice and koji (rice with aspergillus mold). Since no edible alcohol is added, the taste of this type of sake is mostly rich and full-bodied.

本酿造酒指在米和酒曲里加入了少量食用酒精制作而成的清酒。直接压榨酒醪获得的叫作纯米酒，但是本酿造酒在压榨之前会加入少量的食用酒精调整酒的味道。

Honjozo: refers to a type of sake made by adding a small amount of edible alcohol to rice and koji. The sake directly pressed from fermentation mash is called pure rice sake, but Honjozo sake adds a small amount of edible alcohol to adjust the taste before pressing.

吟酿酒指在纯米酒或本酿造酒中，使用精米步合在 60% 以下的高度研磨的大米作为原料，再经过低温发酵制成的具有特别口感的酒。

Ginjo: refers to a type of sake made from highly milled rice with less than 60% semiaibuai as the raw material in Junmai or Honjozo sake, and then fermented at low temperatures to create a unique taste.

三、清酒的酿造要点 (Key Points of Sake Brewing)

日本清酒在品质、工艺和文化传统上都有其独特性。日本清酒制酒三要素：一水、二米、三酵母。此外，酒质也会因为产地的不同而展现出不一样的风味与地理特色。

Japanese sake has its own unique qualities in terms of quality, craftsmanship, and cultural tradition. The three key elements of Japanese sake brewing are: water, rice, and yeast. In addition, the quality of the sake can vary due to different production locations, showcasing unique flavors and geographical characteristics.

（一）水 (Water)

水按照其矿物质含量，特别是钙和镁的含量，可以分为软水和硬水。软水中所含的矿物质成分较少，酵母得到的养分少使得活性低、发酵速度慢，随之糖分转变为酒精的速度也较慢，酿成的酒入口较为甘甜柔和，酒质清淡爽口。硬水的矿物质含量较多，有助于发酵的进行，酵母活跃使得糖分加快成为酒精，造就出浓烈辛辣的口感，酒质相对浓郁醇厚。

Water can be classified as soft or hard based on its mineral content, particularly the levels of calcium and magnesium. Soft water contains fewer minerals, resulting in lower yeast activity and slower fermentation, which leads to a slower conversion of sugars to alcohol. The resulting wine is sweeter and milder in taste, with a light and refreshing quality. Hard water contains more minerals, which are beneficial for the fermentation process, leading to more active yeast and a faster conversion of sugars into alcohol, resulting in a strong and spicy taste, with a richer and fuller body.

（二）米（Rice）

日本盛产稻米，制酒用米也备受重视，并不断地进行品种改良。以优质米作为酿清酒的原料最为地道，虽然一般的食用米也可以用来酿酒，不过要把清酒的魅力与实力发挥到极致的唯有"酒米"——酒造好适米。制作清酒的专用米才能酿造出吟酿酒等高级酒，公认最好的是新潟县产的米，如山田锦、五百万石，其他代表性品种有美山锦、雄町、八反锦等。

Japan is abundant in rice production, and the rice used for brewing is highly valued and continuously improved. High-quality rice is the most suitable ingredient for brewing sake, although regular rice can also be used for winemaking. However, only "rice varieties" — shuzo-kotekimai can bring out the full charm and strength of sake. The specialized rice for sake brewing can produce high-quality wines such as "ginjo" (high-grade sake). The most recognized variety is the rice produced in Niigata Prefecture, such as Yamada Nishiki and Gohyakumangoku, and other representative varieties include Miyama Nishiki, Omachi, and Hattan Nishiki.

酒米的米粒比食用米颗粒大而且柔软，米心部分也相对较大且富含淀粉质，呈现不透明的乳白色泽，称为"心白"，是酿清酒的主要元素。心白的淀粉质组织较粗，即适合发酵的软质米，培养曲时有利于让曲菌渗透米粒核心，帮助淀粉质易于转化成为糖分，让制曲过程更为顺利。酒中杂味的生成基本上是由于蛋白质的缘故，酒米因为蛋白质含量低，因此，酿清酒时容易调配味道，使酒质更显芳醇、口感变化丰富。

The grains of wine rice are larger and softer than edible rice, with a larger and more opaque, milky-white center part known as the "shinpaku", which is the main element for brewing sake. The starch structure of shinpaku is coarse, making it suitable for fermentation and facilitating the penetration of koji molds into the core of the rice grains, aiding in the smooth conversion of starch into sugar during the koji-making process. The generation of off-flavors in the wine is essentially due to proteins, wine rice has a lower protein content, making it easier to adjust the taste during the sake brewing process, resulting in a more fragrant and richer taste.

（三）酵母（Yeast）

酵母可以决定酒类的口感、香气还有品质。专门用来酿造日本酒的酵母称为"清酒酵母"。日本清酒采用的是"并行复式发酵"法，有别于葡萄酒"单行式发酵"法，原料中的米有的是淀粉质而没有足够的糖分，这使制酒过程更趋复杂，需要先将淀粉质转化成糖分，同时酵母生成酒精。换而言之，"糖化"与"发酵"是同步进行的。

Yeast determines the taste, aroma, and quality of the wine. The yeast specifically used for brewing Japanese sake is called " sake yeast. " Japanese sake employs a " parallel multiple fermentation " method, which is different from the " sequential fermentation " method used in winemaking. The rice used in sake brewing contains starch and does not have sufficient sugar content, making the brewing process more complex. It is necessary to first convert the starch into sugar while simultaneously producing alcohol from the yeast. In other words, " saccharification " and " fermentation " occur simultaneously.

赛证直通 Competitions and Certificates

一、基础知识部分（Basic Knowledge Part）

（一）名词解释（Explanation of Professional Terms）

1. 本酿造酒（Honjozo）：

2. 吟酿酒（Ginjo）：

二维码 1-7-4：
知识拓展

（二）思考题（Thinking Question）

说说清酒制酒三要素。

Discussing the three essential elements of sake brewing.

二、技能操作部分（Skill Operation Part）

品饮清酒并描述其特征。

Taste sake and describe its characteristics.

【单元七　反思与评价】Unit 7　Reflection and Evaluation

学会了：＿＿＿＿＿＿＿＿＿＿＿＿＿＿＿＿＿＿＿＿＿

Learned to：＿＿＿＿＿＿＿＿＿＿＿＿＿＿＿＿＿＿＿

成功实践了：＿＿＿＿＿＿＿＿＿＿＿＿＿＿＿＿＿＿＿

Successful practice：＿＿＿＿＿＿＿＿＿＿＿＿＿＿＿＿

最大的收获：＿＿＿＿＿＿＿＿＿＿＿＿＿＿＿＿＿＿＿

The biggest gain：＿＿＿＿＿＿＿＿＿＿＿＿＿＿＿＿＿

遇到的困难：＿＿＿＿＿＿＿＿＿＿＿＿＿＿＿＿＿＿＿

Difficulties encountered：＿＿＿＿＿＿＿＿＿＿＿＿＿＿

对教师的建议：＿＿＿＿＿＿＿＿＿＿＿＿＿＿＿＿＿＿

Suggestions for teachers：＿＿＿＿＿＿＿＿＿＿＿＿＿＿

模块二　茶

单元一　茶及茶具认知

一碗喉吻润，二碗破孤闷。

三碗搜枯肠，唯有文字五千卷。

四碗发轻汗，平生不平事，尽向毛孔散。

五碗肌骨清，六碗通仙灵。

七碗吃不得也，唯觉两腋习习清风生。

——唐·卢仝《走笔谢孟谏议寄新茶》节选

单元导入 Unit Introduction

中国是茶的故乡。一片小小的树叶，蕴藏着大大的学问。自古以来，茶叶不仅是中华民族的文化瑰宝，更是连接东西方文明的桥梁，承载着深厚的文化底蕴与悠久的历史传承。在这一单元中，我们将一同穿越千年时光的长廊，探寻茶叶的起源、演变与传播，感受中国茶文化的魅力。

China is the hometown of tea. A small leaf, contains a great deal of knowledge. Since ancient times, tea is not only a cultural treasure of the Chinese nation, but also a bridge connecting Eastern and Western civilizations, carrying profound cultural deposits and a long history of inheritance. In this unit, we will explore the origin, evolution and spread of tea and feel the charm of Chinese tea culture through the long corridor of thousands of years.

学习目标 Learning Objectives

➢ 知识目标（Knowledge Objectives）

（1）能描述茶的历史起源。

To describe the historical origin of tea.

（2）能描述六大茶类的不同特征。

To describe the different characteristics of six types of tea.

（3）能描述茶具的用途。

To describe the purpose of the tea set.

➢ 能力目标（Ability Objectives）

（1）能根据所学知识进行茶叶分类，并能区分不同的茶具。

To classify tea according to the knowledge, distinguish different tea sets.

（2）能描述六大茶类的代表名茶及其特点。

Can describe the six representative teas and their characteristics.

（3）能描述不同地区的茶俗文化。

Can describe the tea culture in different regions.

➢ 素质目标（Quality Objectives）

（1）探索中国茶文化的深厚底蕴。

Understand the long history of Chinese tea culture.

（2）提升茶文化素养，增强文化自信。

Improve tea cultural accomplishment, generate traditional cultural pride, and enhance national self-confidence.

（3）了解茶叶对于促进世界交流的重要作用。

Understand the important role of tea in promoting world communication.

【微课】
千年传承：
茶的历史及
常见茶具

专题一
千年传承：茶的历史及常见茶具

图 2-1-1 中的茶具是什么？有什么用途？

_____

What is the tea set in the picture (Figure 2-1-1)? What is it used for?

_____

图 2-1-1　茶具（tea set）

一、茶的历史（The History of Tea）

茶也称"茗"，是一种有益于人体健康的饮品，深受世界各国的欢迎。我国发现茶和利用茶至今已有 4 700 多年的历史，是世界上最早发现茶的国家，种茶技术先后传播到世界各国。我国茶叶的生产有三大特点：产茶地区广泛，制茶历史悠久，名茶多。

Tea, also known as "Ming", is a kind of beverage beneficial to human health and is popular all over the world. The use of tea in China has a history of more than 4,700 years. Also, China is the first country to discover tea in the world, tea planting technology has been spread to all countries in the world. The production of tea in China has three characteristics: tea is widely produced in a lot of regions, tea making has a long history, a great many famous teas.

《神农本草经》记载"神农尝百草，日遇七十二毒，得茶解之"（图 2-1-2）。唐代茶圣陆羽在《茶经》中同样提到"茶之为饮，发乎神农氏"。可见，中国是最早饮茶的国家。

图 2-1-2　神农尝百草（**Shennong tasted hundreds of herbs**）

Shennong's Herbal Classic recorded, " Shennong tasted hundreds of herbs and was poisoned many times. He detoxified by accidentally tasting tea " (Figure 2-1-2). Lu Yu, the Sage of Tea in the Tang Dynasty, mentioned in *The Classic of Tea* that " the drinking of tea originated from Shennong ". It can be seen that China is the earliest country to drink tea.

中国茶叶 8 世纪传入日本，16 世纪传入欧洲各国，进而传入美洲。由于各国文化和饮用习惯的差异，中国主要有六大茶类，其总体特征是种类丰富、历史悠久、风味多样。美国和欧洲一些国家则擅长配制茶，其品种繁多，如水果茶、草药茶、花茶等。

Chinese tea was introduced to Japan in the 8th century, spread to European countries in the 16th century and then to the America. Nonetheless, due to the culture and drinking habits of different countries, both China and Japan are good at natural tea, which is characterized by mellow, simple and regional flavor. The United States and some European countries are good at tea blends. They have many varieties, such as fruit tea, herbal tea, scented tea, and so on.

二、常见茶具（Common Tea Sets）

常见的茶具种类繁多，每种茶具都有其独特的功能和用途。以下是一些常见的茶具及其功能的介绍。

（1）盖碗：一种上有盖、中有碗、下有托的茶具，又称"三才碗"或"三才杯"。盖碗用于泡茶。盖为天，碗为人，托为地，三者合一寓意和谐，蕴含着天地人和的哲学思想。在使用时，盖碗可以保持茶叶的香气和温度，让品茶者更好地感受茶的韵味。

（2）茶壶：主要用于泡茶。茶壶的材质多样，如紫砂、陶瓷等，每种材质都能对茶的口感产生不同的影响。例如，紫砂壶具有良好的保温性能，且泡茶不易变味，能够充分展现茶叶的香气和滋味。同时，紫砂壶还具有一定的收藏价值，是茶友们喜爱的茶具之一。

（3）玻璃杯：一种透明的泡茶器具，适合冲泡外形较好的绿茶、黄茶等。以其透明清澈、易于清洁的特性，成为众多茶友们的喜爱之选。它不仅能够展现茶叶在水中舒展、变化的美丽姿态，还便于观察茶汤的色泽，从而准确判断茶叶的冲泡情况。

（4）茶盘：放置茶具和接水的平底盘，用来盛放茶壶、茶杯等茶具，并承接泡茶时溅出的茶水。茶盘不仅具有实用性，还可以作为装饰品，营造品茶的氛围。

（5）公道杯：又称茶海，用于均匀茶汤。在泡茶过程中，可以将泡好的茶汤从茶壶中倒入公道杯，然后再分别倒入各个品茗杯中，确保每位品茶者都能品尝到浓度一致的茶汤。

（6）品茗杯：专门用于品尝茶的杯子，一般较小且精致。品茗杯的材质多为陶瓷或玻璃，可以清晰地观察到茶汤的颜色和变化。在品茶时，品茗杯是不可或缺的茶具。

（7）茶荷：用于盛放干茶叶，便于观赏和取用。

（8）茶盂：主要用于盛放冲泡过程中产生的废茶水及茶渣。其形状各异，大小也根据使用需求有所不同，但主要设计都是为了方便清洁和保持茶桌的整洁。此外，茶盂也是茶道中一种重要的文化符号。它代表着茶人对茶文化的尊重和传承，也体现了茶人追求整洁、优雅的生活态度。

（9）茶巾：用于擦拭茶具上的水渍和茶渍，保持茶具的清洁和干燥。

（10）茶艺六君子：指茶艺表演中常用的六种茶具，它们分别是茶筒、茶匙、茶漏、茶则、茶夹、茶针。茶艺六君子在茶艺表演中各自扮演着重要的角色，是茶文化的重要组成部分。

①茶筒：作为盛放茶艺用品的器皿，其设计通常优雅而实用，用于收纳和保护其他茶具，使得茶艺表演更加整洁有序。

②茶匙：在泡茶的过程中，用茶匙操作能够确保茶叶的用量恰到好处，为茶汤的口感和品质打下基础。

③茶漏：置茶时，茶漏被放置在壶口上，以引导茶叶顺利进入壶中，同时防止茶叶碎末掉落到壶外。这一设计不仅提高了泡茶的便捷性，还保持了茶桌的整洁。

④茶则：盛茶入壶之用具，可以准确地量取适量的茶叶，确保每一次泡茶的投茶量都保持一致，从而保证茶汤的口感和品质的稳定，一般为竹制。

⑤茶夹：外形如同一个夹子。茶夹的功用多样，既可以用来挟取茶渣，将泡过的茶叶从壶中夹出，也可以用来挟着茶杯进行清洗，既卫生又防烫。

⑥茶针：疏通茶壶的内网，当壶嘴被茶叶堵住时用来疏浚以保持水流畅通，或放入茶叶后把茶叶拨匀，碎茶在底，整茶在上。

这些泡茶所用的常见茶具在茶艺表演的过程中各自发挥着重要的作用，共同构成了中国茶文化的独特魅力。

There are many kinds of common tea sets, and each tea set has its unique function and use. Here is an introduction to some common tea sets and their functions.

(1) Cover bowl: a tea set with a cover on the top, a bowl in the middle and a tray under, also known as " three bowls" or " three cups". Tureens are used to make tea. Cover for the day, the bowl for people, for the ground, the integration of the three implies harmony, containing the philosophical thoughts of heaven and earth. When in use, the covered bowl can maintain the aroma and temperature of the tea, so that tea drinkers can better feel the charm of the tea.

(2) Teapot: mainly used for brewing tea. Teapots are made of a variety of materials, such as purple sand, ceramics, etc., and each material can have a different effect on the taste of tea. For example, the purple clay pot has good thermal insulation performance, and the tea is not easy to change taste, which can fully show the aroma and taste of tea. At the same time, the purple clay pot also has a certain collection value and is one of the favorite tea sets of tea friends.

(3) Glass: a transparent tea making apparatus, suitable for brewing green tea, yellow tea and so on. With its transparent, clear and easy to clean characteristics, it has become a favorite choice of many tea friends. It can not only show the beautiful posture of tea stretching and changing in water,

but also facilitate the observation of the color of tea soup, so as to accurately judge the brewing situation of tea.

(4) Tea tray: flat chassis for placing tea sets and receiving water, used to hold tea sets such as teapots and teacups, and to undertake the tea spilled when making tea. The tea tray is not only practical, but also can be used as an ornament to create an atmosphere of tea tasting.

(5) Fair cup: also known as tea sea, used for even tea soup. In the process of brewing tea, the brewed tea can be poured into a fair cup from the teapot, and then poured into each sample tea cup, to ensure that each tea tasting can taste the same concentration of tea.

(6) Sample tea cup: A cup specifically used for tasting tea, which is generally small and delicate. The material of the sample tea cup is mostly ceramic or glass, and the color and change of tea can be clearly observed. When tasting tea, sample tea cup is an indispensable tea set.

(7) Tea lotus: used to hold dry tea, easy to watch and access.

(8) Tea cup: mainly used to hold the waste tea and tea residue generated in the brewing process. Its shapes and sizes vary according to the needs of use, but the main design is to facilitate cleaning and keep the tea table clean. In addition, the cup is also an important cultural symbol in the tea ceremony. It represents the tea people's respect and inheritance of tea culture, and also reflects the tea people's pursuit of neat and elegant life attitude.

(9) Tea towel: used to wipe water and tea stains on the tea set, to keep the tea set clean and dry.

(10) Six gentlemen of tea art: refers to the six kinds of tea sets commonly used in tea art performances, which are tea cans, teaspoons, tea glasses, tea rules, tea clips, tea needles. The six gentlemen of tea art each play an important role in the tea art performance, which is an important part of tea culture.

① Tea tube: As a vessel for holding tea art supplies, its design is usually elegant and practical, which is used to store and protect other tea sets, making the tea art performance more tidy and orderly.

② Teaspoon: In the process of brewing tea, using a teaspoon can ensure that the amount of tea is just right, laying the foundation for the taste and quality of the tea.

③ Tea glass: When placing tea, the tea glass is placed on the spout to guide the tea into the pot smoothly, while preventing the tea dust from falling out of the pot. This design not only improves the convenience of brewing tea, but also keeps the tea table clean.

④ Tea: The utensils for holding tea into the pot can accurately measure the right amount of tea to ensure that the amount of tea thrown into each tea is consistent, so as to ensure the stability of the taste and quality of the tea soup, generally made of bamboo.

⑤ Tea holder: Shaped like a clip. The functions of the tea holder are diverse, both can be used to carry tea residue, the soaked tea from the pot, can also be used to carry the tea cup for cleaning, both health and anti-hot.

⑥ Tea needle: dredge the internal network of the teapot. When the spout is blocked by tea leaves, it is used to dredge to keep the water flowing smoothly. Or, after adding tea leaves, stir the tea leaves evenly, with broken tea on the bottom and whole tea on the top.

These common tea sets used for making tea each play an important role in the process of tea art performance, and together constitute the unique charm of Chinese tea culture.

赛证直通 Competitions and Certificates

二维码 2-1-1：知识拓展

一、基础知识部分（Basic Knowledge Part）

（一）专业名词解释（Explanation of Professional Terms）

1. 盖碗（Covered bowl）：
2. 公道杯（Fair cup）：

（二）思考题（Thinking Question）

谈谈你对中国茶文化的理解。

Talk about your understanding of Chinese tea culture.

二、技能操作部分（Skill Operation Part）

练习不同茶具的使用方法。

Train the use of tea sets.

【微课】品类名优：中国名茶

专题二
品类名优：中国名茶

你可以从视觉上描述图 2-1-3 中茶叶的特点吗？

_____

Can you visually describe the tea leaves in this picture (Figure 2-1-3)?

_____

图 2-1-3 不同的茶叶（different tea leaves）

中国是世界上最早发现和利用茶树的国家。大量的历史资料和近代调查研究材料都证明了中国是茶树的原产地。中国作为产茶大国，各地各类名茶比比皆是。为了便于识别茶叶的品质特征，根据加工方式和发酵程度的不同，一般习惯于把茶叶分为六大类：绿茶、白茶、黄茶、青茶、黑茶、红茶。六个种类的茶叶外观由绿向黄绿、黄、青褐、黑色渐变，茶汤也由绿向黄绿、黄、青褐、红褐色渐变。

China is the first country in the world to discover and utilize tea plants. A large amount of historical materials and modern research data have proven that China is the birthplace of tea plants. As a major tea-producing country, there are numerous types of famous teas from various regions. To facilitate the identification of tea quality characteristics, according to the different processing methods and fermentation degrees, it is generally customary to divide tea into six major categories: green tea, white tea, yellow tea, oolong tea, dark tea, and black tea. The appearance of the six types of tea gradually changes from green to green-yellow, yellow, green-brown, and black, and the tea soup also gradually changes from green to green-yellow, yellow, green-brown, and red-brown.

一、绿茶（Green Tea）

绿茶的典型特征是"三绿"，即干茶绿、茶汤绿、叶底绿。绿茶制法一般经过鲜叶杀青、揉捻、干燥三个工序。其中，杀青是使绿茶保持绿汤、绿叶的关键。杀青是通过高温方式，让鲜叶中多酚氧化酶的活性钝化，从而阻止茶叶氧化变红，也就是通常说的不发酵。根据不同的杀青和干燥方法，绿茶可分为炒青绿茶、晒青绿茶和蒸青绿茶。绿茶一般冲泡水温以85℃为宜，冲泡时间以2～3分钟为好。绿茶越嫩，茶汤滋味就越鲜爽，所以名优绿茶大多采用极嫩的原料制作。绿茶是我国产茶量最多的一类茶叶，产地遍布全国，重要的生产省份有浙江、安徽、湖北、湖南、四川、贵州等，代表名品有西湖龙井、洞庭碧螺春、黄山毛峰、都匀毛尖等。

The typical characteristics of green tea are "three greens", dry tea is green, the tea soup is green, and tea dregs, after soaking, are green. The green tea is usually made by three steps: kill-green fresh leaves, rolling, and drying. Among them, kill-green is the key to keep leaves and soup green. Kill-green is to inactivate the activity of polyphenol oxidase in fresh leaves in high temperature, so as to prevent tea leaves from oxidizing and turning red, we usually call it non-fermentation. According to different methods of kill-green and drying, green tea can be divided into stir-fried green tea, sun-dried green tea and steamed green tea. Green tea is generally brewed with 85 ℃ water for 2-3 minutes. The more tender the green tea, the more refreshing the tea soup tastes, so the most famous green tea is made of extremely tender raw materials. Green tea has the largest amount in China, with production areas all over the country, mainly in Zhejiang, Anhui, Hubei, Hunan, Sichuan, Guizhou, etc. We have well-known green teas like West Lake Dragon Well, Dongting Biluochun, Huangshan Maofeng, and Duyun Maojian, etc.

二、白茶（White Tea）

白茶因茸毛不脱，白毫满身而得名。我国是世界上唯一的白茶产地。白茶的制作工艺最自然，采摘多毫的幼嫩芽叶后，只需自然摊晾和文火烘干，让茶叶中的茶多酚自然氧化即可，属于轻度发酵茶。杀青是通过高温钝化茶叶中的氧化酶活性，茶叶较绿；揉捻是给茶叶造型，让茶汁外流；焙火可以让茶叶干燥，产生更多的香气。而白茶不杀青、不揉捻、不焙火，最大化保留了茶叶原始的营养物质，其茶汤颜色较浅。白茶最著名的产地是福建，知名的白茶有白毫银针、白牡丹、寿眉等。

White tea is famous for being covered with white hairs as no fuzz taking off. China is the only country to produce white tea in the world. It has the most natural process. After picking a few young buds and leaves, we just use natural drying and gentle heat to oxidize naturally the tea polyphenols in leaves, it is a mildly fermented tea. Kill-green is to inactivate the enzyme activity in tea leaves in a high temperature to make leaves greener; rolling is to shape tea leaves and flow tea juice out; roasting can dry leaves to make more aroma. However, white tea does not have kill-green, rolling, and roasting, it retains original nutrients to the maximum, the color of the tea soup is very light. Fujian is the most famous area to produce white tea. Well-known white teas include Baekho tip Silver Needle, White Peony, Shou Mei, etc.

三、黄茶（Yellow tea）

黄茶较为少见，它的初制工艺与绿茶相似，但是加入了闷黄的工序，从而促使部分多酚类物质发生了自动氧化，让黄茶的香气、滋味变醇。和绿茶相比，黄茶更加柔和、回甘。黄茶叶黄、汤黄，茶汤含有大量的消化酶，饮用可助消化，代表茶有君山银针、霍山黄芽、蒙顶黄芽。

Yellow tea is relatively rare. Its preliminary processing is similar to green tea, but adding the step of heaping for yellowing, which promotes partial auto-oxidation of polyphenols, so as to make its aroma and taste more mellow. Compared with green tea, it is softer and sweeter. The yellow tea has yellow leaves and soup, there are a lot of digestive enzymes in the soup which can help digestion. Representative teas include Jun Shan Silver Needle, Huoshan Huangya and Mengding Huangya.

四、青茶（Oolong tea）

青茶又叫作乌龙茶，属于半发酵茶，但是这个"半"并不准确。因为在乌龙茶中，发酵程度最轻的包种茶接近绿茶，发酵程度最重的东方美人茶接近红茶。乌龙茶制作工艺中的做青，是它区别于其他茶类的独特之处，也是形成乌龙茶天然花果香品质的关键。乌龙茶的特点在于先促进茶多酚的氧化，然后突然终止氧化，因此乌龙茶兼具绿茶和红茶的特点，既有绿茶的鲜浓，又有红茶的甜醇。没有哪种茶比乌龙茶更讲究冲泡技巧，因此，乌龙茶又叫作工夫茶。乌龙茶的产区主要分布在福建、广东和台湾，其中福建是乌龙茶的发源地和最大产区。福建所产的乌龙茶按地域可分为闽北乌龙和闽南乌龙，主要代表产品分别为武夷岩茶和安溪铁观音。

Green tea is also called oolong tea, belongs to partially fermented tea, but the "partially" is not very accurate. Because in oolong tea, Pouchong Tea with the lightest degree of fermentation close to green tea, and Oriental Beauty Tea with the highest degree of fermentation close to black tea. The fine manipulation of green tea leaves in the production process of oolong tea is a unique feature that distinguishes it from other teas, and also the key to form natural floral and fruity quality of oolong tea. Oolong tea is characterized by promoting the oxidation of tea polyphenols first, and then stopping oxidation suddenly. Therefore, oolong tea has both features of green tea and black tea, with freshness of green tea and sweetness of black tea. No other tea is more skilled in brewing than Oolong tea, so it is also called Kungfu tea. Oolong tea is mainly produced in Fujian, Guangdong and Taiwan, among which Fujian is the birthplace and the largest producing area of oolong tea. Oolong tea produced in Fujian can be divided into northern Fujian Oolong and southern Fujian Oolong by region. The main representative products are Wuyi Rock Tea and Anxi

Tieguanyin.

五、黑茶（Dark tea）

早期黑茶大多销往边区，作为茶马古道的重要物资，黑茶对于边疆游牧民族而言是生活必需品，需求量也比较大。为了运输方便，大多数黑茶被压制成各种形状的紧压茶。长期以来黑茶都使用较为粗老的原料制作，饮用时经过熬煮而不是冲泡。不过现在也有使用采摘较嫩的原料制作的黑茶，如宫廷普洱茶等，适合冲泡饮用。黑茶如果用的是粗老的原料则可以煮着喝，如果用的是嫩的原料则可以用 100 ℃的沸水冲泡喝。黑茶属于后发酵茶，在加工、储藏和运输过程中，因为微生物胞外酶的作用，产生了一些其他茶类没有的或者说是含量比较低的一些生化活性物质，所以它有助消化、顺肠胃的作用。黑茶中品质相对好的是云南普洱茶。云南普洱茶分为散茶和紧压茶。随着现在消费区域和消费群体的拓展，黑茶的形态有了很大的变化，变得更加方便、时尚且符合人们现在的需求。

Early black tea was mostly sold to border areas as an important material of ancient tea-horse Road. Black tea is a daily necessity for border herders with large demand. For easy transportation, most dark teas are pressed into compressed Tea of various shapes. For long, dark tea has been made by relatively crude and old raw materials, and boiled instead of brewed when drinking. But there are also dark teas made from more tender raw materials, such as Gongting Pu'er tea, which are suitable for brewing. Dark tea can be boiled if it is crude and old raw materials, and be brewed in boiling water at 100 ℃ if it is tender materials. Dark tea belongs to post-fermented tea. In processing, storage and transportation, some biochemical active substances that other teas do not have or have a relatively low content are produced by the effect of microbial extracellular enzymes, so it can help digestion. Relatively good dark tea is Yunnan Pu'er tea, which is divided into loose tea and pressed tea. With the expansion of our current consumption areas and consumer groups, the form of dark tea has changed a lot, more convenient and fashionable to meet people's needs.

六、红茶（Black Tea）

中国是世界上最早生产和饮用红茶的国家。红茶作为一种氧化性发酵茶类，其起源可以追溯到中国的明清时期。如今红茶已经成为国际茶叶市场的大宗产品，全世界有 40 多个国家生产红茶，主要的红茶产茶国包括中国、印度、斯里兰卡等。红茶属于全发酵茶类，萎凋、揉捻、发酵等方式都是为了促进红茶中的茶多酚氧化，在加工过程中形成黄色的茶黄素和红色的茶红素，因此，红茶的干茶色泽和冲泡的茶汤色泽都是以红色为主，在整体上表现出甜醇的风格。我国的红茶产地集中在安徽、福建、广东和云南一带，代表茶有金骏眉、正山小种、滇红等。

China is the earliest country in the world to produce and drink black tea. As an oxidative fermented tea, it can be traced back to the Ming and Qing Dynasties of China. Today, it has become a major product in international tea market. Black tea is produced in over 40 countries include: China, India, Sri Lanka, etc. Black tea is a complete fermented tea. Withering, rolling and fermentation are all to promote the oxidation of tea polyphenols in black tea, forming yellow theaflavins and red thearubigins, so the color of dry tea and brewed tea soup are mainly red, showing a sweet and mellow quality. Producing areas of black tea in China are concentrated in Anhui, Fujian, Guangdong and Yunnan, representative teas include Jin Junmei, Lapsang Souchong and Yunnan Black Tea.

中国茶叶历史和知识博大精深，对于刚接触茶的新手，从主要的六大茶类入手，对日后品茶形成系统的认知很有必要。

The history and knowledge of Chinese tea is extensive and profound. For beginners, it is necessary to start from the main six tea categories to form a systematic understanding of drinking tea.

 ## 赛证直通 Competitions and Certificates

一、基础知识部分（Basic Knowledge Part）

二维码 2-1-2：
知识拓展

（一）名词解释（Explanation of Professional Terms）

杀青（kill-green fresh leaves）：

（二）思考题（Thinking Question）

1. 茶叶分为哪六大类？
What are the six main types of tea?

2. 茶汤颜色和茶叶制作时的发酵程度之间有什么关系？
What is the relationship between the color of the tea soup and the degree of fermentation?

二、技能操作部分（Skill Operation Part）

品饮绿茶并说出其特点。
Taste green tea and describe its characteristics.

【微课】
千姿百态：
国外名茶

专题三
千姿百态：国外名茶

你知道图 2-1-4 中的女孩们是哪国人吗？她们在做什么？

_____

Do you know the nationality of the girls in the picture (Figure 2-1-4)? What are they doing?

_____

在古代相当长的时间里，只有中国饮茶、种茶、制茶并掌握茶

图 2-1-4 女孩们（girls）

的有关知识。随着世界交流、贸易的发展，这片来自东方的神奇叶子，早已超越了地域和文化的界限，成为全球范围内广受欢迎的饮品。在各类饮料市场产品中，茶是仅次于水的饮料，消费者分布最广泛。

For quite a long time in ancient times, only Chinese people drank, planted, produced tea and mastered the relevant knowledge of tea. With the development of global communication and trade, this magical leaf from the East has long transcended geographical and cultural boundaries, becoming a widely beloved beverage across the world. In the market of various beverage products, the tea is only second to the water, with the widest distribution of consumers.

一、东亚茶（East Asian Tea）

中国是茶的故乡和茶文化的发源地，因此，离中国较近的日本和韩国在茶叶和茶文化方面的发展较完善。

China is both the hometown of tea and the birthplace of its culture, Japan and South Korea, which are close to China, have relatively complete development in tea and its culture.

日本产出的茶叶中，有90%都是绿茶，它的分类非常细致。其中最高级的茶品是玉露，据说一百棵茶树里可能都找不到一棵可以生产玉露，可见对茶树的要求之高。玉露的涩味较少，口感鲜爽，香气清雅，茶汤清澈。日本人最常喝的绿茶是煎茶，占日本茶产量的80%。日本的茶道起源于中国，但具有浓郁的日本民族特色。日本有两种茶道文化形式，一为抹茶道，传自我国唐宋时期，采用当时的抹茶法，用蒸青茶碾制成粉状茶叶饮用；一为煎茶道，源于中国明清时期，采用以炒为主加工而成的散状芽条。

Ninety percent of the tea produced in Japan is green tea, with very meticulous classification, in which the most advanced tea is Gyokuro. It is said that one hundred tea trees may not be able to produce Gyokuro, which shows high requirements for the tea tree. Gyokuro is less astringent, fresh in taste, elegant in fragrance and limpid in tea soup. The Japanese drink the Steamed green tea most, which accounts for about 80% of Japanese tea. Japanese tea ceremony originates from China, but owns strong Japanese national characteristics. It is divided into two schools: the first one is matcha which originated from the Tang and song Dynasties period in China and adopted the Maccha method at that time, and the steamed green tea was grinded into powdery tea for drinking; the second one is Steamed green tea which originated from Ming and Qing Dynasties in China, using mainly stir-fried mainly processed from the loose bud.

韩国早在新罗时代便有了茶文化，是韩国传统文化的一部分。韩国最常见的是绿茶，具有清新的香气和淡雅的口感。但韩国的许多传统茶饮其实并不使用茶叶，而是以其他植物、果实、谷物或药材为原料制成，这些茶饮被称为"传统茶"。它们不仅味道独特，还具有丰富的营养价值和保健功效。比较常见的有五谷茶如大麦茶、玉米茶等，大麦茶可以增进食欲、温暖肠胃，很多韩国家庭都以大麦茶代替饮用水。常见的还有水果茶如柚子茶、大枣茶、橘皮茶等，柚子茶可以祛除风寒，添加蜂蜜后更具独特风味，很受欢迎。

South Korea's tea culture occurred in the Silla Era. It is considered as a part of traditional Korean culture. The most common type of tea in Korea is green tea, known for its refreshing aroma and delicate taste. However, many traditional Korean beverages, referred to as " traditional teas," do not actually use tea leaves. Instead, they are made from other plants, fruits, grains, or medicinal herbs. These traditional teas not only have unique flavors but also offer rich nutritional value and health benefits. There is relatively common Grain tea, such as barley tea, corn tea, etc., in which the barley tea can promote appetite and warm the stomach. Many South Korean families replace drinking water with the barley tea. The common tea also includes the fruit tea, such as grapefruit tea, jujube tea, orange peel tea, etc., among them, the grapefruit tea can remove the chill, and owns a more unique flavor after being added honey, which is very popular.

二、南亚茶 (South Asian Tea)

南亚是世界茶叶的重要产区，主产国有印度、斯里兰卡和孟加拉国等国家和地区，其中印度产量居世界第二位。

South Asia is a significant tea-producing region in the world. The major producing countries include India, Sri Lanka, Bangladesh, etc., in which Indian output ranks the second in the world.

印度是世界红茶的主要产地之一，世界四大名茶中的大吉岭红茶和阿萨姆红茶就产自于印度。大吉岭红茶以其独特的幽雅香气被誉为"红茶中的香槟"，其3—4月的初摘一号茶多为青绿色，5—6月的次摘二号茶为金黄色，以二号茶品质最优。其气味芬芳高雅，上品尤其带有麝香葡萄味，滋味纯净浓郁回甘。

India is one of the major producing areas of the world's black tea, produces the Darjeeling Black Tea and Assam Black Tea are among the world's four famous teas. Darjeeling Black Tea is honored as "champagne in black tea" by virtue of its unique quiet and tasteful fragrance. Most of No. 1 tea picked in the beginning of Mar.– Apr. are blue-green and No. 2 tea picked in May– Jun. are golden yellow, in which No. 2 tea has the best quality. Its odor is elegant and fragrant and its top-quality product is equipped with muscat grape flavor, with a pure, strong and sweet taste.

阿萨姆红茶以6—7月采摘的品质最优，茶汤伴有淡淡的麦芽香和玫瑰香。阿萨姆红茶茶叶中含量最高的生物碱——咖啡碱，具有兴奋作用，能够使头脑思维活动迅速清晰，消除睡意，消除肌肉疲劳，具有使感觉更加敏锐和提高运动技能的作用。

Assam Black Tea picked in Jun.–Jul. has the best quality, and the tea soup is accomplished with light malt fragrance and rose scent. The alkaloid — caffeine with the highest content in Assam Black Tea plays an excitation role and can make mind-thinking activity rapidly clear, eliminate drowsiness and muscle fatigue, make feeling more acute and improve sport skills.

斯里兰卡是世界第四大茶叶生产国，其锡兰红茶也是世界四大名茶之一。斯里兰卡的红茶风味强劲、口感浑重，适合泡煮香浓奶茶。斯里兰卡人不仅在产量上做文章，还提高茶叶附加值，开发特色茶，塑造茶叶品牌等，可以说将茶叶运用到了极致。

Sri Lanka, as the fourth largest tea producing country in the world, produces Ceylon Black Tea which is deemed as one of the world's four famous teas. The black tea in Sri Lanka is suitable for brewing fragrant tea with milk by virtue of powerful flavor and strong taste. People of Sri Lanka do not only pay attention to output, but also boost the added value of tea by developing characteristic tea, shaping tea brands, etc, they apply the tea to an extreme.

三、非洲茶（African Tea）

非洲产茶国家共有十余国，主要集中在东非。其中，肯尼亚、马拉维、乌干达等都是非洲主要产茶国。

There are over ten tea-producing countries in Africa which mainly gather together in the East Africa. Kenya, Marawi, Uganda, etc. are the main tea-producing countries in Africa.

肯尼亚是非洲重要的茶叶生产国之一。在非洲的红茶产区中，无论是红茶的数量还是质量，肯尼亚都是居于首位的。肯尼亚横贯赤道，日照丰富，其境内东非大裂谷两侧略带酸性的火山灰土壤最适宜茶树的种植。肯尼亚出产的红茶又叫作红碎茶，凭借茶色优美、茶香浓郁、味道甘醇闻名于世。

Kenya, as the one of the important tea producing countries in Africa, ranks the first in both quantity and quality of black tea in the African black tea producing area. Straddling the equator, Kenya has plenty of sunshine, and its East African Rift Valley is flanked by acidic volcanic ash soil, making it is ideal for growing tea. The black tea produced in Kenya is also called as broken black tea and is world-renowned by virtue of its graceful tea color, strong tea fragrance and mellow taste.

肯尼亚红茶可以冲泡出高品质的茶汤，被认为是世界上最好的饮料。这是因为其茶园主要位于海拔 1 500 ～ 2 700 米的高原地区，周围没有工业污染，加之紫外线强，所以病虫害极少。独特的地理位置形成了肯尼亚红茶无污染、无农药、无化肥的有机生长环境。肯尼亚红茶的喝法颇具多样性，除了冲泡，还可以煮着喝。因为肯尼亚曾经是英国的殖民地，当地人们喝茶深受英国的影响，喝红茶时喜欢加糖、加奶。

Kenya black tea can be brewed into high-quality tea soup, and is considered as the best beverage in the world. Because its tea plantations are mainly located in the plateau area at an altitude of 1 500– 2 700 meters, there is no industrial pollution around it, and ultraviolet light is strong, so there are few diseases and insect pests. Unique geographical location of Kenya forms the pollution-free, pesticide-free and fertilizer-free organic growth environment of black tea. There are many ways to drink Kenyan black tea, besides brewing, it can also be boiled and drunk. Because Kenya was once a British colony, the local people drink tea deeply influenced by the British, so they like adding sugar and milk to black tea.

非洲还有一种非常特别的如意茶（博士茶），是用一种南非的豆荚类植物加工制成的茶。这种植物的生长范围非常有限，只在南非西南好望角附近的山地生长。这种茶树是和普通山茶树完全不相同的植物。如意茶凭借香味深邃和口感浓郁的特点被称为非洲最流行的饮品，其享用方法很多，可热、可冷，可加糖、加奶。

There is a kind of very special Rooibos in the Africa which is also called Rooibos Vanilla and made from legume plants in South Africa. The plant is very restricted in growth range, only grows in the mountain region near Cape of Good Hope in the southwest of South Africa. The Rooibos tree is completely different from common camellia. Rooibos is called the most popular beverage in Africa by virtue of its deep fragrance and strong taste, and can be hot or cold, be added sugar and milk.

四、国际茶文化（International Tea Culture）

茶作为一种全球性的饮品，在国际交流中扮演了多重角色。国际茶文化是独特的世界文化形式，超越了国家、种族、地域等限制，是全人类共同的精神财富之一。21世纪全世界人们已经认识到，经济发展并不是社会发展的全部，协调人际关系、重视品格修行、调和人类与自然的平衡，才是有意义的文明飞跃。

Tea, as a global beverage, plays multiple roles in international exchanges. International tea culture is a unique form of world culture, transcending limitations such as countries, races, and regions. It is one of the spiritual wealth shared by all humanity. In the 21st century, people around the world have recognized that economic development is not the entirety of social development. Only by coordinating interpersonal relationships, valuing character cultivation, and balancing the relationship between humans and nature we can achieve meaningful civilizational leaps.

（一）文化桥梁（Cultural Bridge）

茶叶作为文化桥梁，不仅促进了不同文化之间的交流，也丰富了全球的文化多样性。通过茶叶，人们可以体验和学习到世界各地的不同文化和传统。茶文化作为东方文化的重要组成部分，通过茶文化的国际传播，促进了东西方文化的交流与理解。当今世界各国、各民族的饮茶风俗，都因本民族的传统、地域民情和生活方式的不同而各有所异，然而"客来敬茶"却是古今中外的共同礼俗。

Tea, as a cultural bridge, not only promotes communication between different cultures but also enriches global cultural diversity. Through tea, people can experience and learn about different cultures and traditions from around the world. As an important part of Eastern culture, the international dissemination of tea culture has promoted the exchange and understanding of Eastern and Western cultures. Today, the tea-drinking customs of various countries and ethnic groups differ due to their own traditions, regional customs, and lifestyles. However, the common etiquette of "serving tea to guests" is shared across all times and places.

（二）经济媒介（Economic Media）

茶叶贸易是国际贸易的重要组成部分，对许多国家的经济发展具有重要意义。通过茶叶的进出口，可以增加就业机会，促进经济增长。随着全球贸易的发展，茶叶市场变得更加多元化。茶叶不仅在传统生产国继续发展，也在其他地区如非洲的肯尼亚等地形成了新兴的茶叶产业。此外，当今的茶叶贸易不仅包括传统的茶叶销售，还包括茶馆文化、茶艺表演以及与茶叶相关的旅游和体验活动。

Tea trade is an essential part of international trade and holds significant importance for the economic development of many countries. Through the import and export of tea, employment opportunities can be increased, and economic growth can be promoted. With the development of global trade, the tea market has become more diversified. Tea is not only continuing to develop in traditional producing countries but also forming emerging tea industries in other regions, such as Kenya in Africa. In addition, today's tea trade not only includes traditional tea sales but also teahouse culture, tea art performances, and tea-related tourism and experiential activities.

（三）外交手段（Diplomatic Means）

茶可以作为一种外交手段，促进国家间的友好关系。如通过举行茶会表达敬意和欢迎；通过茶艺表演、茶道展示等活动，向外国友人展示本国的茶文化和艺术，增进文化理解和相互尊重；在国际会议或论坛期间，提供茶歇不仅是为了休息，也是为了让与会者有更多的非正式交流机会，促进信息交流和关系建立；茶叶还可作为国礼赠送给外国政要，表达友好关系的同时也推广了本国的特产和文化。

Tea can serve as a means of diplomacy, promoting friendly relations between countries. For example, holding tea parties can express respect and welcome, while tea art performances and tea ceremony demonstrations showcase the country's tea culture and art to foreign friends, enhancing cultural understanding and mutual respect. During international conferences or forums, providing tea breaks is not only for rest but also to provide more informal communication opportunities for participants, promoting information exchange and relationship building. Tea is also given as a national gift to foreign dignitaries, expressing friendly relations and promoting the country's specialties and culture.

总之，茶不仅是一种饮品，它在促进国际交流、文化传播、经济发展和健康生活等方面都发挥着重要作用。

In summary, tea is not only a beverage, but it also plays an important role in promoting international exchange, cultural dissemination, economic development, and a healthy lifestyle.

赛证直通 Competitions and Certificates

一、基础知识部分（Basic Knowledge Part）

思考题（Thinking Question）

二维码 2-1-3：
知识拓展

1. 非洲的主要产茶国有哪些？

What are the main tea producing countries in Africa?

2. 世界四大名茶是哪些？

What are the four famous teas in the world?

二、技能操作部分（Skill Operation Part）

查阅资料，说说原本不产茶且无饮茶习惯的印度人，因何接触茶叶，又因何开始大规模种植茶叶，一跃成了全球知名的产茶大国？

Refer to the materials and explain why Indians, who originally did not produce tea nor had a tea-drinking habit, came into contact with tea and subsequently began large-scale tea cultivation, eventually becoming a globally renowned tea-producing nation?

❁【单元一 反思评价】Unit 1 Reflection and Evaluation

学会了：_____

Learned to: _____

成功实践了：_____

Successful practice: _____

最大的收获：_____

The biggest gain: _____

遇到的困难：_____

Difficulties encountered: _____

对教师的建议：_____

Suggestions for teachers: _____

单元二　茶的泡饮及创意制作

坐酌泠泠水，看煎瑟瑟尘。

无由持一碗，寄与爱茶人。

——唐·白居易《山泉煎茶有怀》

单元导入 Unit Introduction

茶，不仅是一种饮品，更是中华优秀传统文化的重要组成部分。从绿茶、白茶、黄茶、乌龙茶到红茶及黑茶，每一种茶都承载着丰富的文化内涵和民族情感。在这个单元中，我们将一起探索茶的泡饮技巧与创新制作方法，学习如何正确泡茶，体验不同的茶叶带来的口感变化，并尝试将传统茶文化与现代元素相结合，创作出新颖的茶饮。

Tea is not only a kind of beverage, but also an important part of fine traditional Chinese culture. From green tea, white tea, yellow tea, oolong tea to black tea and black tea, each tea carries rich cultural connotations and national emotions. In this unit, we will explore tea brewing techniques and innovative ways of making tea. We will learn how to brew tea properly, experience the changes in taste brought about by different tea leaves, and try to combine traditional tea culture with modern elements to create novel tea drinks.

🎯 学习目标 Learning Objectives

➢ 知识目标（Knowledge Objectives）

（1）能描述不同茶器的特征。

Can describe the characteristics of different tea ware.

（2）能描述不同茶的品质特征。

Can describe the quality characteristics of different teas.

（3）掌握创意茶饮的配制原理及要求。

Master the principles and requirements of the preparation of creative tea drinks.

（4）掌握茶席设计的构成要素。

Master the components of tea ceremony design.

➢ 能力目标（Ability Objectives）

（1）根据所学知识选用正确的泡茶器具。

Can choose different utensils to make different tea according to the knowledge learned.

（2）根据所学知识掌握六大类茶的冲泡方式。

Can master the brewing method of six kinds of tea according to the knowledge.

（3）根据所学知识掌握点茶方法。

Can master the method of ordering tea according to the knowledge.

（4）根据所学知识自创茶饮。

Can create tea drinks according to the knowledge.

（5）根据所学知识自创茶席。

Can create their own tea room according to the knowledge.

➢ 素质目标（Quality Objective）

（1）培养学生具备良好的职业精神、专业精神和工匠精神。

Train students to have a good professional spirit, professionalism and craftsman spirit.

（2）在冲泡六大类茶的过程中培养学生的创新精神并进行挫折教育及双创教育。

In the process of brewing six kinds of tea, cultivate students' innovative spirit and carry out frustration education and double innovation education.

（3）提升文化自觉，坚定文化自信，增强对传统文化的热爱与尊重，使学生能够深刻理解并认同中华优秀传统文化的价值。

Enhance cultural consciousness, strengthen cultural self-confidence, enhance their love and respect for traditional culture, and enable students to deeply understand and recognize the value of excellent traditional Chinese culture.

专题一
异曲同工：茶的泡饮

你知道图 2-2-1 中的紫砂壶可以用来冲泡什么茶吗？

Do you know what kind of tea can be made in the purple clay pot in the picture above (Figure 2-2-1)?

图 2-2-1 紫砂壶（purple clay teapot）

在泡茶的技艺中，对于用水的选择极为讲究，以"清、轻、甘、冽、活"五种特性为最佳。唐代茶学专家陆羽在其著作《茶经》中明确指出："其水，用山水上，江水中，井水下。"这一观点强调了泡茶用水的来源和质量对茶汤品质的重要性。明代茶人张大复在《梅花草堂笔谈》中进一步阐述："茶性必发于水，八分之茶，遇十分之水，茶亦十分矣；八分之水，试十分之茶，茶只八分耳。"可见，水对于茶的重要性。

In the technology of tea making, the choice of water is extremely exquisite, with five characteristics of " clear, light, sweet, cold and alive " as the best. Lu Yu, a tea expert in the Tang Dynasty, clearly pointed out in his book *The Classic of Tea*: " As for the water, Spring water is the best, river water is the second, and well water is the worst. " This point of view highlights the importance of the source and quality of the water used to make tea to the quality of the tea soup. Zhang Dafu, a tea man in the Ming Dynasty, further elaborated in " Plum Blossom Cottage Essays " : " Tea's nature relies on the water. Common tea meets good water, tea reaches its perfection as well; While fine tea meets common water, the tea is common as well. " It can be seen that water is important to tea.

一、绿茶冲泡（Green Tea Brewing）

（一）冲泡注意事项（Precautions for Brewing）

1. 茶具选择（Tea Set Selection）
饮用绿茶，通常用透明度好的玻璃杯、瓷杯或盖碗冲泡。

When drinking green tea, it is usually brewed in clear glasses, porcelain cups or gaiwan.

2. 茶与水的比例（Ratio of Tea to Water）
一般推荐茶与水的比例控制在 1∶50 ～ 1∶60，即每 1 克茶叶应使用 50 ～ 60 mL 的水。这样的比例能够确保茶叶在热水中充分展开，释放其特有的香气和风味成分，同时避免茶汤过于浓厚或淡薄，使得茶汤的浓淡程度适中，口感鲜爽。

It is generally recommended that the ratio of tea to water be controlled between 1∶50 and 1∶60, that is, 50 to 60 ml of water should be used for every 1 gram of tea. Such a ratio can ensure that the tea is fully unfolded in the hot water, releasing its unique aroma and flavor components, while avoiding the tea soup is too thick or weak, making the tea soup moderate shade and fresh taste.

3. 水温（Water Temperature）

对于冲泡绿茶的水温，建议控制在 80～90 ℃，过高的水温可能会导致茶叶中的茶多酚类物质发生氧化反应，进而使茶汤色泽偏黄，丧失原有的香味。而水温不够时，茶叶则不容易被冲泡开。

For brewing green tea, it is recommended to control the water temperature between 80 ℃ and 90 ℃, too high water temperature may lead to oxidation of tea polyphenols in tea, and then make the tea color yellow, lose the original flavor. Too low temperature may cause the tea not be fully brewed.

（二）冲泡方法（Brewing Method）

绿茶的品种较丰富，根据形状、紧结程度和鲜叶老嫩程度不同，有上投法、中投法和下投法三种冲泡方法。

The variety of green tea is rich, according to the shape, the degree of tightness and the degree of fresh leaves are different, there are three kinds of brewing methods: upthrow method, mid-stroke method, downthrow method.

1. 上投法（Upthrow Method）

在泡茶的过程中，一次性向茶杯中注入七分满的热水，后再投入茶叶。这种方法特别适用于细嫩度极高的绿茶品种，如碧螺春、信阳毛尖等。值得注意的是，茶叶的嫩度越高，对水温的要求则越低。

In the process of brewing tea, pour seven points of hot water into the cup at one time, and then pour the tea leaves. This method is especially suitable for green tea varieties with high tenderness, such as Biluochun, Xinyang Maojian, etc.. It is worth noting that the higher the tenderness of the tea, the lower the requirements for water temperature.

2. 中投法（Mid-stroke Method）

在泡茶的过程中，先向茶杯中注入约三分之一的热水，然后投放茶叶，此举旨在使茶叶吸收水分并得以舒展，待茶叶逐渐展开后，再向茶杯中继续注入热水至七分满。中投法适用于细嫩紧实、扁平的绿茶，如西湖龙井、六安瓜片等。这种泡茶技巧有助于茶叶在热水中均匀受热，释放出其独特的香气和风味。

During the tea brewing process, the teacup is first filled with about one-third of the hot water, and then the tea leaves are poured in, so that the tea leaves absorb the water and stretch, and when the tea leaves are gradually unrolled, the teacup is continued to be filled with hot water until it is seventy percent full. The method is suitable for soft, firm and flat green tea, such as West Lake Longjing, Lu'an melon slices, etc.. This tea brewing technique helps the tea to heat evenly in hot water, releasing its unique aroma and flavor.

3. 下投法（Downthrow Method）

在泡茶的过程中，应先将茶叶投放至杯中，随后一次性向杯内注入七分满的热水。下投法适用于细嫩度相对较低的绿茶。一次性直接注入热水，使得茶叶能够迅速被热水浸润，有助于茶叶中的营养成分和香气充分释放，进而为饮茶者提供一杯香气四溢的茶汤。

In the process of brewing tea, the tea should be first put into the cup, and then injected into the cup at one time seven points full of hot water. The downcasting method is suitable for green tea with relatively low tenderness. A one-time direct injection of hot water, so that the tea can be quickly soaked by hot water, help to fully release the nutrients in the tea and aroma, and then provide tea drinkers with a cup of fragrant tea soup.

二、白茶冲泡（White Tea Brewing）

（一）冲泡注意事项（Precautions for Brewing）

1. 茶具选择（Tea Set Selection）

盖碗是冲泡白茶的万能选择，无论新茶还是老茶都适用。盖碗操作方便，可控性强，能够很好地展现白茶的香气和滋味。对于新茶，可以快速出汤，保持香气鲜爽，滋味甜润；对于老茶，可以让茶叶泡得更久一些，香气独特，滋味甘醇。

Tureens are a versatile choice for brewing white tea, both new and old. The cover bowl is easy to operate and has strong controllability, which can well show the aroma and taste of white tea. For new tea, it can quickly make soup, keep the aroma fresh and sweet taste; For old tea, the tea can be steeped for a longer time, with a unique aroma and sweet taste.

2. 茶与水的比例（Ratio of Tea to Water）

一般 150 毫升的水用 3 ～ 5 克的茶叶。茶水比的范围一般为 1∶20 至 1∶50。

Generally 150 mL of water with 3 to 5 grams of tea. The tea to water ratio generally ranges from 1∶20 to 1∶50.

3. 水温（Water Temperature）

白茶冲泡水温要求 80 ～ 100 ℃。一般来说，白毫银针冲泡水温 80 ～ 90 ℃、白牡丹冲泡水温 90 ～ 95 ℃、贡眉寿眉冲泡水温 95 ～ 100 ℃。

The water temperature for brewing white tea is 80–100 ℃ . Generally speaking, the brewing water temperature of pekoe silver needle is about 80–90 ℃ , the brewing water temperature of white peony is about 90–95 ℃ , and the brewing water temperature of Gongmei Shoumei is about 95–100 ℃ .

（二）老白茶煮饮法（Old White Tea Boiling Method）

白茶有"一年茶，三年药，七年宝"的说法。正常陈化三年以上的寿眉适合煮饮，寿眉叶片大、茶梗粗，内含有益成分多，其煮出的茶汤甘醇，有浓郁的枣香、药香，具有养生保健功效。

White tea has " one year of tea, three years of medicine, seven years of treasure " saying. Normal aging of more than 3 years of Shou Mei suitable for boiling, Shou mei leaves large and thick tea stalk, containing beneficial ingredients. The boiled out of the tea soup is sweet, has a strong jujube aroma, medicine fragrance, which has health care effects.

三、黄茶冲泡（Yellow Tea Brewing）

（一）冲泡注意事项（Precautions for Brewing）

1. 茶具选择（Tea set selection）

黄茶通常用玻璃杯或盖碗冲泡。

Yellow tea is usually brewed in a glass or tureen.

2. 茶与水的比例（Ratio of Tea to Water）

可以根据个人口味放入适量茶叶，通常茶叶投放量与水的比例是 1∶50。

You can add the right amount of tea according to personal taste, usually the ratio of tea to water is 1∶50.

3. 水温（Water temperature）

黄茶不宜用沸水冲泡。黄芽茶比较嫩，用 80 ℃的水冲泡即可，而黄小茶和黄大茶则可以用 90 ℃以上的水来冲泡。

Yellow tea should not be brewed in boiling water. Yellow bud tea is relatively tender and can be brewed with water at 80 ℃, while yellow tea and yellow tea can be brewed with water above 90 ℃.

（二）冲泡方法（Brewing Method）

黄茶第一次冲泡 15 秒左右出汤，之后每次冲泡时间延长 10 秒，可冲泡 3 ～ 5 次。

Yellow tea is brewed for about 15 seconds for the first time, and then the brewing time is extended by 10 seconds each time, and it can be brewed 3 to 5 times.

四、乌龙茶冲泡（Oolong Tea Brewing）

（一）冲泡注意事项（Precautions for Brewing）

1. 茶具选择（Tea Set Selection）

乌龙茶通常用紫砂壶、盖碗等冲泡，它们各有特点，适用于不同的冲泡场景和品鉴需求。紫砂壶因其双孔透气结构和较强的吸附力，能够留住茶香且提升茶汤的香浓度，是冲泡乌龙茶的传统佳选。

Oolong tea is usually brewed in purple clay POTS, covered bowls, etc., which have their own characteristics and are suitable for different brewing scenes and tasting needs. Because of its double-hole breathable structure and strong adsorption force, the purple clay pot can retain the aroma of tea and improve the aroma concentration of tea soup, which is the traditional best choice for brewing oolong tea.

2. 投茶量（Tea consumption）

在泡茶时，茶叶的投放量需根据茶叶的品种和形态来合理调整。根据不同的乌龙茶形态，投茶量也有所不同。对于条形的乌龙茶，投茶量差不多占紫砂壶容量的 1/6 到 1/5 ；而对于球形的乌龙茶，由于其形状特殊，茶叶展开比较慢，通常投茶量会盖过紫砂壶底部大半。

When brewing tea, the amount of tea needs to be adjusted reasonably according to the variety and form of tea. The amount of tea poured varies according to different forms of oolong tea. For bar oolong tea, the amount of takes up about 1/6 to 1/5 of the capacity of the purple sand pot. But for spherical oolong tea, because of its special shape, tea leaves unfold slowly, usually over the bottom half of the teapot.

3. 水温（Water Temperature）

乌龙茶内含特定的芳香物质，在冲泡乌龙茶时，需使用沸腾的水以确保这些芳香物质得以充分激发，从而展现乌龙茶独特的香气特质。

Oolong tea contains specific aromatic substances, and boiling water is used when brewing Oolong tea to ensure that these aromatic substances are fully stimulated to reveal the unique aroma characteristics of Oolong tea.

（二）冲泡方法（Brewing Method）

1. 浸泡时间（Soaking time）

在冲泡闽南乌龙茶时，应严格控制冲泡时间。首泡时，建议将茶叶浸泡约 45 秒，以确保茶叶中的香气和风味物质得到初步释放。在接下来的冲泡中，每次浸泡时间应适当增加，第二次冲泡约 60 秒，之后每次冲泡时间可在此基础上稍加延长数十秒，以逐步释放茶叶中的深层风味。而对于闽北和广东的乌龙茶，由于其特有的制作工艺和茶叶特性，出汤速度相对较快。在冲泡时，首泡的浸泡时间应控制在 15 秒左右，以迅速提取茶叶中的香气和滋味，避免过度浸泡导致茶汤苦涩；后续冲泡可根据茶叶的展开程度和茶汤的浓淡程度，适当调整浸泡时间。

When brewing Oolong Tea from southern Fujian, the brewing time should be strictly controlled. For the first time, it is recommended to soak the tea for about 45 seconds to ensure that the aroma and flavor substances in the tea are initially released. In the next brewing, each soaking time should be appropriately increased, the second brewing time is about 60 seconds, and then each brewing time can be slightly extended for tens of seconds on this basis to gradually release the deep flavor in the tea. For Oolong Tea from northern Fujian and Guangdong, due to its unique production technology and tea characteristics, the soup speed is relatively fast. When brewing, the first soaking time should be controlled at about 15 seconds to quickly extract the aroma and taste of the tea, and avoid excessive soaking leading to bitter tea soup. Subsequent brewing can be adjusted according to the degree of tea spread and the depth of the tea soup, the appropriate soaking time.

2. 冲泡次数（Number of Brewing Times）

乌龙茶因其独特的制作工艺，具备能多次冲泡的特点，素有"三泡四泡是精华，七泡有余香"之赞誉。乌龙茶在冲泡过程中，第三泡和第四泡所呈现的茶汤品质尤为突出，为茶叶风味的精华所在。在正确的冲泡方法下，乌龙茶甚至能够持续冲泡七次以上，每次冲泡后仍有余香缭绕，充分展现了乌龙茶的香气和滋味。

Oolong tea because of its unique production process, with the characteristics of multiple brewing, known as " three bubbles four bubbles is the essence, seven bubbles have lingering aroma " praise. Which means that in the brewing process of oolong tea, the quality of the tea soup presented by the third and fourth time is particularly prominent, which is the essence of tea flavor. Under the correct brewing method, Oolong Tea can even be brewed more than seven times, and there is still lingering incense after each brewing, fully demonstrates the aroma and taste of oolong tea.

五、红茶冲泡（Black Tea Brewing）

1. 茶具选择（Tea Set Selection）

白瓷盖碗作为冲泡红茶的首选茶具与红茶的鲜艳汤色相得益彰，形成了鲜明的视觉对比，从而更加凸显红茶的色泽魅力。此外，盖碗的设计充分考虑了茶叶的浸泡需求，便于精准控制茶叶的浸泡时间和温度，确保冲泡好的红茶口感醇厚。

As the preferred tea set for brewing black tea, the white porcelain covered bowl and the bright color of black tea complement each other, forming a sharp visual contrast, thus more highlighting the color charm of black tea. In addition, the design of the covered bowl fully considers the tea's soaking needs, which is convenient to accurately control the tea's soaking time and temperature, and ensure that the brewed black tea has a mellow taste.

2. 茶与水的比例（Ratio of Tea to Water）

一般来说，红茶与水的比例在 1∶50 到 1∶60，即每克红茶对应 50 到 60 mL 的水。高品质的红茶可能需要适当增加茶叶的用量，以充分展现其丰富的内含物质和醇厚的口感。对于白瓷盖碗冲泡红茶，投入 4 到 5 克的茶叶是一个常见的选择。

In general, the ratio of black tea to water is between 1∶50 and 1∶60, that is, each gram of black tea corresponds to 50 to 60 mL of water. High quality black tea may require a moderate increase in the amount of tea leaves to fully express its rich contents and mellow taste. For white porcelain bowl brewing black tea, 4 to 5 grams of tea is a common choice.

3. 水温（Water temperature）

通常用 90～95 ℃的热水冲泡红茶，泡出来的茶汤清澈，香气馥郁，滋味醇厚，叶底明亮。

Usually use 90-95 ℃ hot water brewing black tea, the tea out of the clear, fragrant, mellow taste, bright leaf bottom.

六、黑茶的冲泡（Dark Tea Brewing）

（一）冲泡注意事项（Precautions for Brewing）

1. 茶具选择（Brewing method）

通常新茶和散茶用盖碗泡茶，老茶和紧压茶用紫砂壶为宜，紫砂壶吸附性强，可以有效地清除黑茶的异味。

Usually, new tea and loose tea are made in a covered bowl, and old tea and pressed tea are made in a purple clay pot, which has strong adsorption and can effectively remove the odor of black tea.

2. 茶与水的比例（Tea to water ratio）

在冲泡黑茶时，茶水比的选择是确保茶汤品质的关键因素。通常建议的茶水比范围在 1∶30 至 1∶50，即每克茶叶对应的水量在 30～50 毫升。

When brewing dark tea, the choice of tea to water ratio is a key factor to ensure the quality of tea soup. The recommended ratio of tea to water is usually between 1∶30 and 1∶50, that is, the amount of water corresponding to each gram of tea is between 30 ml and 50 mL.

3. 水温（Water temperature）

一般需要用 95～100 ℃的水冲泡黑茶，粗老的紧压茶需要煎煮才能充分提取其内含物质。

It is generally necessary to brew black tea with water of 95-100 ℃, and the crude old pressed tea needs to be decocted to fully extract its contents.

（二）泡饮方法（Soaking method）

1. 冲泡与煮饮（Brewing and boiling）

黑茶需要润茶，起到醒茶和去除茶叶表面杂质和灰尘的作用。一般可冲泡 3～5 次。在冲泡过程中，可以采用"凤凰三点头"的方式，即高冲低斟，使茶叶在壶中翻滚，有利于茶汁的浸出。此外，黑茶还适合煮饮。取茶叶 10～15 克，加水烧开，放入茶叶，待水再次烧开后，炖煮两分钟左右，然后停火滤渣，趁热饮用。

Dark tea needs to moisten the tea, to wake up the tea and remove impurities and dust on the surface of the tea. Generally, it can be brewed 3 to 5 times. In the brewing process, the way of "phoenix three nods" can be used, that is, high flushing and low pouring, so that the tea leaves roll in the pot, which is conducive to the leaching of tea juice. In addition, black tea is also suitable for boiling. Take 10–15 grams of tea leaves, add water to boil, add tea leaves, wait for the water to boil again, simmer for about two minutes, and then cease the filter residue, drink while hot.

2. 奶茶饮法（Milk tea drinking method）

按传统煮饮法煮好后，可按奶、茶汤1∶5的比例调制后可以直接饮用，也可以加适量盐或其他配料等，即成特色奶茶。

After cooking according to the traditional cooking method, it can be directly drunk according to the ratio of milk and tea soup 1∶5, or it can be added with an appropriate amount of salt or other ingredients, which is a special milk tea.

赛证直通 Competitions and Certificates

一、基础知识部分（Basic Knowledge Part）

思考题（Thinking Question）

冲泡乌龙茶最适宜用什么茶具？为什么？

What is the most suitable tea set for brewing Oolong Tea? Why?

二维码 2-2-1：知识拓展

二、技能操作部分（Skill Operation Part）

实操六大类茶的冲泡。

Practice the brewing of six kinds of tea.

【微课】
温文儒雅：
宋代点茶

专题二
温文儒雅：宋代点茶及茶百戏

宋代点茶的步骤有哪些（图2-2-2）？

_____

What were the steps of dian cha in Song Dynasty (Figure 2-2-2)?

_____

盛行于宋代的点茶技艺，凭借其深厚的文化内涵和独特的

图 2-2-2　宋代点茶
（dian cha in Song Dynasty）

艺术魅力，被广泛喜爱。通过点茶这一传统技艺，人们不仅能够深入体验中国传统文化的精髓，还能学习到独特的泡茶方式，从而在繁忙生活中找到一丝宁静，回归本真，收获内心的喜悦与满足。

The technique of dian cha prevailing in the Song Dynasty, with its profound cultural connotation and unique artistic charm, has been widely loved. Through the traditional skill of dian cha, people can not only deeply experience and understand the essence of traditional Chinese culture, but also learn the unique way of making tea, so as to find a little peace in the busy life of the city, return to the truth, and harvest the joy and satisfaction of the heart.

一、宋代点茶（Tea in Song Dynasty）

（一）点茶概述（The Introduction of Dian Cha）

点茶，是宋代特有的一种烹茶方式，它不仅是一种饮茶方法，更是一门精致的艺术。其由唐代煎茶法演变而来，至宋代达到鼎盛。宋代社会的经济繁荣与文化昌盛，为点茶艺术的蓬勃发展奠定了坚实的基础，使茶事活动从日常饮用上升为雅集品鉴，尤其在文人墨客之间备受推崇。

Dian Cha is a unique way of brewing tea in the Song Dynasty. It is not only a method of drinking tea, but also a delicate art. The method of decocting tea evolved in Tang Dynasty and reached its peak in Song Dynasty. The economic and cultural prosperity of the Song Dynasty laid a solid foundation for the vigorous development of tea ordering art which made tea activities rise from daily drinking to elegant tasting, especially among scholars and writers.

（二）七汤点茶法（Seven Orders of Dian Cha）

宋徽宗赵佶所著的《大观茶论》中，七汤点茶法被详细阐述，这是古代中国茶艺的精髓，展现了宋代茶艺的高超技艺与深厚文化底蕴（图2-2-3）。

图 2-2-3　点茶工具（tool of dian cha）

第一汤，称为"量茶受汤，调如融胶"。将沸水注入茶粉中，将茶粉调成黏稠状，形成胶质物。此时，茶粉颗粒必须全部溶解，茶粉与茶粉之间咬合在一起，为后续的冲泡打下坚实基础。

第二汤，称为"击拂既力，珠玑磊落"。通过快速而有力的击拂，茶面上会迅速产生大泡泡和小泡泡，这些泡泡如同珠玑般磊落，彰显出茶面的活力与动感。

第三汤，则是"击拂轻匀，粟文蟹眼"。注水量减少，使用茶筅的速度要均匀而稳定，将大泡泡匀速击碎成小泡泡，形成如粟粒般细密且均匀的茶面，同时茶面上会出现蟹眼般的小气泡，这是茶汤品质优良的标志。

第四汤，名为"稍宽勿速，轻云渐生"。注水要更加谨慎，量要少而均匀，茶筅转动的幅度要大而慢。随着击拂的进行，云雾般的茶沫渐渐从茶面升起，营造出一种轻盈飘逸的氛围。

第五汤，则是"乃可稍纵，茶色尽矣"。再次注入少量水，击拂的动作可以稍微放松一些。随着茶汤的逐渐融合，茶色达到最佳状态，呈现出清澈透亮、色泽均匀的茶汤。

第六汤，称为"以观立作，乳点勃然"。继续注水并击拂茶汤，目的是将底部未被打散的茶粉继续打上来，使乳面更加厚实。同时，要注意观察茶汤的变化，当乳点勃然兴起时，即达到了理想的冲泡效果。

第七汤，称为"乳雾汹涌，溢盏而起"。这是点茶法的最后一步，也是最为壮观的一步。在中上部快速而有力地击打茶汤，直到乳雾汹涌而起，溢满整个茶盏。此时，茶汤表面形成一层厚厚的乳沫，如同云雾般缭绕在茶盏之上，展现出点茶法的最高境界——"咬盏"。

通过这七个步骤的精心操作，宋代的茶艺师们能够冲泡出一杯色香味俱佳的茶汤，让人在品味中感受到茶文化的博大精深与独特魅力。

In the "Treatise on Tea" written by Emperor Huizong of the Song Dynasty,the seven-decoction tea method is elaborated, which is the essence of ancient Chinese tea art, showing the superb skill and profound cultural heritage of tea art in the Song Dynasty (Figure 2-2-3).

The first soup is called "tea by soup, such as melt glue". The boiling water is injected into the tea powder, and the tea powder is adjusted into a sticky shape to form a colloidal substance. At this time, the tea powder particles must all dissolve, and the tea powder and the tea powder bite together to lay a solid foundation for the subsequent brewing.

The second soup is called "blow both force, pearls and aboveboard". Through quick and powerful strokes, the tea surface will quickly produce large and small bubbles, which are like pearls, highlighting the vitality and dynamics of the tea surface.

The third soup is "stroke gently, millet crab eyes". The water injection is reduced, the speed of using the tea whisk should be uniform and stable, the large bubbles should be broken into small bubbles at a uniform speed, forming a fine and even tea surface like a millet grain, and there will be small bubbles like crab eyes on the tea surface, which is a sign of good quality tea soup.

The fourth soup, called "a little wider not fast, light clouds gradually born." Water injection should be more careful, the amount should be small and uniform, and the amplitude of the tea whisk rotation should be large and slow. As the stroke progresses, the cloud of tea droplets gradually rises from the tea surface, creating a light and airy atmosphere.

Fifth soup, is "can be a little longitudinal, brown done." Inject a small amount of water again, and the stroke can be slightly relaxed. With the gradual fusion of the tea soup, the tea color reaches the best state, showing a clear and bright, uniform color of the tea soup.

The sixth soup is called "to observe and stand, milk point is booming". Continue to inject water and stroke the tea soup, the purpose is to continue to hit the tea powder that has not been broken at the bottom, so that the milk surface is thicker. At the same time, pay attention to the changes in the tea soup, when the milk point rises, the ideal brewing effect is achieved.

The seventh soup is called "milk mist surging, overflowing and rising". This is the last and most spectacular step in the tea-ordering process. Beat the tea quickly and vigorously in the upper middle until the milk mist rises and overflows the whole cup. At this time, a thick layer of milk is formed on the surface of the tea soup, which is like cloud and mist curling around the tea cup, showing the highest realm of the tea method-"biting the cup".

Through the careful operation of these seven steps, the tea artists in the Song Dynasty were able to brew a cup of tea soup with excellent color and flavor, so that people can feel the profound and unique charm of tea culture in the taste.

（三）点茶的文化意义（The Cultural Significance of Dian Cha）

在宋代，点茶超越了茶单纯的饮用功能，成为文人间交流互动的重要媒介。它体现了宋代人对生活精致化的追求，也反映了当时社会文化的繁荣。点茶过程中所展示的宁静、专注和精湛的技艺，都被认为是修身养性的重要表现。同时，茶会也是文人抒发情感、切磋诗文、交流思想的场所，许多文学佳作都诞生于这些充满诗意的茶会之中。

In the Song Dynasty, dian cha went beyond the simple function of drinking, and became an important medium of communication and interaction among scholars. It reflects the Song Dynasty people's pursuit of life refinement, but also reflects the prosperity and elegance of social culture at that time. The serenity, concentration and craftsmanship displayed during the process of dian cha are all considered to be important manifestations of self-cultivation. At the same time, the tea party is also a place for literati to express their feelings, exchange poems and texts, and exchange ideas. Many great literary works are born in these poetic tea parties.

（四）点茶的影响与传承（The Influence and Inheritance of Dian Cha）

宋代点茶艺术影响了后世茶文化的发展。随着时代的变迁，虽然点茶艺术逐渐淡出了日常生活，但其精神内核——劳动精神、专业精神、工匠精神等，依然在现代茶艺中得到体现和传承。点茶是中国茶艺史上的瑰宝，它不仅展现了古人对生活品质的追求，也折射了当时社会的文化成就。通过深入学习点茶艺术，我们能够更好地理解和欣赏中国传统文化的内涵，同时也能获得对现代生活美学的启发和思考。

Dian cha art in Song Dynasty influenced the development of tea culture in later generations. With the changes of the times, although dian cha art gradually faded out of daily life, its spiritual core-the pursuit of harmony and refinement is still reflected and inherited in modern tea art. Dian cha is a treasure in the history of Chinese tea art, which not only shows the high requirements of ancient people for the quality of life, but also reflects the cultural achievements of the society at that time. Through in-depth study of dian cha, we can better understand and appreciate the connotation of traditional Chinese culture, and also get inspiration and reflection on modern life aesthetics.

二、茶百戏（Cha Baixi）

茶百戏有时也被称作分茶、水丹青、汤戏等，这种艺术形式始见于唐代，尤其在宋代闽北武夷山一带十分流行。

Cha Baixi is sometimes called sorting tea, water painting, soup show, etc. This art form began in the Tang Dynasty, especially in the Song Dynasty in the Wuyi Mountain area of northern Fujian very popular.

（一）茶百戏的历史背景（Historical Background of Cha Baixi）

茶百戏起源于唐代，当时已有使用茶汤作画的记载，至宋代在闽北武夷山一带达到鼎盛。

作为一种生活与艺术的结合体，茶百戏体现了中国古代文人对生活细节的审美化追求。它不仅是一种饮茶方式，更是文化交流和艺术展示的平台。

Cha Baixi originated in the Tang Dynasty, when there were records of painting with tea soup, and reached its peak in the Wuyi Mountain area in northern Fujian in the Song Dynasty. As a combination of life and art, Cha Baixi embodies the aesthetic pursuit of the details of life of ancient Chinese literati. It is not only a way to drink tea, but also a platform for cultural exchange and artistic display.

（二）茶百戏的技艺特点（The Technical Characteristics of Cha Baixi）

（1）原料纯粹：茶百戏仅使用茶和水作为原料，不添加其他物质。
（2）技艺高超：通过特殊的斟茶和调汤技巧，使茶汤表面形成各种文字和图案。
（3）形式多样：能够变幻出山水、花鸟或抽象图案等，展现出丰富的视觉效果。
（4）风格独特：每一幕茶百戏都是独一无二的即兴创作，具有很高的艺术价值。

（1）Pure raw materials: Cha Baixi only uses tea and water as raw materials, without adding any other substances.

（2）Excellent skills: through special tea pouring and soup skills, various characters and patterns are formed on the surface of the tea soup.

（3）Diverse forms: it can conjure landscapes, flowers and birds or abstract patterns, etc., showing rich visual effects.

（4）Unique style: each scene of Cha Baixi is a unique improvisation, with high artistic value.

（三）茶百戏的文化价值（The Cultural Value of Cha Baixi）

（1）艺术融合：茶百戏将书法、绘画艺术与茶艺相结合，展现了中国文化的多样性和深度。
（2）文化传承：作为一种传统的艺术形式，茶百戏承载了中国传统文化的精髓，是非物质文化遗产的重要组成部分。
（3）精神享受：在品茶之余观赏茶百戏，为日常生活增添乐趣。

（1）Art integration: Cha Baixi combines calligraphy and painting art with tea art, showing the diversity and depth of Chinese culture.

（2）Cultural inheritance: as a traditional art form, Cha Baixi carries the essence of traditional Chinese culture and is an important part of the intangible cultural heritage.

（3）Spiritual enjoyment: Enjoy tea drama in addition to tea tasting, which adds fun to everyday life.

（四）茶百戏的保护与传承（Protection and Inheritance of Cha Baixi）

在现代社会，茶百戏虽然曾一度濒临失传，但经过非遗传承人的不懈努力，这项技艺已得到了恢复和重现。茶百戏不仅是茶文化的重要组成部分，也是中国古代文化创造力的体现。深入了解和学习茶百戏，不仅能够领悟到古代人的智慧，还能够感受到中国传统文化的深远影响和魅力。

In modern society, although Cha Baixi was once on the verge of being lost, but now through the unremitting efforts of non-genetic successors, this skill has been restored and reproduced. Cha

Baixi is not only a precious part of tea culture, but also the embodiment of ancient Chinese cultural creativity. Through in-depth understanding and learning of Cha Baixi, we can not only appreciate the wisdom and artistic achievements of the ancients, but also appreciate the far-reaching influence and charm of traditional Chinese culture.

赛证直通 Competitions and Certificates

一、基础知识部分（Basic Knowledge Part）

（一）专业名词解释（Explanation of Professional Terms）

1. 茶筅（Tea whisk）：
2. 茶百戏（Cha Baixi）：

（二）思考题（Thinking Question）

宋代点茶的文化意义及其影响与传承。

The cultural significance, influence and inheritance of tea in Song Dynasty.

二、技能操作部分（Skill Operation Part）

在实训室进行点茶实操。

Perform dian cha exercises in the training room.

二维码 2-2-2：
知识拓展

专题三
推陈出新：创意茶饮制作

【微课】
推陈出新：
创意茶饮制作

你知道创意茶饮的调制原理吗（图 2-2-4）？

_____

Do you know how creative tea is prepared (Figure 2-2-4)?

_____

在各式各样的茶饮中，将传统的原叶茶融入新的冲泡方式，探索茶的另一面，通过创意茶饮的制作，满足广大消费者对新鲜、健康饮品的选择和需求。

图 2-2-4　创意茶饮（creative tea）

In a variety of tea drinks, we integrate traditional original leaf tea into new brewing methods, explore the other side of tea, through the production of creative tea drinks, to meet the majority of consumers' choice and demand for fresh and healthy drinks.

一、调饮茶概述（Overview of Mixing Tea）

　　调饮茶指的是以茶叶为主要原料，辅以其他一种或多种茶料（可称为辅料），经过精心调和而成的茶饮。这类茶饮品不仅满足了社会和生活的多样化需求，还承载了深厚的精神理念。在调饮茶的制作过程中，通常配以精巧的茶具，遵循规范的冲泡技能，从而赋予调饮茶丰富的文化内涵和艺术雅趣。

　　Mixing tea refers to tea as the main raw material, supplemented by one or more other tea ingredients (called accessories), after careful coordination of tea drink. This tea beverage fully which not only meets the diverse needs of society and life, but also carries a profound spiritual concept. In the process of making tea, delicate tea utensils are usually used and standardized brewing skills are followed, thus giving mixing tea rich cultural connotation and artistic elegance.

二、常见的调饮方法（Common Methods of Drinking）

（一）冰茶调饮法（Mixing Method of Ice Tea）

　　冰茶调饮法是一种主要在北美地区（如美国、加拿大等）流行的茶饮调制方法。该方法首先将速溶茶粉倒入清洁的玻璃杯中，随后加入适量的凉白开进行冲泡。为了丰富口感，可以进一步添加糖、柠檬片或其他柠檬风味的调味品，以及冰块。通过充分的搅拌或摇动，使所有成分充分融合，最终得到一杯清爽宜人的冰茶。这种调饮法制作的茶饮适合在炎热的夏季饮用，具有解暑降温的功效。

　　Iced tea is a popular tea preparation method mainly in North America (such as the United States, Canada, etc.). The method first pour the instant tea powder into a clean glass, and then add an appropriate amount of cold water for brewing. To enrich the taste, further add sugar, lemon slices or other lemon-flavored flavorings, and ice cubes. By stirring or shaking enough, so that all the ingredients are fully integrated, the final cup of refreshing and pleasant iced tea. This method is suitable for drinking in hot summer and has the effect of relieving summer heat and cooling down.

（二）薄荷茶调饮法（Mixing Method of Mint Tea）

　　薄荷茶调饮法是一种在非洲各国广泛流行的茶饮调制方式，其中以摩洛哥最具代表性。在街头，常常可以看到身背茶桶的零售者，他们提供茶汤供路人购买饮用。在这些地区，饮用糖茶与粮食同等重要，被视为日常生活中不可或缺的一部分。西非国家的薄荷茶以其独特的口感而广受欢迎，其甜爽的滋味和清新的口气使人倍感舒适。制作薄荷茶时，首先将茶叶冲泡成茶汤，再加入适量的糖和薄荷叶进行调味，使茶汤既具有茶叶的醇厚口感，又带有薄荷的清新香气。这种茶饮不仅味道独特，而且具有提神醒脑、解暑降温的功效，深受当地人的喜爱。

　　The mint tea blending method is a widely popular tea blending method in African countries, among them, Morocco is the most representative. It is common to see sellers carrying tea buckets on the streets, they provide tea soup for passers-by to buy and drink. In these areas, drinking sugar tea is as important as food and is seen as an indispensable part of daily life. The West African country's mint tea is popular for its unique taste, which makes people feel comfortable with its sweet and refreshing taste and fresh breath. When making mint tea, the tea leaves are first brewed into tea soup, and then the appropriate amount of sugar and mint leaves are added for seasoning, so that the tea soup has both the mellow taste of tea and the fresh aroma of mint. This kind of tea not only has a unique taste, but also has the effect of refreshing the mind, relieving the summer heat and cooling down, which is deeply loved by the local people.

（三）牛奶红茶调饮法（Method of Milk Black Tea）

牛奶红茶调饮法在国外，尤其是在欧洲、南亚、大洋洲以及北美等地区广泛流行，其中以英国最具代表性，通常被称为英式饮茶法。这种饮茶方式深受当地人的喜爱，许多人习惯从早到晚多次品饮，其中以"下午茶"最为隆重。在享用下午茶时，人们不仅品饮红茶，还搭配各种点心，同时进行交流，成了一种便捷的社交方式。在俄罗斯的寒带地区，人们也常采用俄式茶炊煮水泡茶的方式，以抵御严寒。在调制红茶时，他们会在茶汤中加入果酱、蜂蜜、奶油或甜酒等调味品，不仅丰富了茶汤的口感，还增加了热量，有助于御寒保暖。这种饮茶方式体现了俄罗斯人对红茶的独特喜爱和创意调配。

The method of milk black tea is widely popular abroad, especially in Europe, South Asia, Oceania and North America, among them the United Kingdom is the most representative, and is often referred to as the British tea method. This way of drinking tea is loved by the local people, many people are used to drinking from morning to night, of which " afternoon tea " is the most solemn. When enjoying afternoon tea, people not only drink black tea, but also match various snacks, and communicate and chat at the same time, which has become a convenient way to socialize. In the cold regions of Russia, people often use Russian samovars to brew water and tea to resist the cold. When brewing black tea, they will add jam, honey, cream or dessert wine and other condiments to the tea soup, which not only enriches the taste of the tea, but also increases the heat and helps to keep warm. This way of drinking tea reflects the Russian unique love for black tea and creative mixing.

（四）奶茶（Milk Tea）

奶茶的饮用习惯在印度源远流长，其起源可追溯到中国的西藏地区。由于印度人的口味较重，他们在传承奶茶制作工艺的基础上进行了独特的创新。在印度，人们习惯将鲜奶与茶叶同煮，以此增强奶茶的醇厚口感。更为特别的是，印度人还会在奶茶中加入生姜、豆蔻、肉桂、槟榔等多种香辛料，这些添加物不仅使得奶茶的味道更为香烈，还具有一定的保健作用。这种独特的奶茶调制方式，体现了印度人对茶饮文化的热爱与创意。

The drinking habit of milk tea has a long history in India, its origin can be traced back to the Tibetan region of China. Due to the strong taste of Indians, they have carried out unique innovations on the basis of inheriting milk tea. In India, it is customary to boil fresh milk and tea directly together to enhance the mellow taste of milk tea. What is more special is that Indians also add ginger, cardamom, cinnamon, betel nut and other spices to milk tea, which not only makes the taste of milk tea more fragrant, but also has certain health are effects. This unique way of making milk tea reflects the Indian people's love and creativity for tea drinking culture.

三、创意调饮茶的制作原理及要求（Principles Requirements for Making Creative Tea）

（一）创意调饮茶的配制原理（Principle for Preparation of Creative Tea）

在调饮茶的制作过程中，必须充分了解茶叶的属性以及辅料的功效。这是因为不同的茶叶和辅料在搭配时会产生不同的味道和效果。例如，某些辅料可能会与茶叶中的某些成分产生化学反应，影响茶叶的口感和功效。因此，在制作调饮茶时，需要注意茶品与辅料之间的搭配关系，避免产生不良的影响。同时，也应该注意到，每个人的体质都是不同的，在饮茶时也需要因人而异。一般来说，遵循"寒者热之、热者寒之、实则泻之、虚则补之"的原则，可以更好地选择适合自己的茶饮。此外，在配置调饮茶时，我们还需要兼顾所调茶品的色、香、味、形，使其达到最佳的口感和视觉效果。

In the process of making tea, we must fully understand the properties of tea and the property flavors and efficacy of excipients. This is because different tea leaves and accessories will produce different tastes and effects when combined. For example, certain excipients may react chemically with certain components in the tea, affecting the taste and efficacy of the tea. Therefore, when configuring tea, it is necessary to pay attention to the matching relationship between tea and accessories to avoid adverse effects. At the same time, we should also note that everyone's physique is different, so it also needs to vary from person to person when drinking tea. Generally speaking, following the principle of " cold disease should be treated by warm therapy， warm disease should be treated by cold therapy， deficiency syndrome should be treated by tonifying therapy， excess syndrome should be treated by purgation therapy" can help us better choose the tea that suits us. In addition, when configuring tea, we also need to take into account the color, aroma, taste and shape of the tea to achieve the best taste and visual effect.

（二）创意调饮茶配制要求（Requirements for Preparation of Creative Tea）

在配制调饮茶时，需要遵循以下六项核心原则，以确保茶品的品质与饮用体验。

（1）显著的茶味：调饮茶的首要原则是要保持显著的茶味。茶作为基础成分，其特有的香气和口感是调饮茶的核心。无论加入何种配料，都不应掩盖或削弱茶的原味。

（2）性质相宜的配料：调饮茶中应加入一至数种与茶性质相宜的配料。这些配料应与茶的味道、香气等相互衬托，形成独特的口感和风味，而不是相互冲突。

（3）明确的数量规定：每种茶料在调饮茶中的数量应有明确的规定。合理的数量配比可以确保茶品口感的均衡和稳定，避免因过量或不足而影响品质。

（4）合理的操作程序：配制调饮茶应遵循合理的操作程序。这包括配料的准备、混合、冲泡等步骤，每一步都应精确控制，以确保茶品的品质。

（5）科学的泡饮方法：泡饮方法是调饮茶品质的关键因素之一。这包括泡茶的时间、水温以及茶汤的颜色等。科学的泡饮方法可以最大限度地发挥茶和配料的品质，使茶品达到最佳口感和风味。

（6）口味适宜的茶汤与意境：调饮茶应具有口味适宜的茶汤，并在品尝时能够营造出一定的意境。这不仅包括茶品的口感和风味，还包括品茶时的环境、氛围等。综合这些因素，调饮茶可以为品茶者带来愉悦的品饮体验。

The following six core principles should be followed when preparing tea to ensure the quality and drinking experience of the tea.

（1）Significant tea flavor: the first rule of tea mixing is to maintain a significant tea flavor. Tea as the basic ingredient, its unique aroma and taste are the core of tea. No matter what ingredients are added, they should not cover up or weaken the taste of the tea.

（2）Suitable ingredients: one to several ingredients suitable for tea should be added to the tea. These ingredients should complement the taste and aroma of the tea to form a unique taste and flavor, rather than conflict with each other.

（3）Clear quantity regulation: the quantity of each tea ingredient in tea mixing should be clearly regulated. A reasonable quantity ratio can ensure the balanced and stable taste of tea, and avoid affecting the quality because of excessive or insufficient.

（4）Reasonable operating procedures: the preparation of tea should follow reasonable operating procedures. This includes the preparation of ingredients, mixing, brewing and other steps, each step should be precisely controlled to ensure the quality of the tea.

（5）Scientific brewing method: the brewing method is one of the key factors of tea quality. This includes the time of brewing, the water temperature, and the color of the tea. Scientific brewing methods can maximize the quality of tea and ingredients, so that tea to achieve the best taste and flavor.

（6）Suitable taste of tea soup and artistic conception: tea should have a suitable taste of tea soup, and can create a certain artistic conception when tasting. This includes not only the taste and flavor of the tea, but also the environment and atmosphere of the tea. By combining these factors, it can bring pleasant drinking experience to tea drinkers.

在制作调饮茶的过程中，需格外注重配料与茶本身的协调与融合。以红茶为例，红茶在口感上通常带有轻微的"涩"感，这种特性要求我们在选择配料时需进行细致的考量。为了取得口感上的平衡，建议选择酸甜度适中的水果作为配料。当这些水果与红茶混合时，其酸甜的口感可以中和红茶的"涩"感，使得整体口感更为和谐、平衡。这一原则不仅适用于红茶与水果的搭配，也为其他茶类与不同配料的搭配提供了有益的参考。

In the process of making mixing tea, special attention should be paid to the coordination and integration of ingredients and tea itself. For example, black tea often has a slight "astringent" taste, a characteristic that requires careful consideration when selecting ingredients. In order to achieve a balance on the taste, it is recommended to choose fruits with moderate sweetness and sour as ingredients. When these fruits are mixed with black tea, their sour and sweet taste can neutralize the "astringent" feeling of black tea, making the overall taste more harmonious and balanced. This principle not only applies to the collocation of black tea and fruit, but also provides a useful reference for the collocation of other teas with different ingredients.

（三）创意调饮茶的冲泡原则（The Brewing Principle of Creative Tea）

创意调饮茶的冲泡原则具体如下：

（1）冲泡方法：创意调饮茶的冲泡方法需参照基本茶类的冲泡法，确保每一步骤都符合茶叶本身的特性和要求。

（2）茶具选配：在冲泡过程中，茶具的选配需合理得当。根据茶叶种类、冲泡方式及饮用习惯，选择合适的茶壶、茶杯等茶具，以确保茶味能够充分展现。

（3）茶叶用量：在冲泡果茶、奶茶等调饮茶时，茶叶的用量需科学合理。通常，冰茶类的茶叶用量应加倍，以保证茶汤有足够的浓度和茶味。

（4）冰奶茶的冲泡技巧：在调制冰奶茶时，需添加奶、奶酪、冰块等配料。茶叶冲泡后，应放入冰柜进行冷却，这样可以使茶汤不易结块或呈豆腐花状，保持清爽的口感。

（5）辅料混合冲泡：若茶与几种辅料混合冲泡，可将细碎的辅料放置在滤茶器中，大朵或大块的辅料则可置于外层。这样的布局有利于茶与辅料在水中充分舒展，使茶味与辅料香味更好地融合，达到理想的冲泡效果。

The brewing principles of creative tea are as follows:

（1）Brewing method: the brewing method of creative tea should refer to the brewing method of basic tea to ensure that each step meets the characteristics and requirements of the tea itself.

（2）Tea set selection: in the brewing process, the selection of tea sets should be reasonable and proper. According to the type of tea, brewing method and drinking habits, choose the right teapot, tea cup and other tea sets to ensure that the tea taste can be fully displayed.

（3）Tea dosage: when brewing fruit tea, milk tea and other tea, the amount of tea should be scientific and reasonable. In general, the amount of tea used in iced tea should be doubled to ensure that the tea soup has sufficient concentration and tea flavor.

（4）Brewing skills of iced milk tea: when making iced milk tea, it is necessary to add milk, cheese, ice and other ingredients. After brewing, the tea should be put into the freezer to cool, so that the tea soup is not easy to cake or tofu flower shape, maintaining a refreshing taste.

（5）Auxiliary materials mixing brewing: if the tea is mixed with several auxiliary materials, the finely broken auxiliary materials can be placed in the tea filter, and the large auxiliary materials can be placed on the outer layer. Such a layout is conducive to the tea and accessories in the water fully stretched, so that the tea flavor and accessories flavor better integration, to achieve the ideal brewing effect.

（四）常用辅料介绍（Introduction to Commonly Used Auxiliary Ingredients）

在调制创意茶饮的过程中，辅料的选择和搭配对最终茶品的口感、营养价值和健康效益起着至关重要的作用。辅料可根据其来源和性质分为以下四大类。

（1）植物类辅料。这一类别涵盖了植物的多个部位，如根、茎、叶、花和果实等。常见的植物类辅料包括金银花、薄荷、甘草、陈皮和枸杞等。这些植物辅料不仅为调饮茶增添了独特的香气和口感，同时也使其具备一定的营养价值和健康功效。

（2）动物类辅料。动物类辅料主要包括乳酪、牛奶等。这些辅料在调饮茶中起到增香、增味和补充营养的作用。例如，蜂蜜能增添茶品的甜味和香气，牛奶和乳酪则能丰富茶品的口感和营养价值。

（3）矿物类辅料。矿物类辅料主要以盐为代表。在特定的调饮茶品中，盐能起到平衡口感、增强风味的作用。同时，矿物类辅料也为调饮茶提供了一定的矿物质元素。

（4）其他类辅料。除了上述三类辅料外，还有一些其他类别的辅料也被广泛应用于调饮茶的制作中，如酒、冰、果冻等。这些辅料不仅能丰富调饮茶的口感和风味，还能满足消费者的不同需求和喜好。例如，加入冰块可以制作冰茶，增添清凉感；加入果冻可以增加茶品的口感层次。

In the process of making creative tea, the selection and collocation of auxiliary materials play a crucial role in the taste, nutritional value and health benefits of the final tea. Excipients can be divided into the following four categories according to their source and nature.

（1）Plant accessories. This category covers many parts of the plant, such as roots, stems, leaves, flowers and fruits. Common plant accessories include honeysuckle, mint, licorice, orange peel and wolfberry. These plant accessories not only add unique aroma and taste to tea, but also have certain nutritional value and health effects.

（2）Animal accessories. Animal accessories mainly include, cheese, milk and so on. These auxiliary materials play the role of enhancing aroma, flavor and nutritional supplement in tea mixing and drinking. For example, honey can add sweetness and aroma to tea, and milk and cheese can enrich the taste and nutritional value of tea.

（3）Mineral auxiliaries. Mineral auxiliaries are mainly represented by salt. In the specific tea, salt can play a role in balancing taste and enhancing flavor. At the same time, mineral excipients also provide certain mineral elements for tea mixing.

（4）Other types of accessories. In addition to the above three types of accessories, there are some other types of accessories are also widely used in the production of tea, such as wine, ice, jelly, etc. These accessories can not only enrich the taste and flavor of tea, but also meet the different needs and preferences of consumers. For example, ice can be added to make iced tea, adding a cool feeling; adding jelly can increase the taste level of tea.

赛证直通 Competitions and Certificates

一、基础知识部分（Basic Knowledge Part）

（一）专业名词解释（Explanation of Professional Terms）

1. 调饮茶（mixing tea）:
2. 奶茶（milk tea）:

二维码 2-2-3：
知识拓展

（二）思考题（Thinking Question）

如何做好创意茶饮设计?

How to do creative tea design?

二、技能操作部分（Skill Operation Part）

以六大类茶为基础进行创意茶饮制作。

Make a creative tea drink based on six kinds of tea.

专题四
标新立异：创意茶席设计

你知道创意茶席设计的构成要素吗（图 2-2-5）？

Do you know what constitutes a creative tea banquet design (Figure 2-2-5)?

图 2-2-5　创意茶席（creative tea banquet）

茶席设计是以茶作为精髓，以茶具作为核心载体，在特定的空间格局中，与多种艺术形式交融，共同打造出一个拥有独特主题的茶道艺术整体。

The essence of tea banquet design is to take tea as the essence and tea set as the core carrier. In a specific spatial pattern, it blends with various art forms to create a unique theme of tea art as a whole.

一、创意茶席设计元素（The Elements of Creative Tea Banquet Design）

（一）茶品（Tea Products）

茶席设计的核心灵魂，常常成为整个设计构思的主线与精髓。在茶席的设计过程中，茶品占据着无可替代的核心地位，它不仅是品茗的对象，更是整个设计构思的灵魂和主要线索。茶品的特性、风味、品质，以及与之相匹配的冲泡方式，都深深影响着茶席的整体氛围和感受，是茶席设计中不可忽视的重要元素。

The core soul of tea banquet design often becomes the main line and essence of the entire design concept. In the design process of tea banquets, tea products occupy an irreplaceable core position. It is not only the object of tea tasting, but also the soul and main clue of the whole design concept. The characteristics, flavor and quality of tea products, as well as the matching brewing method, have a deep impact on the overall atmosphere and feeling of the tea banquet, and are important elements that cannot be ignored in the design of the tea banquet.

（二）茶具组合（Tea Set Combination）

作为茶席设计的基石与核心要素，茶具的选择与组合直接决定了茶席的整体风貌。各种质地、形状和颜色的茶具，应依据不同主题和形式的茶席设计进行精心搭配。这些茶具不仅要具备实用性，满足泡茶、品茶的基本需求，同时还应展现出独特的艺术性，为茶席增添一份雅致与品味。因此，茶具组合的基本特征即是实用性与艺术性的完美融合。

As the cornerstone and core elements of tea banquet design, the choice and combination of tea sets directly determine the overall style of the tea banquet. Tea sets of various textures, shapes and colors should be carefully matched according to different themes and forms of tea banquet design. These tea sets not only need to be practical, to meet the basic needs of tea brewing and tea tasting, but also should show unique artistic, to add a tasteful tea banquet. Therefore, the basic feature of tea set combination is the perfect integration of practicality and artistry.

（三）铺垫 (Mat)

作为茶桌上用于摆放茶器或其他物件的底部衬垫，铺垫涵盖了布艺类及其他各种质地的材料。铺垫的首要功能在于防止茶桌上的器物直接接触桌面，保持器物的清洁与卫生。此外，铺垫自身的特征，如质地、款式、大小、色彩及花纹等，能够辅助器物共同塑造和强化茶席设计的主题与立意。选取铺垫时，应根据茶席设计的整体主题与具体立意进行精心选择，以确保铺垫与茶席的和谐统一。

As the bottom liner for placing tea utensils or other objects on the tea table, it covers fabrics and other materials of various textures. The primary function of the mat is to prevent the utensils on the tea table from directly touching the table top and keeping the utensils clean and hygienic. In addition, the unique characteristics of the mat, such as texture, style, size, color and pattern, can assist the utensils to jointly shape and strengthen the theme and intention of the tea banquet design. When selecting the mat, we should carefully choose according to the overall theme and specific intention of the tea banquet design to ensure the harmonious unity of the mat and the tea banquet.

（四）插花艺术 (Flower Arranging Art)

用自然界的鲜花和叶草作为原材料，经过精心的艺术加工，将线条与造型的变幻融入情感和思考，创造出花卉的崭新形象。在茶席中，插花不仅仅是一种装饰，更是茶文化和茶席精神的体现。它追求的是一种崇尚自然、返璞归真、朴实秀雅的艺术风格，旨在通过花卉的灵动与美丽，进一步升华茶席的氛围，使品茶者能够在宁静与和谐中感受到茶文化的独特魅力。

It uses natural flowers and leaf grasses as raw materials, through careful artistic processing, the changes of lines and shapes into emotions and thinking, to create a new image of flowers. In the tea banquet, flower arranging is not only a kind of decoration, but also the embodiment of tea culture and tea banquet spirit. It pursues an artistic style of advocating nature, returning to nature, simple and elegant, aiming to further sublimate the atmosphere of the tea banquet through the agility and beauty of flowers, so that tea drinkers can feel the unique charm of tea culture in peace and harmony.

（五）焚香仪式 (Incense Burning Ceremony)

焚香仪式是一种古老而高雅的艺术形式，它源于对自然界中动物和植物所提取的天然香料的精心加工。经过精妙的处理，这些香料散发出各种不同的芬芳。人们在不同的场合焚香，旨在追求嗅觉上的极致享受。在我国，焚香与点茶、插花、挂画一同被誉为茶文化的"四艺"。在茶席中，焚香不仅作为一种艺术形式融入其中，其散发出的迷人香气更是弥漫在茶席四周的空间，为品茶者带来嗅觉上的愉悦与舒适，使人们在品茗的同时，也能享受到精神上的宁静与和谐。

Incense burning ceremony is an ancient and elegant art form, which originates from the careful processing of natural spices extracted from animals and plants in nature. After delicate treatment, these spices emit a variety of different fragrances, and people burn incense on different occasions, aiming to pursue the ultimate enjoyment of smell. In China, burning incense, dian cha, arranging flowers, and hanging pictures are known as the "four arts" of tea culture. In the tea banquet, incense burning is not only integrated into it as an art form, but also diffuses the charming aroma in the space around the tea banquet, bringing olfactory pleasure and comfort to tea drinkers, so that people can enjoy spiritual peace and harmony while drinking tea.

（六）挂画艺术（Hanging Pictures Art）

挂画艺术也称为挂轴，是茶席中不可或缺的一部分，主要指悬挂在茶席背景环境中，用以装饰和营造氛围的书法和绘画作品。其中，书法作品以汉字书法为主，绘画作品则多为中国画，这两者的结合，为茶席增添了深厚的文化底蕴和艺术气息。

Painting art also known as the hanging axis, is an indispensable part of the tea banquet, mainly refers to the calligraphy and painting works hanging in the background environment of the tea banquet to decorate and create the atmosphere. Among them, the calligraphy works are mainly Chinese calligraphy, while the painting works are mostly Chinese painting. The combination of the two adds profound cultural heritage and artistic flavor of the tea banquet.

（七）配套工艺品（Supporting Crafts）

配套工艺品在茶席设计中扮演着不可或缺的辅助角色。这些精心挑选的工艺品能够有效地衬托和强化茶席的主题，甚至在特定条件下，还能进一步深化茶席所要传达的意境。虽然数量上无需过多，但它们的摆放位置却至关重要，通常被巧妙地安置在茶席的侧边、下方或背景处，默默地为茶席的主体元素增添光彩，共同营造出一种和谐而富有层次感的艺术氛围。

Supporting crafts plays an indispensable auxiliary role in the design of tea banquets. These carefully selected handicrafts can effectively set off and strengthen the theme of the tea banquet, and even further deepen the artistic conception of the tea banquet under certain conditions. Although the number does not need to be too much, their placement is crucial, and they are usually cleverly placed on the side, below or in the background of the tea banquets, silently adding luster to the main elements of the tea banquets, and jointly creating a harmonious and hierarchical artistic atmosphere.

（八）茶食（Tea Snacks）

"茶食"涵盖了饮茶时与之相配的各种小食，如茶点、茶果等。这些茶食通常具备几个显著特点：分量轻盈、体积小巧、制作精细且样式雅致。在挑选时，应充分考虑不同的茶品、季节变换、茶席的主题以及品茶者的身份等因素，以确保茶食与整体品茶体验相得益彰，营造出更加和谐的品茶氛围。

This term covers the various snacks that go with drinking tea, such as tea, tea fruits and tea food.

These tea snacks usually have several distinctive characteristics: light weight, small size, delicate workmanship and elegant style. In the selection, factors such as different tea products, seasonal changes, the theme of the tea banquet and the identity of the tea drinker should be fully considered to ensure that the tea snacks and the overall tea tasting experience complement each other, and create a more harmonious tea tasting atmosphere.

（九）背景设置（Background Setting）

背景设置是为了确保观赏者能够准确地理解和感受到茶席主题所要传达的思想及特定的视觉效果而精心设计的艺术形态。背景设置主要分为室外背景设置和室内背景设置两种形式，它们作为茶席的衬托，旨在增强茶席的艺术效果和主题深度。

Background setting is a carefully designed art form to ensure that viewers can accurately understand and feel the ideas and specific visual effects to be conveyed by the theme of the tea banquet. The setting of background is mainly divided into outdoor background and indoor background, which serve as a foil to the tea banquet, aiming to enhance the artistic effect and theme depth of the tea banquet.

（十）主题设定（Theme Setting）

在茶席设计作品中，主题设定是至关重要的一环。它是对茶席设计所蕴含思想的提炼与概括，可以在设计过程开始前明确，也可以在创作过程中逐步明确。通过围绕特定主题进行创作，茶席设计作品能够更有效地传达其思想内涵和艺术特色，引发观者的深思与共鸣，从而获得更为丰富的艺术体验与感悟。

Theme setting is a crucial part of tea banquet design works. It is a distillation and summary of the ideas contained in the design of the tea banquet, which can be clarified before the design process begins, or it can be gradually clarified during the creation process. By creating around a specific theme, the tea banquet design works can more effectively convey its ideological connotation and artistic characteristics, and trigger the viewer's reflection and resonance, so as to obtain a richer artistic experience and sentiment.

二、创意茶席设计题材选择（Creative Tea Mat Design Theme Selection）

（一）茶品的选择（The Choice of Tea Products）

茶品的选择是茶席设计中的重要环节。不同产地、不同制作方法的茶品展现出各自独特的特性与特征。这些差异性不仅丰富了茶席设计的题材选择，也体现了茶文化的多样性。对于茶席设计而言，深入了解不同地域的茶文化，以及茶产地的自然景观、人文风情、风俗习惯、制茶手艺、饮茶方式、品茗意趣、茶典志录等，都是不可或缺的素材和灵感来源。这些元素不仅为茶席设计提供了丰富的题材，还能使茶席设计更具深度和内涵，更好地传承和弘扬茶文化。

The selection of tea products is an important link in the design of tea banquets, The tea products of different origins and different production methods will show their unique characteristics and characteristics. These differences not only enrich the theme selection of tea banquet design, but also reflect the diversity of tea culture. For tea banquet design, an in-depth understanding of the tea culture, as well as the natural landscape, cultural customs, customs, tea-making technology, tea-drinking methods, tea taste, tea ceremony records, etc. are indispensable materials and sources of inspiration. These elements not only provide rich themes for tea banquet design, but also make tea banquet design more profound and connotation, and better inherit and carry forward tea culture.

（二）茶事的选择（The Choice of Tea Events）

生活与历史事件历来都是艺术作品不可或缺的表现主题。在茶文化的发展历程中，那些具有特别影响的事件、重大的茶文化历史事件，以及茶席设计者基于自身审美感知对某些茶事的理解和解读，都可以作为茶席设计的选题内容。这些茶事不仅反映了茶文化的历史变迁和发展脉络，也为茶席设计提供了丰富的素材和灵感来源。茶席设计者可以通过深入挖掘这些茶事背后的文化内涵和艺术价值，将其融入茶席设计中，使茶席作品更具历史深度和文化底蕴。

Life and historical events have always been indispensable themes in artistic works. In the course of the development of tea culture, those events with special influence, major historical events of tea culture, and tea table designers' understanding and interpretation of certain tea things based on their own aesthetic perception can be selected as the content of tea banquet design. These tea events not only reflect the historical changes and development of tea culture, but also provide a rich source of material and inspiration for tea banquet design. Tea banquet designers can deeply explore the cultural connotation and artistic value behind these tea events, and integrate them into the tea banquet design, so that the tea banquet works have more historical depth and cultural heritage.

（三）茶人的融入（The Integration of Teaman）

在茶文化的广阔领域中，那些热爱茶、对茶有突出贡献并以茶品作为自己品质象征的人，我们称之为"茶人"。茶席设计可将茶人作为重要的题材，通过茶席的布局、元素选择等手法，深入展现茶人的精神风貌、光辉事迹及其独特的贡献，以此彰显茶人行事之高雅。茶人的融入，不仅丰富了茶席设计的文化内涵，也为观赏者提供更多了解茶文化、感受茶人精神的机会。

In the broad field of tea culture, those who love tea, have outstanding contributions to tea and take tea as a symbol of their quality, we call " teaman ". The design of tea banquet can take teaman as an important subject. Through the layout of tea banquet, the selection of elements and other techniques, the spiritual style, glorious deeds and unique contributions of teaman can be deeply displayed, so as to highlight the elegance. The integration of tea people not only enriches the cultural connotation of tea banquet design, but also provides viewers with more opportunities to understand tea culture and feel the spirit of tea people.

三、创意茶席设计题材表现形式（Creative Tea Mat Design Theme Expression）

在茶席设计中，以茶人、茶事、茶品为题材的茶席作品，其表现形式通常分为具象的物态语言和抽象的感觉语言两种。具象的物态语言主要通过具体的、可感知的茶具、茶器、装饰物等元素，直观地展现茶人、茶事、茶品的形象与特征。这些元素的选择与搭配，能够直接反映茶席的主题和内涵，使观赏者能够直观地感受到茶文化的魅力和茶席设计的艺术价值。抽象的感觉语言则更注重通过色彩、线条、形态等视觉元素，以及氛围的营造和情感的表达，传递茶人、茶事、茶品所蕴含的精神和情感。这种表现形式更加注重观赏者的心理感受和情感共鸣，能够引导观赏者深入体会茶文化的深邃内涵和茶席设计的独特魅力。

In the design of tea banquet, the tea banquet works with teaman, tea events and tea products as the theme are usually divided into two forms of expression: figurative material language and abstract feeling language. The figurative state language mainly displays the images and characteristics of teaman, tea events and tea products intuitively through specific and perceptible elements such as tea sets, tea utensils and decorative objects. The selection and collocation of these elements can directly reflect the theme and connotation of the tea banquet, so that the viewer can intuitively feel the charm of tea culture and the artistic value of the tea banquet design. Abstract sensory language pays more attention to convey the spirit and emotion contained in teaman, tea events and tea products through visual elements such as color, line and form, as well as the creation of atmosphere and the expression of emotion. This form of expression pays more attention to the viewer's psychological feelings and emotional resonance, and can guide the viewer to deeply understand the profound connotation of tea culture and the unique charm of tea banquet design.

在茶席设计中，具象的表现形式通过对物态形式的精准把握体现主题。当需要表现特定人物时，设计师需精心挑选能够准确反映该人物特性的特殊物品或象征物。同样，在表现历史事件时，设计师应仔细挑选能够典型反映该事件特殊性质的物品及象征物。以宋代斗茶与分茶为例，设计师应选用反映当时时代风貌的茶具，如宋代的建盏等，以还原和表现当时斗茶与分茶的盛况，使观赏者能够直观地感受到历史文化的魅力。

In the design of tea banquet, the figurative expression reflects the theme through the accurate grasp of the physical state form. When it comes to representing a particular person, designers need to carefully select special objects or symbols that accurately reflect that person's characteristics. Similarly, when representing historical events, designers should carefully select objects and symbols that typically reflect the special nature of the event. Taking tea fighting and Fencha in Song Dynasty as an example, designers should choose tea sets that reflect the style of the time, such as Jian Zhan in Song Dynasty, etc., in order to restore and express the grand situation of tea fighting and Fencha at that time, so that viewers can intuitively feel the charm of history and culture.

抽象的表现形式，在于通过人的视觉、听觉、嗅觉、味觉、触觉及心理感受，对某一事物形成深刻印象后，选择并运用那些最能反映这种印象与感觉的形态具体体现。例如，若要表现"快乐"这一主题，设计师可以运用音乐中跳跃的节奏和欢快的旋律，结合茶席中色彩明快、充满活力的器物以及自由洒脱的摆置结构，共同营造一种愉悦、轻松的氛围，使观赏者能够深刻感受到"快乐"这一情感。这种抽象的表现方式，不仅丰富了茶席设计的艺术手法，也提升了茶席作品的艺术感染力和表现力。

The abstract form of expression is that after forming a deep impression on something through people's vision, hearing, smell, taste, touch and psychological feelings, they choose and use those forms that can best reflect this impression and feeling to embody it concretely. For example, in order to express the theme of happiness, the designer can use the jumping rhythm and cheerful melody in the music, combined with the bright colors and energetic objects in the tea table and the free and unrestrained arrangement structure to create a pleasant and relaxed atmosphere, so that the audience can deeply feel the emotion of happiness. This abstract expression not only enrich the artistic technique of tea banquet design, but also enhances the artistic appeal and expression of tea banquet works.

总之，在茶席设计中，需要充分且巧妙地运用各个设计元素，确保它们之间和谐共生、相互映衬，共同营造一种高雅的艺术氛围。同时，还应注重调动茶席客人的审美情趣和参与性，通过他们的感官体验与心理共鸣，进一步提升茶席主题的表现张力。只有如此，才能打造出完整且富有表现力的茶席主题表现形式，让茶席文化得以充分展现和传承。

In a word, in the design of tea banquet, we need to fully and skillfully use various design elements to ensure that they coexist harmoniously and set off each other, and jointly create an elegant artistic atmosphere. At the same time, we should also pay attention to mobilizing the tea guests' aesthetic taste and participation, through their sensory experience and psychological resonance, to further enhance the performance tension of the tea table theme. Only in this way can we create a complete and expressive expression of the theme of the tea banquet, so that the tea banquet culture can be fully displayed and passed on.

赛证直通 Competitions and Certificates

一、基础知识部分（Basic Knowledge Part）

二维码 2-2-4:
知识拓展

（一）专业名词解释（Explanation of Professional Terms）

1. 茶品（Types of tea）：
2. 茶之"四艺"（"Four arts" of tea）：

（二）思考题（Thinking Question）

如何做好创意茶席设计？

How to do a creative tea table design?

二、技能操作部分（Skill Operation Part）

以表现贵州绿茶的品质特性为主题进行茶席设计。

The tea banquet design was carried out with the theme of showing the quality characteristics of Guizhou green tea.

【单元二 反思与评价】Unit 2 Reflection and Evaluation

学会了：_____

Learned to: _____

成功实践了：_____

Successful practice: _____

最大的收获：_____

The biggest gain: _____

遇到的困难：_____

Difficulties encountered: _____

对教师的建议：_____

Suggestions for teachers: _____

单元三 中国茶艺表演

活水还须活火烹，自临钓石取深清。

大瓢贮月归春瓮，小杓分江入夜瓶。

——北宋·苏轼《汲江煎茶》节选

单元导入 Unit Introduction

茶艺是饮茶活动中特有的文化现象，它包括茶叶品评技法和艺术操作手段的鉴赏及对品茗美好环境的领略等。茶艺表演是在茶艺的基础上产生的，通过各种茶叶冲泡技艺的形象演示，科学、生活化、艺术地展示泡饮过程，使人们在精心营造的优雅环境氛围中，得到美的享受和情操的熏陶。这个单元将带你见识绿茶、红茶、乌龙茶的茶艺表演。

Tea art is a special cultural phenomenon in tea drinking activities. It includes the appreciation of tea-tasting techniques and artistic operation means and the appreciation of the beautiful environment for tea-drinking. Tea art performance is derived from the foundation of tea art. It demonstrates the process of brewing various types of tea through a visual presentation of brewing skills, showcasing the process in a scientific, life-like, and artistic manner. This allows people to enjoy the beauty and cultivate their sentiments in a meticulously created elegant environment. This unit will take you through different types of tea art performances.

🎯 学习目标 Learning Objectives

➤ 知识目标（Knowledge Objectives）

（1）掌握绿茶茶艺表演步骤。

Master the steps of green tea art performance.

（2）掌握红茶茶艺表演步骤。

Master the steps of black tea art performance.

（3）掌握乌龙茶茶艺表演步骤。

Master the steps of oolong tea art performance.

➤ 能力目标（Ability Objectives）

（1）能进行绿茶茶艺表演。

Capable of performing green tea art performances.

（2）能进行红茶茶艺表演。

Capable of performing black tea art performances.

（3）能进行乌龙茶茶艺表演。

Capable of performing oolong tea art performances.

（4）能在茶艺表演中进行创意设计。

Able to incorporate creative design in tea art performances.

➤ 素质目标（Quality Objectives）

（1）领略茶艺中的中国传统美学观点。

Appreciate the Chinese traditional aesthetic perspectives in the art of tea.

（2）领略茶事技艺所蕴含的精神寄托。

Comprehend the spiritual sustenance embodied in the skills of tea ceremony.

（3）理解中国人"天人合一"的哲学观念。

Understand the Chinese philosophical concept of "harmony between humanity and nature".

专题一
碧波浮翠：绿茶茶艺表演

【微课】
碧波浮翠：
绿茶茶艺表演

你能说出图 2-3-1 中茶具的名字吗？

_____

Can you identify the name of the tea set in the picture (Figure2-3-1)?

_____

图 2-3-1　茶具（tea set）

一、茶艺的内容（Content of Tea Art）

人们在饮茶过程中追求享受，对水、茶、器具、环境都有较高的要求，同时以茶培养、修炼自己的精神道德。在各种茶事活动中协调人际关系，求得自己思想的自信、自省，也沟通彼此的情感，以茶雅志，以茶会友。茶本身存在着一种从形式到内容，从物质到精神，从人与物的直接关系到成为人际关系的媒介，逐渐形成传统东方文化瑰宝——中国茶文化。

In the process of tea drinking, people seek enjoyment and have high demands on water, tea, utensils, and the environment; they also cultivate and refine their spiritual morals through tea. In various tea events, they coordinate interpersonal relationships, gain self-confidence and self-reflection, communicate emotions, express themselves and elegantly through tea to make friends. The tea itself embodies a transformation from form to content, from material to spirit, from the direct relationship between humans and objects to becoming a medium of interpersonal relations, gradually evolving into a treasured aspect of traditional Eastern culture—Chinese tea culture.

就形式而言，茶艺包括选茗、择水、烹茶技术、茶具艺术、环境的创设等内容。背景中景物的形状，色彩的基调，书法、绘画和音乐的形式及内容，都是茶艺风格形成的影响因子。茶艺是形式和精神的结合，其中包含着美学观点和人的精神寄托。因此，茶艺既包含着我国古代朴素的辩证唯物主义思想，又包含了人们主观的审美情趣和精神寄托。

Formally, tea art includes selecting tea, choosing water, brewing techniques, the art of teaware, and the creation of the environment. The shapes of objects in the background, the basic tones of colors, the forms and contents of calligraphy and painting, and music are all factors influencing the formation of tea art style. Tea art is the combination of form and spirit, which contains aesthetic views and the spiritual aspirations of people. Therefore, in the tea art, it encompasses both the ancient Chinese dialectical materialist thought and people's subjective aesthetic taste and spiritual aspirations.

茶艺主要包括以下内容：

Tea art mainly includes the following contents:

第一，茶叶的基本知识。学习茶艺，首先要了解和掌握茶叶的分类、主要名茶的品质特点和制作工艺，以及茶叶的鉴别、贮藏、选购等。这是学习茶艺的基础。

Firstly, the basic knowledge of tea. To learn tea art, first understand and master the classification of tea, the quality characteristics of major famous teas and the production process, as well as the identification, storage and purchase of tea. This is the foundation of learning tea art.

第二，茶艺的技术。这是指茶艺的技巧和工艺，包括茶艺术表演的程序、动作要领、讲解的内容，茶叶色、香、味、形的欣赏，茶具的欣赏与收藏等。这是茶艺的核心部分。

Secondly, the technology of tea art. It refers to the skills and techniques of tea art, including the procedures of tea art performances, the key points of actions, the content of explanations, the appreciation of tea in terms of color, aroma, taste, and shape, and the appreciation and collection of teaware. This is the core part of tea art.

第三，茶艺的礼仪。这是指服务过程中的礼貌和礼节，包括服务过程中的仪容仪表、迎来送往、互相交流与彼此沟通的要求与技巧等。

Thirdly, the etiquette of tea art. It refers to politeness and etiquette in the service process, including the appearance and demeanor in the service process, welcoming and seeing off, the requirements and skills of mutual communication and understanding.

第四，茶艺的规范。茶艺要真正体现出茶人之间平等互敬的精神，因此对宾客都有规范的要求。作为客人，要以茶人的精神与品质要求自己，投入地品赏茶。作为服务者，也要符合待客之道，尤其是茶艺馆，其服务规范是决定服务质量和服务水平的一个重要因素。

Fourthly, the norms of tea art. Tea art should truly reflect the spirit of equality and mutual respect among teaman, so there are normative requirements for guests. As a guest, one should embody the spirit and qualities of a teaman and immerse oneself in appreciating tea. As a service provider, one should also conform to the etiquette of hospitality, especially in a teahouse, where service norms are a crucial factor determining the quality and level of service.

第五，悟道。道是指一种修行，一种生活的道路和方向，是人生的哲学，属于精神内容。悟道是茶艺的一种最高境界，通过泡茶与品茶感悟生活、感悟人生，探寻生命的意义。

Fifthly, the path to enlightenment (Wu Dao). Dao refers to a practice, a way of life and direction, the philosophy of life, and it belongs to the content of the spirit. Enlightenment is the highest state of tea art, achieved through brewing and tasting tea to perceive life, understand life, and explore the meaning of existence.

二、绿茶茶艺表演步骤（Steps of Green Tea Art Performance）

绿茶，作为中国受欢迎的茶类之一，有着悠久的历史和丰富的品种。绿茶以嫩叶为主要原料，经过采摘、杀青、揉捻、干燥等工艺制成，具有清爽、鲜爽、滋味爽口等特点。

Green tea, one of the most popular tea categories in China, boasts a long history and a rich variety. Green tea is made primarily from tender leaves through processes such as plucking, fixing, rolling, and drying, characterized by its refreshing, brisk, and smooth flavors.

（一）备具（Preparing the Utensils）

泡茶前，准备好茶叶和茶具。观察干茶的外形、嫩度、色泽和匀净度，根据个人喜好进行选择。茶具对于绿茶茶艺的呈现也非常重要，人们常说，"水为茶之母，器为茶之父"。名优绿茶通常需要用透明度较高的茶具来展现其美感，如玻璃杯、水晶杯等。

Before brewing, prepare the tea leaves and utensils. Observe the shape, tenderness, color, and evenness of the dry tea and choose according to personal preference. The tea utensils are also crucial for the presentation of green tea art. It is often said that "water is the mother of tea and utensils are the father". High-quality green tea usually requires transparent utensils, such as glass or crystal cups, to showcase its beauty.

（二）烫杯（Warming the Cups）

绿茶鲜嫩，对水温要求比较高，一般在 80 ℃左右。泡茶前需要预热茶壶和茶杯。预热的水可以倒入茶盘中，以便清洗茶具，确保杯子干净且温度适宜。

Green tea is delicate and requires a relatively high water temperature, usually around 80 ℃.

Prior to brewing, preheat the teapot and cups. The preheating water can be poured into a tea tray for cleaning the utensils, ensuring they are clean and at the right temperature.

（三）投茶（Adding Tea）

根据茶具大小和个人喜好，将适量绿茶倒入茶杯。一般而言，绿茶的茶水比常为 1∶50。用玻璃杯冲泡绿茶时，投茶可分为上投法、中投法、下投法。上投法是先投水后投茶，中投法是先投水再投茶再投水，下投法是先投茶后投水。也可以根据季节选择冲泡方法，有"夏上投、秋中投、冬下投"的说法。

Pour an appropriate amount of green tea into the cup according to the utensil size and personal preference. Generally, the tea-to-water ratio for green tea is 1∶50. When brewing in a glass cup, tea can be added using the top-pour, middle-pour, or bottom-pour method. The top-pour method involves adding water first then tea, the middle-pour method adds water, then tea, then more water, and the bottom-pour method adds tea first then water. The method can also be chosen according to the season, with sayings like "top-pour in summer, middle-pour in autumn, bottom-pour in winter".

（四）洗茶（Washing the Tea）

第一次冲泡时，注水后稍等片刻，常为三秒，然后将茶汤倒掉，这一步称为洗茶，有助于唤醒茶叶并去除杂质。

During the first brewing, pour water and wait for a moment, usually three seconds, then discard the liquid. This step, known as washing the tea, helps awaken the tea and remove impurities.

（五）冲泡（Brewing）

洗茶之后，再次注入热水，冲水至杯的七分满，用玻璃杯冲泡可以看到茶在水中缓缓舒展、流动、变换的美妙景象。把握好冲泡水温，名优绿茶芽叶较为细嫩，水温应在 70～80℃；普通绿茶水温掌握在 85～90℃ 即可，第一泡浸泡时间在 30 秒至 1 分钟，可根据个人口感喜好进行调整。

After washing, pour in hot water again, filling the cup to about 70% full. Brewing in a glass cup allows you to witness the wonderful sight of the tea slowly unfurling, flowing, and transforming in the water. Control the brewing temperature; high-quality green tea with tender leaves requires water between 70℃ and 80℃; regular green tea can be brewed at 85℃ to 90℃. The first infusion time is typically between 30 seconds and 1 minute, adjustable according to personal taste.

（六）奉茶（Serving the Tea）

将泡好的茶汤缓缓倒入准备好的杯中，倒七分满即可，然后细细品饮。品饮绿茶，可以先赏形、观色，透过晶莹清亮的茶汤观赏茶的沉浮和舒展，茶汤应以浅绿色、浅黄绿色为佳；然后端起茶杯，闻茶香清新淡雅；最后再细细品味，小口品尝，含在口中，感受茶汤在口舌间生出的清鲜甘甜。

Slowly pour the brewed tea into the prepared cup to about 70% full, then savor it. When drinking green tea, first appreciate its form and color, observing the tea leaves ups and downs and unfolding through the clear tea soup, which should ideally be light green or light yellow-green. Then, lift the cup, smell the fresh and subtle fragrance of the tea, and finally, savor it by sipping, letting the tea soup linger on the tongue to experience its fresh sweetness.

（七）续泡（Reinfusing）

第一泡后的茶叶还可以再泡 2 ～ 3 次，每次泡茶的时间可以逐渐延长，口感也会逐渐变化，可以根据自己的口感进行调整。

The tea leaves after the first brewing can be re-brewed 2-3 more times, with the brewing time gradually increasing and the taste evolving each time. Adjust according to your preferred tea strength.

（八）洁具（Cleaning the Utensils）

泡完茶后，要及时清洗茶具。将茶壶、茶杯等放入茶盘中，用热水清洗干净，然后用茶巾擦干。清洗茶具的目的是保持茶具的清洁卫生，以便下次使用。

After tea drinking, promptly clean the tea utensils. Place the teapot, cups, etc., in a tea tray, clean with hot water, and then dry with a tea towel. The purpose of cleaning the utensils is to maintain hygiene for the next use.

三、绿茶茶艺表演展示（Demonstration of Green Tea Art Performance）

（一）冰心去尘凡（Clean in Addition to Dirt）

茶是至清至洁，天涵地育的灵物，泡茶要求所用的器皿也需至清至洁。用开水再烫洗一遍茶杯，做到一尘不染。

Tea is a divine and pure creation, nurtured by heaven and earth. The utensils used to brewing tea also tend to be pure and spotless. Re-rinse the tea cups with boiling water to ensure they are completely free from any contamination.

（二）玉壶养太和（Teapot Keep Water）

绿茶属芽茶类，茶芽细嫩，若用滚烫的开水直接冲泡，则会破坏茶中的维生素并造成熟汤失味。将水温降至 80 ℃左右再进行冲泡，才能恰到好处，让茶色、香、味俱佳。

Green tea, being a bud tea, features delicate buds. If brewed directly with boiling water, it would destroy the vitamins in the tea and result in a spoiled taste. Reducing the water temperature to around 80℃ before brewing allows for the perfect balance, ensuring the tea's color, aroma, and taste are at their best.

（三）清宫迎佳人（Pure Palace Welcomes Beauties）

苏东坡有诗云："戏作小诗君勿笑，从来佳茗似佳人"。他把优质的茶比喻成让人一见倾心的绝代佳人。"清宫迎佳人"即用茶匙将茶叶倾置入晶莹剔透的玻璃杯中。

Su Dongpo once wrote, " Please don't laugh at my little poem, for a fine tea is like a beauty." He metaphorically compared high-quality tea to an enchanting beauty that captivates at first sight. " pure palace welcomes beauties" refers to using a tea spoon to transfer the tea leaves into crystal clear and pristine glass cups.

（四）甘露润莲心（Sweet Dew Moistens the Lotus Heart）

好的绿茶外观嫩如莲心，清代乾隆皇帝曾把茶叶称为"润心莲"。"甘露润莲心"即是在开泡前向杯中注入少许热水，起到润茶的作用。

Good green tea appears tender like a lotus heart. Emperor Qianlong of the Qing Dynasty once referred to tea as the "moist heart lotus". The phrase "sweet dew moistens the lotus heart" refers to pouring a little hot water into the cup before brewing to moisten the tea.

（五）凤凰三点头（The Phoenix Three Nods）

冲泡绿茶时讲究高冲水。在冲水时水壶有节奏地三起三落，犹如凤凰向各位嘉宾点头致意。

When brewing green tea, it is important to pour the water from a high position. As the water is poured, the kettle rhythmically rises and falls three times, resembling a phoenix nodding in greeting to the guests.

（六）碧玉沉清江（Emerald Jade Sinks into the Clear River）

冲入热水后，绿茶先是浮在水面，而后慢慢沉入杯底，将这一景象称之为"碧玉沉清江"。

After pouring in hot water, the green tea initially floats on the surface of the water before slowly sinking to the bottom of the cup. This scene is referred to as "emerald jade sinking into the clear river."

（七）观音捧玉瓶（Guanyin Holds the Jade Flask）

佛教故事中传说观世音菩萨常捧着一个白玉净瓶，净瓶中的甘露可消灾祛病，救苦救难。现将泡好的茶敬奉给各位，意在"祝福好人一生平安"。

In Buddhist stories, it is said that the Bodhisattva Guanyin often holds a white jade flask, from which the sweet dew inside can alleviate suffering, cure diseases, and rescue those in distress. Now, the brewed tea is presented to you, symbolizing "a wish for peace and blessings throughout your life".

（八）碧波展旗枪（Unfurling Flags on Green Waves）

品绿茶要一看、二闻、三品味。杯中的热水如春波荡漾，在热水的浸泡下，茶芽慢慢地舒展开，千姿百态的茶芽在杯中随波晃动，栩栩如生，宛如春兰初绽，又似有生命的绿精灵在舞蹈，十分生动有趣。

To appreciate green tea, one must observe, smell, and taste. The hot water in the cup ripples like a spring wave. Under the immersion of the hot water, the tea buds gradually unfold, displaying a myriad of postures. These tea buds sway in the cup, life-like, as if spring orchids are just blooming, or like green sprites dancing with vitality, making it incredibly vivid and interesting.

（九）慧心悟茶香（Wisdom in Perceiving Tea Fragrance）

绿茶的茶香清幽淡雅，需要用心感悟，才能闻到绿茶带有的春天气息，以及清纯悠远难以言传的生命之香。

The fragrance of green tea is delicate and subtle, requiring one's heart to perceive. Only then can one smell the spring essence carried by the green tea, along with its pure and distant, almost indescribable life-like fragrance.

（十）品茗乐无穷（Endless Joy in Tea Tasting）

绿茶的茶汤清纯甘鲜，淡而有味。只要用心品味，就一定能从这淡淡的绿茶汤中品出天地间至清、至醇、至真、至美的韵味。

The tea soup of green tea is pure, sweet, and subtly flavorful. With a thoughtful approach, one can certainly taste the utmost purity, mellowness, authenticity, and beauty from this simple green tea soup.

品茶有三乐：一曰"独品得神"，一个人面对着青山绿水或置身于一个高雅的茶室，通过品茗，心驰宏宇，神交自然，物我两忘，此为一乐；二曰"对品得趣"，即两个知心的朋友相对品茗，无需多言即心有灵犀一点通，或推心置腹倾诉衷肠，此亦一乐也；三曰"众品得慧"众人相聚品茗，相互沟通，相互启迪，可以学到书本中所学不到的知识，这同样是一大乐事。在品完头道茶后，邀请来宾亲自实践，从茶事活动中感受修身养性，品味人生的无穷乐趣。

Tea tasting brings three joys: first, "tasting alone enlivens the spirit." Facing the green hills and waters or being in an elegant tea room, through tea tasting, one's heart soars and communes with nature, reaching a state of forgetting oneself and the world, which is one joy. Second, "tasting together sparks interest." That is, two close friends tasting tea together, no words are needed as they connect heart to heart, or reveal their innermost thoughts, which is another joy. Third, "tasting in a group enlightens wisdom." Gathering to taste tea, communicating and inspiring each other, one can learn knowledge that cannot be found in books, which is also a great joy. After tasting the first infusion of tea, invite guests to engage in the tea ceremony themselves, experiencing the endless joy of self-cultivation and life reflection through the tea activities.

赛证直通 Competitions and Certificates

一、基础知识部分（Basic Knowledge Part）

思考题（Thinking Question）

绿茶玻璃杯冲泡法有哪些投茶方式？

What are the green tea pouring methods in the glass brewing method?

二维码 2-3-1：
知识拓展

二、技能操作部分（Skill Operation Part）

进行绿茶玻璃杯冲泡茶艺表演实操。

Please do tea art performance of brewing green tea in a glass.

专题二
气韵流朱：红茶茶艺表演

你喝过奶茶吗？你知道大多数奶茶使用的是什么茶叶吗？

_____

Have you ever had milk tea? Do you know what kind of tea is used in most milk tea?

_____

一、红茶茶艺表演步骤（Steps of Red Tea Art Performance）

（一）备具（Preparing the Utensils）

红茶属于高香型茶叶，若要从香气、滋味方面品鉴茶汤，建议优先选择盖碗冲泡。盖碗具有快速注水、快速出汤的优点，更能客观真实地体现红茶的品质特点。

Red tea is a type of tea with a high aromatic profile. To truly appreciate its aroma and taste, it is recommended to use a gaiwan for brewing. The gaiwan allows for quick water infusion and fast pouring of the tea, which more objectively and truly reflects the quality characteristics of the red tea.

（二）烫杯（Warming the Cups）

将沸水倾入盖碗、茶壶、公道杯、闻香杯、品茗杯等茶具中进行冲洗，既是为了卫生清洁，同时也能给茶具预热，使得茶的味道更香。

Pour boiling water into the gaiwan, teapot, fair mug, fragrance-smelling cups, and tea-sipping cups to rinse them. This not only serves a hygienic purpose but also preheats the teaware, enhancing the fragrance of the tea.

（三）投茶（Adding Tea）

投茶也称"请茶"，即将茶叶投放到茶具中。该过程比较简单，但名字中的"请"字体现了茶道中对客人的尊重。

Adding tea also known as "qiang cha", this involves putting tea leaves into teaware. Although this process is simple, the term "qing" reflects the respect for guests in the tea ceremony.

（四）洗茶（Washing the Tea）

将沸水倒入茶具中，让水和茶叶适当接触，然后又迅速倒出，有助于激发茶香并去除茶叶表面的杂质。

Pour boiling water into teaware, allowing the water and tea leaves to interact briefly before quickly pouring it out. This helps to stimulate the tea's aroma and remove impurities from the surface of the tea leaves.

（五）冲泡（Brewing）

红茶在冲泡的时候，采用定点高冲的手法，有利于茶汤的香气更好地被激发出来。注完水即可以出汤，出汤快有利于茶汤的香气更高扬。

When brewing red tea, use a high-angle pouring technique at a fixed point, which helps to bring out the aroma of the tea soup more effectively. After pouring in the water, the tea can be poured out immediately; quick pouring enhances the high-pitched aroma of the tea soup.

（六）拂盖（Skimming）

冲水时，沸水要高出壶口，用壶盖拂去漂浮的茶沫儿，防止客人喝到，影响口感。

When pouring water, the boiling water should rise above the spout and use the lid to skim off the floating tea foam, preventing guests from drinking it and affecting the taste.

（七）封壶（Sealing the Pot）

盖上壶盖以保存茶壶里茶叶冲泡出来的香气；用沸水遍浇壶身也是这个目的。

Cover the pot to preserve the aroma of the tea leaves that have been brewed inside, and pouring boiling water over the entire pot serves the same purpose.

（八）回壶（Returning the Pot）

轻轻将壶中茶水倒入公道杯，使每个人都能品到色、香、味一致的茶。这个步骤也是体现茶道"公正、公平、公道"的文化。

Gently pour the tea from the pot into the fair mug, ensuring that everyone can taste the tea with consistent color, aroma, and taste. This step also reflects the culture of tea ceremony, which emphasizes "fairness, equality, and justice".

（九）奉茶（Serving the Tea）

把茶汤均匀地倒入每个客人的闻香杯中，一般而言，斟茶只需七分满，之后双手将杯子递到客人面前，以茶奉客。

Pour the tea evenly into each guest's fragrance-smelling cup. Generally, the tea should be filled to about 70% full. Then, with both hands, present the cups to the guests as a gesture of serving tea.

（十）品茗（Tea Appreciation）

一闻香，二观色，三尝味。轻嗅闻香杯中的余香，闻香过后，观察茶汤的颜色和明亮度，然后便可品茗。品茗前，需用三指拿起品茗杯，轻啜慢饮。

First, smell the aroma, then observe the color, and finally taste the flavor. Lightly sniff the remaining fragrance in the fragrance-smelling cup. After appreciating the aroma, observe the color and brightness of the tea, and then it's time to taste the tea. Before tasting, use three fingers to pick up the tea cup and sip slowly.

（十一）洁具（Cleaning the Utensils）

泡完茶后，要及时清洗茶具。将茶壶、茶杯等放入茶盘中，用热水清洗干净，然后用茶巾擦干。清洗茶具的目的是保持茶具的清洁卫生，以便下次使用。

After tea is brewed, it is important to promptly clean the tea utensils. Place the teapot, cups, etc., on the tea tray and wash them clean with hot water, then dry them with a tea towel. The purpose of cleaning the utensils is to maintain their cleanliness and hygiene for the next use.

二、红茶茶艺表演展示（Demonstration of Black Tea Art Performance）

（一）备具列器（Preparing Utensils）

首先，将所需的器皿按要求陈列在相应的位置，为茶艺表演做好准备。

Firstly, the required utensils are arranged in their designated positions to prepare for the tea art performance.

（二）清泉初沸（Spring Water Begins to Boil）

将所备之水煮沸，准备用于冲泡茶叶。微沸的壶中上浮的水泡，仿佛"蟹眼"已生。

The prepared water is brought to a boil, ready for brewing the tea. The small bubbles that rise in the slightly boiling pot are reminiscent of "crab eyes" being born.

（三）温热壶盏（Warming the Pot and Cups）

用初沸之水，注入瓷壶及杯中，给壶、杯升温。

Using the freshly boiled water, pour it into the porcelain pot and cups to warm them.

（四）宝光初现（The Emergence of Treasured Light）

红茶外形条索紧秀，锋苗较好。其实红茶干茶的色泽并非人们通常认为的红色，而是呈现乌黑润泽，红茶干茶独特的颜色称为"宝光"。

The black tea has a tight and elegant texture, with good tips. The color of the dried black tea is not the red that people usually think of, but a glossy black. This unique color of black tea is referred to as "treasured light."

（五）王子入宫（Prince Enters the Palace）

红茶也被誉为"王子茶"。"王子入宫"指的是用茶匙将茶荷或赏茶盘中的红茶轻轻拨入壶中。

Black tea is also hailed as "Prince Tea". The term "Prince Enters the Palace" refers to gently scooping the black tea from the tea caddy or tea tray with a tea spoon and placing it into the pot.

（六）悬壶高冲（Pouring High from the Pot）

刚才初沸的水，此时已是"蟹眼已过鱼眼生"，处于完全沸腾的状态，正好用于冲泡。高悬壶，斜冲水，将开水徐徐注入盖碗或壶中，让茶叶在水的激荡下，充分浸润，以利于色、香、味的充分发挥，这是冲泡红茶最关键的一步。

The water, which has just started to boil, is now at the stage of "the crab eyes have passed and the fish eyes are emerging", indicating a full rolling boil, perfect for brewing. Hang the pot high and pour the water obliquely, slowly injecting the boiling water into the gaiwan or teapot. This allows the tea leaves to be fully immersed and stirred by the water, facilitating the full expression of color, aroma, and taste, which is the most crucial step in brewing black tea.

（七）平分秋色（Sharing the Autumn Colors）

浸润茶叶 2～3 分钟后，将红艳明亮的茶汤平分入品茗杯中。

After steeping the tea leaves for 2-3 minutes, divide the bright red and clear tea soup evenly into tea-sipping cups.

（八）敬献佳茗（Offering the Finest Tea）

双手将茶敬献给客人，请客人先闻一闻茶汤的香气，然后品一口茶汤，体验红茶的醇厚悠长。

Holding the tea with both hands, present it to the guests. Ask the guests to first smell the aroma of the tea soup, then take a sip to experience the rich and long-lasting flavor of the black tea.

（九）三品得趣（Experiencing Delights Through Three Times）

红茶通常可冲泡三次，三次的口感各不相同，细饮慢品，徐徐体味茶之真味。三泡之后，方得茶之真趣。

Black tea can typically be brewed three times, with each steep offering a different taste. Slowly sipping and savoring, one gradually discerns the true essence of the tea, and only after the third steeping can truly grasp the tea's true charm.

（十）收杯谢客（Closing Cups and Thanking Guests）

茶艺表演者以红茶为媒，将一腔独特内质、隽永情怀舒展开来，让饮者更能领略红茶的温润醇和，感受茶主彬彬有礼的茶礼茶习，让饮者真正喜茶乐茶。

The tea artist, using black tea as the medium, unfolds a unique inner quality and enduring emotions, allowing drinkers to better appreciate the gentle and mellow nature of black tea, and to feel the courteous tea etiquette. This allows drinkers to truly enjoy and appreciate tea, blending the guest-host experience with the fragrance and elegance of tea.

赛证直通 Competitions and Certificates

一、基础知识部分（Basic Knowledge Part）

思考题（Thinking Question）

红茶冲泡三要素指的是什么？

What are the three elements of black tea brewing?

二维码 2-3-2：
知识拓展

二、技能操作部分（Skill Operation Part）

请操作：红茶盖碗冲泡茶艺表演。

Please do this operation: tea art performance of brewing black tea in gaiwan.

【微课】
炉火纯青：乌
龙茶茶艺表演

专题三
炉火纯青：乌龙茶茶艺表演

图片中展示的是乌龙茶茶艺表演的哪道程式（图2-3-2）？

_____

Which routine of the oolong tea art show is shown in the picture (Figure 2-3-2)?

_____

图 2-3-2　茶艺（tea art）

在中国广东、福建、台湾等地区，品茗文化深厚，人们热衷于使用小杯细品茶汤，这种独特的品茗方式被称为"工夫茶"。工夫茶不仅仅是一种饮茶的方式，更是一种深厚的文化传统和茶艺的体现。在泡茶的过程中，每一步骤都充满了讲究和仪式感。

In China, Guangdong, Fujian, Taiwan and other regions, tea-tasting culture is profound, people are keen to use small cups of fine tea soup, this unique way of drinking tea is called "kung fu tea". Kung fu tea is not only a way of drinking tea, but also a profound cultural tradition and the embodiment of tea art. In the process of brewing tea, every step is full of exquisite and ritualistic sense.

一、乌龙茶茶艺表演步骤（Procedure of Oolong Tea Art Performance）

（一）备器（Preparing Utensils）

备器是饮茶前的重要步骤，涉及对茶壶、茶盘、茶杯等茶具的清洁和预热。通过沸水淋洗，旨在去除茶具的杂质和异味，提升茶具温度，并为后续的泡茶和品茶打下良好基础。

Preparing utensils is an important step before drinking tea, involving the cleaning and preheating of tea sets such as teapots, tea trays and cups. Through boiling water washing, it aims to remove impurities and odors of tea sets, improve the temperature of tea sets, and lay a good foundation for subsequent tea brewing and tea tasting.

（二）整形（Shaping）

整形是泡茶前的一个重要步骤，主要是针对乌龙茶。它涉及将茶叶倒入茶荷中，通过轻轻抖动使茶叶自然分散，形成层次分明的状态。随后，使用专用的竹匙将较粗的茶叶和细末部分分开，以确保茶叶的均匀分布和冲泡时茶汤口感的纯净。这一步骤不仅提升了泡茶效果，还体现了茶艺中对细节和品质的把控。

Shaping is an important step before brewing tea, mainly for oolong tea. It involves pouring tea leaves into a tea holder and dispersing tea leaves naturally by gently shaking them to form a well-defined state. Then, a special bamboo spoon is used to separate the thicker tea leaves from the fine parts to ensure an even distribution of tea leaves and a pure taste of the tea. This step not only improves the effect of tea brewing, it also reflects the attention to detail and quality in the tea art.

（三）置茶（Adding Tea）

置茶，又称为"乌龙入宫"，是茶艺中的关键步骤。将适量的乌龙茶叶放入茶壶中，确保茶叶在壶内均匀分布。这一步骤对于茶叶的展开和冲泡效果至关重要，是茶艺中不可或缺的一环，为后续的冲泡和品茗做好了充分的准备。

Adding tea, also known as "oolong into the palace", is a key step in the tea art. In this step, an appropriate amount of oolong tea leaves is placed into the teapot to ensure that the tea leaves are evenly distributed in the teapot. This step is crucial to the development and brewing effect of the tea, and is an indispensable part of the tea art, it is fully prepared for the subsequent brewing and tea tasting.

（四）冲水（Flushing）

冲水，在茶艺中被称为"悬壶高冲"，是泡茶过程中的重要步骤。执行此步骤时，需将盛水壶置于高处，沿茶壶边缘缓缓冲入热水。水流需保持稳定且缓慢，使茶叶在壶中翻滚、打转，形成旋转的圈。这一步骤有助于茶叶的充分展开和香气释放，同时使茶汤更加均匀，提升品茗体验。

Flushing, known as "hanging pot high flushing" in tea art, is an important step in the tea brewing process. To perform this step, place the kettle on a high position and slowly pour in the hot water along the edge of the teapot. The water flow needs to be steady and slow so that the tea leaves tumble and swirl in the pot, forming a rotating circle. This step helps to fully unfold the tea leaves and release the aroma, while making the tea soup more uniform and enhancing the tea-tasting experience.

（五）刮沫（Scraping）

刮沫，这一步骤在茶艺中通常被称为"春风拂面"。在执行刮沫操作时，首先需要将沸水冲入茶壶中，直至沸水满出茶壶，溢出壶口。此时，茶壶内的茶叶会因热水的冲击而翻滚，茶汤表面会浮起一层细腻的泡沫。接着，使用壶盖轻轻地在茶汤表面掠过，将浮沫轻轻刮去。这一步骤的目的在于去除茶汤表面的浮沫，使茶汤更加清澈，同时也有助于茶叶香气的散发。通过这一步骤，可以体验到茶艺中的细腻与精致，感受到茶艺文化的魅力。

Scraping is often referred to as "spring breeze brushing" in tea art. To perform the scraping operation, you first need to pour boiling water into the teapot until the boiling water fills the teapot and overflows the spout. At this time, the tea leaves in the teapot will roll due to the impact of hot water, and the surface of the tea soup will float a layer of delicate foam. Then, use the lid to gently brush over the surface of the tea, gently scraping the foam away. The purpose of this step is to remove the foam on the surface of the tea, make the tea more clear, and also help the aroma of the tea. Through this step, we can experience the delicacy and refinement of tea art and feel the charm of tea art culture.

（六）斟茶（Pouring Tea）

斟茶是茶艺中展示细致操作和品质控制的重要步骤。在斟茶时，采用低斟的方式，缓缓将茶汤注入杯中，以确保茶汤的均匀分配。通过轮流注入茶汤至各杯，先注入一半，再来回倾入，直至每杯茶汤达到七分满，这种操作手法被称为"关公巡城"，形象地展现了茶汤在杯中流转的过程。最后，将茶壶中剩余的几滴浓茶分别注入各杯，这一步骤被称为"韩信点兵"，既展现了茶艺的细致严谨，也确保了茶汤的充分利用。

Pouring tea is an important step in tea art to demonstrate meticulous manipulation and quality control. When pouring tea, the way of low pouring is used to slowly inject the tea soup into the cup to ensure the uniform distribution of the tea soup. By taking turns to inject tea soup to each cup, first inject half, and then pour back and forth until each cup of tea soup reaches 70% full, this operation is called "Duke Guan Patrols the City", which vividly shows the process of tea soup in the cup. Finally, the remaining drops of strong tea in the teapot are injected into each cup, a step known as "Han Xin Selects the Soldier", which shows the meticulous and rigorous tea art, but also ensures the full use of the tea soup.

（七）品茗（Tasting Tea）

品茗是茶艺中的核心环节，旨在深入体验茶的口感和香气。首先，通过双手轻轻搓动闻香杯，释放茶香，以鼻细嗅，品味茶的香气。接着，采用"三龙护鼎"的品茗方式，即用拇指、食指和中指夹住品茗杯，其余手指轻托杯底或自然伸展，形成优雅的兰花指姿态。在品茗过程中，建议分多次饮用，每次品尝都能更深入地感受茶中蕴含的各种滋味与高雅意趣，甚至感悟人生百味。这种品茶方式，不只是对茶这份饮品的珍视，更是一种对生活细致入微的感知与表达。

Tasting tea is the core part of tea art, designed to experience the taste and aroma of tea in depth. First, gently rub the fragrance-smelling cup with both hands to release the aroma of tea, and taste the smell of tea with your nose. Then, use the "three dragons to protect the tripod" way of tasting tea, that is, the thumb, index finger and middle finger hold the tea-slipping cup, and the rest of the fingers lightly support the bottom of the cup or naturally stretch to form orchid fingers posture. In the process of tasting tea, it is recommended to drink it several times, every time you taste tea, you can deeply feel the various tastes and elegant interests contained in tea, and even feel the taste of life. This way of tea tasting is not only the value of tea as a drink, but also a subtle perception and expression of life.

二、乌龙茶茶艺表演展示（Demonstration of Oolong Tea Art Performance）

（一）孔雀开屏（Peacock Spreading Its Tail）

向客人展示茶具（图 2-3-3），需具体介绍。

To show the tea set (Figure 2-3-3) to the guests, specific introduction is required.

图 2-3-3　茶具展示（show the tea set）

（二）嘉叶共赏（Jiaye Co-appreciation）

苏东坡在《叶嘉先生传》中以拟人的手法赞美了茶叶高尚的品德，书中所谓"嘉叶共赏"，意指赏茶（图 2-3-4）。图中展示的是铁观音，产自福建省安溪县，是乌龙茶的一种，是中国十大名茶之一。

Su Dongpo praised the noble virtue of tea with the personification in " Biography of Mr. Ye Jia ". Then so-called " Jiaye co-appreciation" means to admire tea (Figure 2-3-4). The picture shows Tie Guanyin Tea, produced in Anxi County, Fujian Province, is a kind of oolong tea and one of the top ten famous teas in China.

图 2-3-4　嘉叶共赏
（Jiaye Co-appreciation）

（三）高山流水（Lofty Mountain and Flowing Water）

用开水温壶烫盏，在乌龙茶茶艺中称为"高山流水"，目的是提升茶具的温度，使茶叶在冲泡时能更好地发挥色、香、味、型的特点（图 2-3-5）。

It is called " lofty mountain and flowing water " in the oolong tea art by warming the teapot with boiled water. Its purpose is to raise the temperature of the tea set and so that tea brewing can better play the color, aroma, taste, type characteristics (Figure 2-3-5).

图 2-3-5　高山流水
（Lofty Mountain and Flowing Water）

（四）乌龙入宫（Oolong Enters the Palace）

将紫砂壶比作宫殿，以此衬托铁观音茶的精美，将投茶的过程称为"乌龙入宫"（图 2-3-6）。

To compare the purple clay teapot to a palace, so as to serve as a foil to the delicacy of Tie Guanyin tea. The process of throwing tea is called " oolong entering the palace " (Figure 2-3-6).

图 2-3-6　乌龙入宫
（Oolong Enters the Palace）

（五）悬壶高冲（Highly Hanging Teapot to Flush）

使茶叶充分翻动（图2-3-7）。冲泡铁观音讲究高冲水、低斟茶，逆时针旋转冲泡茶叶。

To make the tea leaves turn fully (Figure 2-3-7). When brewing Tie Guanyin tea, strive for high water filling and low tea pouring, and brewing tea leaves by rotating counterclockwise.

图 2-3-7　悬壶高冲
（Highly Hanging Teapot to Flush）

（六）春风拂面（A Sping Breeze Stroking the Face）

用壶盖轻轻刮去壶口处的茶沫（图2-3-8）。

Use the lid to gently scrape off the tea foam spout (Figure 2-3-8).

（七）乌龙入海（Oolong Enters the Sea）

品工夫茶讲究"头泡水，二泡茶，三泡四泡是精华"。头泡冲出的茶水一般不喝，注入茶海，因茶汤呈琥珀色，从壶口流向茶海好似蛟龙入海，故称为"乌龙入海"（图2-3-9）。

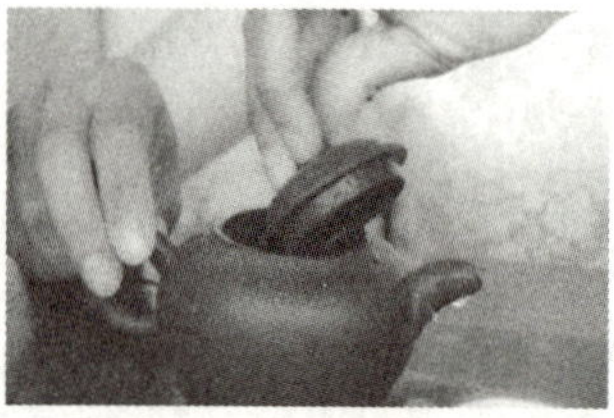

图 2-3-8　春风拂面
（A Spring Breeze Stroking the Face）

Kungfu tea-tasting strives for the principle of "first soaking in water, second soaking in tea, third soaking and fourth soaking is the essence". The tea water rushed out of the first soaking is generally not drunk and poured into the tea beach. As the tea soup is amber in color, it flows from the spout to the tea beach just like a dragon into the sea. Therefore, it is called "oolong enters the sea" (Figure 2-3-9).

图 2-3-9　乌龙入海
（Oolong Enters the Sea）

（八）若琛听泉（Ruochen Listening to the Spring）

若琛是清代的制杯大师，后人将小的瓷质品茗杯称作"若琛杯"。"听泉"，意在赏水（图2-3-10）。

Ruochen was a cup-making master in the Qing Dynasty. Later generations called the small porcelain sample tea cup "Ruochen Cup". "Listen to the spring" is intended to admire the water (Figure 2-3-10).

图 2-3-10　若琛听泉
（Ruochen Listening to the Spring）

（九）关公巡城，韩信点兵（The Duke Guan Patrols the City, Han Xin Selects the Soldiers）

向杯中循环分茶，七分满即可，称为"关公巡城"；以点斟的手法向杯中倒茶，意在向嘉宾点头致意，称为"韩信点兵"（图2-3-11）。手法：需低、快、匀、尽。茶汤均匀，充分体现了"和"的中国茶道精神。

Circularly pour the tea soup into the cup until it is 70% full, which is called "Duke Guan patrolling the city". Pour tea into the cup by spot-pouring, which is intended to nod to the guests. It is called "Han Xin selecting soldiers" (Figure 2-3-11). Manipulation: It needs to be low, fast, even and thorough. Even tea soup fully embodies the spirit of "harmony" in Chinese tea ceremony.

图2-3-11　关公巡城
（**The Duke Guan Patrols the City**）

（十）龙凤呈祥，斗转星移（Prosperity Brought by the Dragon and the Phoenix, the Stars Change in Positions）

将品茗杯扣在闻香杯上，称"龙凤呈祥"；将两个杯子翻转过来称为"斗转星移"（图2-3-12）。

We fastened the tea-sipping cup to the fragrance-smelling cup and called it "prosperity brought by the dragon and the phoenix". Turn the two cups upside down and called it "the stars change in positions" (Figure 2-3-12).

图2-3-12　斗转星移
（**Stars change in positions**）

（十一）空谷幽兰，三龙护鼎（Secluded Orchid in the Hollow Valley, three Dragons Guarding Triod）

轻轻旋转取出闻香杯，揉搓杯身闻茶香（图2-3-13）。拿杯姿势：拇指、食指夹杯沿，中指托住杯底，三手指压为龙，杯如鼎，称为"三龙护鼎"。女士可以翘起兰花指表示雅观，男士则手指往里收，表示稳重。

Gently rotate and take out the fragrance-smelling cup and rub the cup body and smell the tea aroma (Figure 2-3-13). Cup-holding posture: use the thumb, index finger to clamp the cup edge, and use the middle finger to hold the bottom of the cup, and three fingers are like dragons, while a cup is like a tripod. It is called "Three Dragons Guarding Tripod". Ladies can show elegance by raising their orchid fingers like this, while men put their fingers inward to show stability.

图2-3-13　空谷幽兰
（**Secluded Orchid in the Hollow Valley**）

（十二）共品佳茗（Taste Good Tea Together）

"品"字分为三个"口"，所以一杯茶分三口来饮用（图2-3-14）。

图2-3-14　共品佳茗
（**taste good tea together**）

The character "Pin" is divided into three characters "Kou", so a cup of tea is divided into three mouthfuls to drink (Figure 2–3–14).

赛证直通 Competitions and Certificates

一、基础知识部分（Basic Knowledge Part）

二维码 2-3-3：
知识拓展

（一）专业名词解释（Explanation of Professional Terms）

1. 关公巡城（Duke Guan Patrols the City）：
2. 三龙护鼎（Three Dragons Protect the Tripod）：

（二）思考题（Thinking Question）

乌龙茶茶艺表演中"关公巡城、韩信点兵"蕴含的中国茶道精神是什么？

What is the Chinese tea ceremony spirit contained in the "Guan Gong touring the city and Han Xin ordering soldiers" in the Oolong tea art performance?

二、技能操作部分（Skill Operation Part）

练习乌龙茶茶艺表演。

Practice the oolong tea art show.

【单元三 反思与评价】Unit 3 Reflection and Evaluation

学会了：_____

Learned to: _____

成功实践了：_____

Successful practice: _____

最大的收获：_____

The biggest gain：_____

遇到的困难：_____

Difficulties encountered: _____

对教师的建议：_____

Suggestions for teachers：_____

单元四　茶的品鉴及茶文化

小楼一夜听春雨，深巷明朝卖杏花。

矮纸斜行闲作草，晴窗细乳戏分茶。

——宋·陆游《临安春雨初霁》节选

单元导入 Unit Introduction

茶，是中国的传统饮品，承载着深厚的文化底蕴。一杯好茶，犹如一段故事，等待着人们细细品味。在这个单元，我们将探索茶的品鉴之道，学习如何欣赏茶的色泽、品味茶的香气、感受茶的口感以及茶带给人们的宁静与愉悦。同时也深入了解茶文化的丰富内涵，从茶的起源、发展，到茶与礼仪的完美结合，感受茶文化的博大精深。

Tea is a traditional Chinese drink, carrying profound cultural connotations. A good cup of tea, like a story, waiting for people to savor. In this unit, we will explore the way of tea tasting, learn how to appreciate the color of tea, taste the aroma of tea, feel the taste of tea, the peace and pleasure tea brings us. At the same time, we will also have an in-depth understanding of the rich connotation of tea culture, from the origin and development of tea to the perfect combination of tea and etiquette, and feel the breadth and depth of tea culture.

学习目标 Learning Objectives

➢ 知识目标（Knowledge Objectives）

（1）掌握茶品鉴的方法。

Master the principle of tea tasting.

（2）掌握茶品鉴的相关术语。

Terminology related to tea tasting.

➢ 能力目标（Ability Objectives）

（1）能根据所学知识品鉴不同类型的茶。

Can taste different types of tea according to the knowledge.

（2）掌握茶礼仪。

Master tea etiquette.

➤ 素质目标（Quality Objectives）

（1）在品鉴茶的过程中感受中华茶文化的博大精深。

In the process of tasting tea, feel the broad and profound Chinese tea culture.

（2）培养学生的文化自觉和文化自信，增强对中华优秀传统文化的热爱。

Cultivate students' cultural consciousness and self-confidence, and enhance their love for excellent traditional Chinese culture.

专题一

沁人心脾：茶的品鉴

你知道铁观音最主要的香气特征有哪些吗（图2-4-1）？

_____

Do you know what are the main aroma characteristics of Tieguanyin (Figure 2-4-1)?

【微课】
沁人心脾：
茶的品鉴

饮茶，其精髓在于"品"。所谓的"品茶"，不仅是对茶叶品质的鉴别与赏析，同时也蕴含了对茶文化的深刻感受与理解。在繁忙的生活节奏中，冲泡一壶茶，选择一个静谧雅致的环境，悠然自得地斟饮，不仅能有效缓解疲劳、清除杂念、增强思维的条理性，有效地提升人的精神活力。

图 2-4-1　铁观音（Tie Guanyin）

Drinking tea, its essence lies in "tasting". The so-called "tea tasting" is not only the identification and appreciation of tea quality, at the same time, it also contains a deep feeling and understanding of tea culture. In the busy pace of life, brew a pot of tea, choose a quiet and elegant environment, and drink leisurely, which can not only effectively relieve fatigue, clear distractions, enhance the organization of thinking and effectively enhance people's spiritual vitality.

一、品鉴主要内容（The Main Contents of the Tasting）

通常可以从多个方面评判茶叶的品质：首先是观察茶叶的外观，其次是品闻茶叶的香气，接着是品尝茶汤的味道，最后是检查冲泡后茶渣的质地。通过这些步骤，可以更全面地评估茶叶的优劣。

Usually, we can judge the quality of tea from a number of aspects: first by observing the appearance of the tea, next by smelling the tea, then by tasting the taste of the tea, and finally by distinguishing the texture of the tea leaves after brewing. Through these steps, we can evaluate the merits and demerits of tea comprehensively.

（一）观茶（Watching Tea）

查看茶叶的过程包括观察干茶以及茶叶冲泡后的形态变化。其中，"干茶"指的是未经冲泡的茶叶状态，而"茶叶开汤"是指把干茶用开水冲泡后，让茶叶蕴含的物质充分释放出来。通过观察这两个阶段的茶叶形态，对茶叶的整体品质有一个初步了解。

The process of examining tea leaves includes observing the morphology of dried tea and tea leaves after brewing. Among them, " dry tea " refers to the state of unbrewed tea, the " tea opening soup " refers to the dry tea with boiling water after brewing, so that the substances contained in the tea are fully released. By observing the shape of tea in these two stages, we can have a preliminary understanding of the overall quality of tea.

在观察干茶时，首先要关注其干燥程度。若发现茶叶有些回软，建议不要购买。其次，应仔细检查茶叶叶片的清洁度。若茶叶中含有过多的叶梗、黄片、渣沫或其他杂质，则可能品质不佳，至少不属于上等茶叶。此外，观察干茶的条索外形也是重要的一环。条索即茶叶经过揉捻后形成的形态，各类茶叶都有其独特的形态规格。例如，龙井茶呈扁平状，冻顶乌龙茶被揉成半球形，铁观音茶紧结成球状。然而，仅凭观察干茶，只能对其品质作出有限的判断，大概能掌握30%的茶叶情况，无法直接确定其是否为优质茶叶。

When looking at dry tea, we should first pay attention to its degree of dryness. If the tea is found to be soft, it is recommended to avoid buying. Secondly, the cleanliness of the tea leaves should be carefully checked. If the tea contains too many leaf stalks, yellow leaves, residue or other impurities, it may be of poor quality, at least not belong to the first-class tea. In addition, observing the shape of the dry tea rope is also an important part. The rope is the shape of tea after rolling, and all kinds of tea have their unique shape specifications. For example, Longjing tea is flat in shape, Dongding oolong tea is kneaded into a semi-spherical shape, Tieguanyin tea is tightly formed into a ball. However, just by looking at dry tea, we can only make a limited judgment about its quality, we know about 30% of the tea, can not directly determine whether it is high-quality tea.

（二）察色（Observe the Color）

品茶观色，即观茶色、汤色和底色。

The color of tea tasting, that is, the color of tea, soup and infused leaves.

1. 茶色（Tea Color）

茶叶基于颜色可分为绿茶、黄茶、白茶、青茶、红茶和黑茶六大类。这些分类主要基于其干茶状态时的色泽差异，如红与绿、青与黄、白与黑的区别。即使是同种茶叶，其色泽也会因茶树品种、生长环境和采摘季节的不同而有所变化。例如，高档绿茶色泽多样；红茶色泽红艳明亮或乌润显红；乌龙茶如武夷岩茶、铁观音、凤凰水仙和冻顶乌龙等，则各自拥有独特的色泽特征，这些特征是判断乌龙茶品质优劣的关键。

Tea can be divided into six categories based on color: green tea, yellow tea, white tea, green tea, black tea and black tea. These classifications are mainly based on the color differences of its dry tea state, such as the difference between red and green, green and yellow, white and black. Even the same tea, its color will vary depending on the variety of tea tree, growing environment and picking season.

For example, high-grade green tea has various colors, black tea can be bright red or dark red, and oolong teas such as Wuyi Rock tea, Tie Guanyin, Fenghuang Shui Hsien and Dongding Oolong have their own unique color characteristics, which are the key to judge the quality of oolong tea.

2. 汤色（Soup Color）

冲泡茶叶后，茶叶内的成分在沸水中溶解形成的颜色被称为"汤色"。不同茶类的汤色存在显著差异，即使是同一茶类中的不同品种或级别，其汤色也会有所不同。一般来说，优质茶叶的汤色明亮且具有光泽。绿茶的汤色通常为浅绿或黄绿色，清澈明亮；红茶的汤色橙红明亮，若边缘带有金黄色油环，则品质上乘；乌龙茶的汤色以清澈明亮为优质；白茶的汤色则微微发黄，黄中带绿，有光泽。这些汤色特征是判断茶叶品质的重要依据。

After brewing tea leaves, the color formed by the dissolution of the ingredients in tea in boiling water is called " soup color ". There are significant differences in the color of different teas, even in different varieties or grades of the same tea, its soup color will be different. Generally speaking, the soup color of high-quality tea is bright and shiny. Green tea soup color is usually light green or yellow-green, clear and bright; the soup color of black tea is orange-red bright, if the edge with golden oil rings, it is good quality; the soup color of oolong tea is clear and bright for high quality; white tea is slightly yellow, yellow with green, shiny. These characteristics are important basis for judging the quality of tea.

观察茶汤的过程需迅速且及时，因为茶多酚类物质在热水中溶解后与空气接触会迅速氧化变色。绿茶的汤色氧化后会变黄，红茶的则会变暗。长时间放置，茶汤可能变得混浊并出现沉淀物。红茶在温度降至20℃以下时，可能出现"冷后浑"现象，即凝乳状混浊，这是色素与咖啡碱结合产生的。若"冷后浑"现象出现早且呈浅褐色，表明红茶茶味浓郁；若呈暗褐色，则茶味平淡。随温度降低，茶汤颜色逐渐加深。在相同条件下，红茶、大叶种茶、嫩茶和新茶的汤色变化更为显著。为了准确观察茶汤原色，建议在冲泡后10分钟内观察，并确保比较的茶叶属于同种类别。茶汤的颜色因发酵程度和焙火轻重而异，但无论深浅，都应保持清澈透明。浑浊或灰暗都不是好茶汤的表现，清澈透明是评判茶汤质量的基本标准。

The process of observing the tea soup needs to be quick and timely, because the tea polyphenols will be in contact with the air after being dissolved in hot water and will be rapidly oxidized and discolored. The soup color of green tea will turn yellow after oxidation, and black tea will darken. Left for a long time, the tea may become cloudy and precipitate. When the temperature of black tea drops below 20℃, it may appear " cream down " phenomenon, that is, curdy turbidity, which is produced by the combination of pigment and caffeine. If the phenomenon of " cream down " is early and the soup color is light brown, it indicates that the black tea has a rich taste; if it is dark brown, the tea tastes bland. As the temperature decreases, the color of the tea gradually darkens. Under the same conditions, the color changes of black tea, large leaf tea, tender tea and new tea were more significant. In order to accurately observe the primary color of the tea soup, it is recommended to observe it within 10 minutes after brewing, and ensure that the compared tea leaves belong to the same category.The color of the tea soup varies according to the degree of fermentation and the severity of roasting, but it should be clear and transparent regardless of its depth. Cloudy or gray is not a good performance of tea, clear and transparent is the basic standard to judge the quality of tea.

3. 叶底（The Color of Infused Leaf）

所谓"叶底"，即指茶叶冲泡后去掉茶汤所留下的茶渣。在品鉴过程中，除了观察叶底所显现的颜色外，还应当注意其老嫩程度、表面的光滑或粗糙感，以及整体的匀净度等特征。这些方面的综合考量，有助于更全面、准确地评估茶叶的品质。

The so-called "infused leaf" refers to the bottom of the tea leaves left by the tea soup after tea brewing. In the process of tasting, in addition to observing the color of the infused leaf, we should also pay attention to its aging degree, smoothness or roughness of the surface, and the overall evenness and other characteristics. The comprehensive consideration of these aspects helps us to evaluate the quality of tea more comprehensively and accurately.

（三）赏姿（Admiring Posture）

在茶叶冲泡过程中，茶叶吸水展开，呈现出千姿百态。有的如春笋挺拔，有的似雀舌灵巧，有的宛若优雅兰花，有的仿佛墨菊深邃。茶叶的舒展还伴随着动态的视觉效果，如太平猴魁活泼翻跃，君山银针笔直挺立，西湖龙井生机盎然。这些美景在清澈的茶水中映衬，令人陶醉，仿佛茶未醉人，人已自醉。

In the process of tea brewing, tea leaves absorb water and unfold, showing a variety of positions. Some are tall and straight like bamboo shoots, some are dexterous like tongue of sparrow, some are elegant like orchids, and some are deep like blackish chrysanthemum. The stretching of tea leaves is also accompanied by dynamic visual effects, such as Taiping Houkui Tea lively somersaulting, Junshan silver needle standing upright, West Lake Longjing full of vitality. These beautiful scenery set off in the clear tea, intoxicating, as if the tea is not intoxicating, people have drunk themselves.

（四）闻香（Smell the Fragrance）

鉴赏茶香包括三个步骤：首先是"干闻"，即闻干茶的香气；接着是"热闻"，即在茶叶冲泡后闻其充分释放的本香；最后是"冷闻"，即在茶汤稍冷后闻香，以评估香气的持久度。这三个步骤有助于全面品鉴和评估茶叶的香气品质。

Appreciation of tea aroma consists of three steps: first, "dry sniffing", that is, smell the aroma of dry tea; followed by the "hot sniffing", that is, after the tea brewing smell of its full release of the original fragrance; finally, there is "cold sniffing", that is, sniffing the tea after it is slightly cold to assess the durability of the aroma. These three steps help to fully appreciate and evaluate the aroma quality of tea.

"干闻"是品鉴茶叶香气的重要步骤。通过闻干茶，可以初步判断茶叶的香气品质。不同茶叶具有不同的香气特征，如绿茶的清新爽口、红茶的浓烈纯正、花茶的芬芳扑鼻、乌龙茶的浓郁甘润。若茶叶香气低沉或伴有焦、烟、酸、霉、陈等不良气味，则品质可能不佳。在"干闻"时，可将少量干茶置于器皿中或抓一把茶叶放在手心，仔细嗅闻其香气，并留意是否有异味或杂味，以全面评估茶叶的香气品质。

"Dry sniffing" is an important step to appreciate the aroma of tea. By smelling dried tea, we can judge the aroma quality of tea. Different tea leaves have different aroma characteristics, such as the fresh and refreshing green tea, the strong and pure black tea, the fragrant scented tea, rich and sweet. If the tea aroma is low or accompanied by coke, smoke, acid, mold, stale and other bad odors, the quality may be poor. In the "dry sniffing", you can put a small amount of dry tea in a vessel or grab a handful of tea leaves in the palm, carefully smell its aroma, and pay attention to whether there is any odor or miscellaneous taste, in order to comprehensively evaluate the aroma quality of tea.

湿闻法，即在茶叶冲泡后，浸泡 1 ～ 3 分钟，将茶杯送至鼻端嗅闻茶汤表面散发的香气。若使用带盖茶杯，可闻盖香和茶汤香；若用闻香杯，还可闻杯香和茶汤香。闻香过程分为"热闻""温闻"和"冷闻"三个阶段："热闻"主要辨别香气是否正常，香气类型及强弱；"温闻"鉴别茶香的香型，即品质优劣；"冷闻"判断香气持久性。

Wet sniffing method, that is, after the tea is brewed and soaked for 1 to 3 minutes, and bring the tea cup to your nose and smell the aroma of the surface of the tea soup. If use a cup with a lid, you can smell the fragrance of the lid and tea. If use a fragrance-smelling cup, you can also smell the cup and tea. The smelling process is divided into three stages: " hot sniffing " " warm sniffing " and " cold smell ". " Hot sniffing " mainly identifies whether the aroma is normal, type and strength; " warm sniffing " to identify tea flavor type, that is, quality is good or bad; " cold sniffing " to determine the durability of the aroma.

"热闻"的操作主要包括三种方法：从升腾的水汽中品闻香气、闻取杯盖上的残留香气、使用闻香杯嗅闻杯底留香。各种茶叶的香气特征各异，如安溪铁观音的天然花香、红茶的甜香和果味香、绿茶的清香以及花茶的复合花香。茶叶的香气品质与原料鲜嫩程度和加工技术水平密切相关，鲜嫩茶叶含有更丰富的芳香物质，香气更浓郁。"冷闻"则在茶汤冷却后进行，有助于辨识被掩盖的其他气味。

There are three main methods of " hot sniffing ": tasting the aroma from rising water vapor, smelling the residual aroma on the lid of the cup, and smelling the bottom of the cup with the fragrance-smelling cup. The aroma characteristics of various teas are different, such as the natural fragrance of Anxi Tieguanyin, the sweet and fruity aroma of black tea, the fragrance of green tea and the complex fragrance of scented tea. The aroma quality of tea is closely related to the freshness and tenderness of raw materials and the level of processing technology. The fresh and tender tea contain more abundant aromatic substances and the aroma is more intense. " Cold sniffing " is done after the tea has cooled, helping to identify other odors that have been masked.

（五）尝味（Tasting）

尝味是品味茶汤口感的过程。茶汤的滋味由多种味道成分如甜、苦、涩、酸、鲜等共同构成。当这些味道成分的比例和含量达到平衡时，茶汤会显得鲜醇可口、回味悠长。优质的茶汤滋味通常表现为微苦中带有甘甜，口感甘醇浓稠，饮用后喉部会有持久的甘润与舒适感。绿茶的口感应当鲜醇且爽口，红茶的滋味则应浓厚、强烈且带有鲜爽感，而乌龙茶的醇厚与回甘则是其品质上乘的重要标志。由于舌头的不同部位对滋味的感受存在差异，因此，在品味茶汤时，应让茶汤在舌面上循环滚动，以便更准确、全面地体验和分辨茶味的细微差别。

Tasting is the process of tasting the taste of tea soup. The taste of tea soup is composed of a variety of taste components such as sweet, bitter, astringent, sour, and fresh. When the proportion and content of these flavor components are balanced, the tea soup will appear fresh and delicious, with a long aftertaste. The taste of high-quality tea soup is usually slightly bitter with sweet, the taste is mellow and thick, and the throat will have lasting sweetness and comfort after drinking. Because different parts of the tongue have different feelings of taste, therefore, when tasting tea, it should be allowed to circulate on the tongue surface to experience and distinguish the nuances of tea taste more accurately and comprehensively.

品味茶汤的最佳温度范围应保持在 40 ～ 50 ℃。过高的温度（超过 70 ℃）可能烫伤味觉器官，影响对茶汤滋味的正常评判；过低的温度（低于 30 ℃）则会降低味觉对茶汤滋味的敏感度，并可能导致茶汤口感不够醇厚。因此，为确保准确品味和评判茶汤，应将茶汤温度控制在适宜范围内。

The best temperature range for tasting tea should be kept between 40 ℃ and 50 ℃. Excessive temperature (more than 70 ℃) may burn the taste organs, affecting the normal evaluation of the taste of tea; too low a temperature (below 30 ℃) will reduce the sensitivity of the palate to the taste of the tea, and may cause the taste of the tea is not mellow enough. Therefore, in order to ensure accurate taste and evaluation of tea soup, the temperature of tea soup should be controlled within this suitable range.

在品味茶汤的过程中，应关注其浓淡、强弱、爽涩、鲜滞及纯异等特质。为确保准确品味茶叶的本质味道，建议品茶前避免食用辣椒、葱、蒜、糖果等有强烈气味食物，同时避免吸烟，以保持味觉和嗅觉的敏感度。理想的茶汤应带来喉咙的软甜、甘滑感，且口齿留香，令人回味无穷。

In the process of tasting tea, we should pay attention to its denseness and lightness, strong and weak, cool and astringent, fresh and sluggish, and pure and different characteristics. In order to ensure the accurate taste of the essential taste of tea, it is recommended to avoid eating strong smell foods such as chili, onion and garlic, candy, and avoid smoking before tea tasting, so as to maintain the acuity of taste and smell. The ideal tea soup should bring a soft sweet throat, sweet and smooth feeling, the mouth remains fragrant, and endless aftertaste.

二、各类茶的品鉴（Tasting All Kinds of Tea）

（一）高级细嫩绿茶的品饮（Tasting of High-grade Delicate Green Tea）

高级细嫩绿茶在色泽、香气、口感和形态上均展现出独特的魅力，深受人们喜爱。在品茶过程中，首先应通过晶莹透亮的茶汤，欣赏茶叶在水中的沉浮与舒展，以及其优美的姿态。接着，观察茶汁如何浸出并渗透水中，同时注意汤色的变化。之后，端起茶杯，先闻其散发的香气，再轻呷一口，让茶汤含在口中，缓慢在口舌之间回旋品味。通过这样反复的品赏过程，可以更深入地体验高级细嫩绿茶带来的美妙感受。

High-grade delicate green tea has its unique charm in color, aroma, taste and form, and is loved by people. In the process of tea tasting, first through the crystal clear tea soup, appreciate the rise and fall of tea in the water and stretch, as well as its beautiful posture. Next, observing how the tea juice is leached and permeated into the water, and note the change in the color of the soup. After that, pick up the teacup, smell its aroma first, then sip, let the tea soup in the mouth, slowly rotate between the mouth and taste. Through this repeated appreciation process, we can more deeply experience the wonderful feelings brought by high-grade delicate green tea.

（二）乌龙茶的品饮（Tasting of Oolong Tea）

乌龙茶的品饮侧重于闻香和尝味，品形则不是重点。在不同地区，闻香和尝味的重视程度有所不同。例如，在潮汕地区，强调热品的方法。具体来说，就是将茶倒入杯中，然后用拇指和食指按住杯沿，中指抵住杯底，慢慢将杯子由远及近地移至嘴边，使杯沿接触嘴唇，杯面朝向鼻子，先闻其香气。接着，将茶汤含在口中回旋品味，再徐徐咽下，通常三小口即可饮尽杯中茶，之后再嗅闻杯中留存的茶香。

The tasting of oolong tea focuses on the smell and taste, but the shape is not the focus. In different regions, the importance attached to smell and taste varies. For example, in the Chaoshan area, emphasis is placed on the hot product approach. Specifically, the tea is poured into the cup, and then the thumb and index finger hold the rim of the cup, the middle finger against the bottom of the cup, slowly move the cup from far to close to the mouth, so that the rim of the cup touches the lips, the cup face the nose, smell the aroma first. Then, the tea soup is contained in the mouth, and then slowly swallowed, usually three small bites can drink the tea in the cup, and then smell the tea retained in the cup.

(三) 红茶品饮 (Tasting of Black Tea)

红茶被誉为"迷人的茶饮"，这不仅得益于其红艳油润的色泽和甘甜可口的口感，更因为红茶的品饮方式多样，除了可以清饮外，还可以根据个人口味进行调饮。例如，可以加入柠檬增加清爽口感，或者添加肉桂以丰富香气，亦或是加入砂糖增添甜味，甚至可以加入奶酪使其口感更加润滑。这种多样化的品饮方式，使红茶的魅力倍增。品饮红茶时，关键在于全面感受其香气、滋味与汤色。通常使用壶泡法，然后分倒入杯中。品饮时，先嗅香、再观色、后品味，需细心斟酌、细细品尝，以充分领略红茶的精髓与乐趣。

Black tea is known as "charming tea drink", which not only benefits from its red and oily color and sweet taste, but also because of the variety of ways to drink black tea, in addition to clear drinking, but also according to personal taste. For example, lemon can be added for a refreshing taste, or cinnamon for a rich aroma, or granulated sugar for a sweet taste, or even cheese for a smoother taste. This variety of drinking methods makes the charm of black tea double. When drinking black tea, the key is to fully appreciate its aroma, taste and color. It is usually soaked in a pot and then divided into cups. When drinking, first smell, then look at the color, after taste, need to carefully consider and savor, in order to fully appreciate the essence and fun of black tea.

(四) 细嫩白茶与黄茶品饮 (Tasting of Delicate White Tea and Yellow Tea)

白茶属于轻微发酵的茶类，其制作工艺一般包括将鲜叶萎凋后直接烘干，因此其茶汤颜色和口感相对较为清淡。而黄茶的特点在于其黄汤、黄叶，这主要是因为黄茶在制作过程中通常不进行揉捻，导致茶汁较难被浸出，从而形成了其独特的品质。

White tea is a slightly fermented tea, and its production process generally includes the direct drying of fresh leaves after withering, so its tea color and taste are relatively light. Yellow tea is characterized by its yellow soup and yellow leaves, mainly because yellow tea is usually not rolled during the production process, resulting in tea juice being difficult to be leached, thus forming its unique quality.

白茶和黄茶，特别是白毫银针和君山银针，因其独特的冲泡特点，成为以观赏为主的茶品。它们以淡雅茶香、清澈茶汤和微妙甘醇为特点，同时视觉上也极具魅力。品饮前，可欣赏其如银针般的干茶外形。冲泡时，推荐用直筒无花纹的玻璃杯和70℃热水，观看茶芽在杯中舞动并竖直于水中的美景。浸泡约10分钟后，茶香和汤色最佳，此时品尝其甘醇滋味。这类茶品注重观赏，品饮方法别具一格。

White tea and yellow tea, especially Backho silver needle and Junshan silver needle, have become ornamental tea products because of its unique brewing characteristics. They are characterized by light tea flavor, clear tea soup and subtle sweetness, while also being visually appealing. Before tasting, you can enjoy its dry tea shape like a silver needle. When brewing, it is recommended to use a straight cylinder unmarked glass and 70℃ hot water to watch the beauty of tea buds dancing in the cup and standing upright in the water. After about 10 minutes, the tea aroma and soup color is the best, at this time to taste its sweet taste. This kind of tea focuses on watching and drinking methods are unique.

赛证直通 Competitions and Certificates

一、基础知识部分（Basic Knowledge Part）

（一）专业名词解释（Explanation of Professional Terms）

1. 热闻（Hot sniffing）：
2. 冷闻（Cold sniffing）：

（二）思考题（Thinking Question）

如何做好茶的品鉴？

How to taste tea well?

二维码 2-4-1：
知识拓展

二、技能操作部分（Skill Operation Part）

品鉴贵州都匀毛尖并写出其品质特征。

Tasting Guizhou Duyun Maojian and writing its quality characteristics.

【微课】
知书达理：茶
礼仪与茶文化

专题二
知书达理：茶礼仪与茶文化

你知道在茶事活动中需要注意哪些礼仪吗（图 2-4-2）？

_____

Do you know what etiquette need to pay attention to during tea events (Figure 2-4-2)?

_____

用茶招待客人是中国数百年的传统待客之道，形成了独特的茶桌礼仪。这不仅是一种生活习惯，更是深厚的文化传统。主人

图 2-4-2 茶礼仪
（tea etiquette）

以茶示敬，洗风尘叙友情，象征着友谊交流、情感表达。茶文化与礼仪文化紧密相连，成为日常生活中高雅的礼节，体现了纯洁的美德。

Serving tea to guests is a traditional way of hospitality in China for hundreds of years, which has formed a unique tea table etiquette. This is not only a living habit, but also a profound cultural tradition. The host shows respect with tea and washes away the wind and dust, renew friendship, symbolizing friendship and communication, emotional expression. Tea culture is closely linked with etiquette culture, which has become an elegant etiquette in daily life and embodies the virtue of purity.

一、茶礼仪（Tea Etiquette）

（一）职业礼仪（Professional Etiquette）

1. 得体的着装（Appropriately Dress）

茶艺师的服装应与茶艺表演风格一致，选择简洁优雅的服饰，避免花哨与鲜艳。女性泡茶师应淡妆且着装得体，男性则应避免前卫或怪异的装扮。整体而言，泡茶师应保持整洁外表，行为得体，与环境和茶具相协调，展现文化素养与礼貌。

Tea art specialists' clothing should consistent with the style of tea art performance, choose simple and elegant clothing, avoid gaudy and bright. Female tea brewers should wear light makeup and dress appropriately, while men should avoid edgy or bizarre outfits. On the whole, tea art specialists should maintain a clean appearance, behave appropriately, coordinate with the environment and tea sets, and demonstrate cultural literacy and politeness.

2. 优美的手型（Nice Hands）

身为茶艺师，拥有一双修长而纤细的手是至关重要的。茶艺师应当重视手部的日常保养，确保其始终保持干净卫生。这样不仅有利于精湛茶艺的演绎，更是对茶艺工作的尊重与热爱。同时，不建议喷洒气味浓烈的香水，因为这不仅可能掩盖了茶叶本身的香气，还会影响客人品茶时的整体感受。

As a tea master, having a pair of long and tenuous hands is essential. Tea artists should pay attention to the daily maintenance of hands to ensure that they are always clean and hygienic. This is not only conducive to the exquisite interpretation of tea art, but also respect and love for tea art work. At the same time, it is not recommended to spray perfume with a strong smell, because it may not only cover up the aroma of the tea itself, but also affect the overall feeling of the tea taste.

3. 优雅的举止（Elegant Manners）

泡茶不仅彰显了茶艺师的高超技艺，还显露出其独特个性。在泡茶过程中，调整姿态与动作，能展现和提升个人气质。泡茶时，要注重动作的韵律之美，与客人保持自然流畅的交流，让泡茶变成一种将艺术与社交完美融合的优雅体验。这一过程既呈现了茶艺师的个性与品位，又加强了与客人的沟通，因此，优雅的举止和得体的言谈显得极为重要。

Making tea not only shows the superb skills of tea artisan, but also reveals its unique personality. In the process of making tea, adjusting posture and movement can show and enhance personal temperament. When making tea, we should pay attention to the beauty of the rhythm of the movement, maintain natural and smooth communication with the guests, and make tea into an elegant experience that perfectly integrates art and social interaction. This process not only shows the personality and taste of the tea master, but also strengthens the communication with the guests, so elegant manners and decent speech are extremely important.

（二）服务礼仪（Service Attitude）

1. 走姿（Walking Posture）

女性茶艺师在行走时应保持稳定直线步伐，上身稳定不摇摆，双肩放松，下颌微收，双眼平视，双手可交叉于胸前以保持优雅。男性茶艺师则可自然小幅摆动双臂。在接近客人时，茶艺师应稍倾身以示尊重。结束后，应面向客人后退两步并微微倾身转弯离开，展现对客人的敬意和礼貌。

Female tea art specialist should keep a steady straight pace when walking, the upper body is stable and not swaying, the shoulders are relaxed, the jaw is slightly closed, the eyes are flat, and the hands can be crossed in front of the chest to maintain elegance. Male tea art specialist can naturally swing their arms slightly. When approaching the guests, the tea art specialist should lean slightly to show respect. When the end is over, you should face the guest and take two steps back and turn slightly to leave, showing respect and courtesy to the guest.

2. 站姿（Standing Posture）

在泡茶过程中，若茶桌较高，茶艺师可选择站立冲泡以确保操作的顺畅。无论是坐下前还是冲泡时，茶艺师的站立姿势都至关重要，因为它将形成客人的"第一印象"，影响后续的交流和品茶体验。站立时，茶艺师应保持双腿并拢、身体笔直、双肩放松、双眼平视前方。女性茶艺师应双手交叉于身前，右手叠放左手之上，展现优雅气质；男性茶艺师则同样双手交叉，左手叠放右手之上，双脚微开呈外八字，保持稳定与力量。这些动作应自然流畅，避免生硬，以展现茶艺师的专业素养和优雅风度。

In the tea brewing process, if the tea table is higher, the tea art specialist can choose to stand brewing to ensure smooth operation. Whether it is before sitting down or when brewing, the standing posture of the tea art specialist is crucial, because it will form the " first impression " of the guest, affecting the subsequent communication and tea tasting experience. When standing, the tea art specialist should keep legs together, body straight, shoulders relaxed, and eyes straight ahead. The female tea art specialist should cross her hands in front of her body, with her right hand folded on top of her left hand, showing elegance; the male tea art specialise also crosses his hands, puts his left hand on top of his right hand, and opens his feet slightly to maintain stability and strength. These movements should be natural and smooth, avoid stiff, in order to show the professional quality and elegant demeanor of the tea art specialist.

3. 坐姿（Sitting Posture）

当茶艺师坐在椅子上时，应放松身心，稳坐椅子中央，确保重心平衡，保持整体稳定；双腿并拢，上身直立，以展现专业与沉静；头部微仰，下颌内收，鼻尖指向腹部。女性茶艺师可将双手重叠放在茶桌上，展现优雅；男性茶艺师则可以双手分开如肩宽，半握拳轻搭前方桌沿。这种摆放方式既不会显得过于拘谨，又能够随时准备进行泡茶操作，展现出茶艺师的稳重和专业。

When the tea art specialist sits on the chair, he should relax, sit firmly in the center of the chair, ensure the balance of the center of gravity, and maintain overall stability. Keep legs together, upper body upright to show professionalism and calmness. The head is slightly tilted, the jaw is drawn in, and the tip of the nose is pointed towards the abdomen. Female tea art specialist can overlap their hands on the tea table to show elegance; the male tea art specialist can spread his hands shoulder-width apart, with a half-clenched fist lightly resting on the edge of the front square table. This way of placing will not appear too stiff, but can be ready to make tea at any time, showing the steady and professional of tea art specialist.

（三）言谈礼仪（Speech Etiquette）

在茶艺活动中，茶艺师与客人初次见面时，应以自然大方且不失礼节的态度进行自我介绍，简要介绍自己的姓名和即将提供的服务。在泡茶前，茶艺师应简要地介绍茶叶的种类、文化背景、产地、品质特色及冲泡要点。泡茶过程中，茶艺师应对每个步骤进行简要的解说，尤其是有特定含义的操作，以增加茶艺的吸引力。当泡茶流程结束，茶艺师如需离开座位，应礼貌地征询客人意见，以确保茶艺体验的连贯性。总之，在茶艺操作的全程中，茶艺师需使用简洁明了、准确无误的语言，并保持亲切友善的语气，让客人能够深切领略饮茶的高雅韵味，尽情享受其中的美好。这种沟通方式不仅能提升茶艺活动的品质，更能彰显茶艺的文化底蕴和审美价值。

In the tea art activities, when the tea are specialist meets the guest for the first time, he should introduce himself in a natural and generous manner, and briefly introduce his name and the services to be provided. Before brewing tea, tea art specialist should briefly and accurately introduce the types, cultural background, origin, quality characteristics and brewing points of tea. During the tea brewing process, the tea art specialist should give a brief explanation of each step, especially operations that have a specific meaning, in order to increase the appeal of the tea art. When the tea brewing process is over, if the tea art specialist needs to leave the seat, he/she should politely consult the guests to ensure the continuity of the tea art experience. In short, in the whole process of tea art operation, tea art specialist need to use simple, clear and accurate language, and maintain a friendly and friendly tone, so that guests can deeply appreciate the elegant charm of drinking tea, enjoy the beauty of it. This way of communication can not only improve the quality of tea art activities, but also highlight the cultural heritage and aesthetic value of tea art.

二、茶文化（Tea Culture）

（一）茶文化核心要素（The Core Elements of Tea Culture）

"茶文化"是一个历史悠久且丰富的概念，它不仅仅涉及饮茶这一行为，还包括了与之相关的各种文化表现形式和社会活动。以下是茶文化的核心要素：

"Tea culture" is a concept with a long history and richness, which not only involves the behavior of drinking tea, but also includes various cultural expressions and social activities related to it. Here are the core elements of tea culture:

茶道：这是泡茶和饮茶的艺术，包括泡茶的技艺、品茶的礼仪以及与之相关的哲学思想。

Tea ceremony: it refers to the art of making and drinking tea, including the technology of making tea, the etiquette of tasting tea and the philosophy related to it.

茶德：这强调的是饮茶过程中应保持的道德品质，如诚信、和谐和尊重。

Tea virtues: this emphasized the moral qualities that should be maintained during tea drinking, such as integrity, harmony and respect.

茶精神：通常与茶文化中的宁静、清雅和内省有关，它体现了人们对美好生活的追求。

Tea spirit: usually associated with tranquility, elegance and introspection in tea culture, it reflects people's pursuit of a better life.

茶联、茶书、茶诗、茶画：这些是茶文化在文学和艺术领域的体现，通过文字和绘画传达对茶的热爱和赞美。

Tea couplets, tea books, tea poems, tea paintings: these are the embodiment of tea culture in the field of literature and art, conveying the love and praise for tea through words and painting.

茶学：这是对茶叶种植、加工、品鉴等方面的系统研究。

Tea science: this is the systematic study of tea planting, processing, tasting and other aspects.

茶故事：关于茶的历史轶事、传说或现代故事，反映了茶在人们生活中的地位和影响。

Tea stories: historical anecdotes, legends or modern stories about tea that reflect the place and influence of tea in people's lives.

茶艺：它是指泡茶和饮茶的技巧和艺术，包括茶具的选择、水温的控制、泡茶的动作等。

Tea art: it refers to the skills and art of brewing tea and drinking tea, including the choice of tea sets, the control of water temperature, the action of making tea, etc.

茶文化不仅是对茶的认识和饮用方式的具体表现，它还是一种生活方式，一种精神追求，以及一种文化的传承和发展。

Tea culture is not only the understanding of tea and the specific performance of drinking, it is also a way of life, a spiritual pursuit, and a cultural inheritance and development.

（二）历史上的重要茶文化事件（Important Tea Culture Events in History）

历史上的茶文化事件不胜枚举，其中有几个重要的节点值得一提。例如，唐代的"斗茶"活动，这是一种品评茶品质的社交活动，对茶艺的发展产生了深远影响。宋代的茶文化进一步发展，出现了茶馆和茶楼，成为人们社交和休闲的场所。明朝时期，废除了团茶的制作方法，改为制作散茶，这一变革使得茶叶的品种和风味更加多样化。而在近代，随着茶文化的国际化发展，国际茶博会、世界茶文化节等活动的举办，进一步促进了茶文化的交流和茶产业的发展。

There are numerous tea culture events in history, among which several important nodes are worth mentioning. For example, the "tea competition" in the Tang Dynasty, a social activity to evaluate the quality of tea, had a profound impact on the development of tea art. Tea culture in the Song Dynasty developed further, with the emergence of teashops and tea houses, which became places for people to socialize and relax. In the Ming Dynasty, the method of making group tea was abolished and loose tea was made, which made the varieties and flavors of tea more diversified. In modern times, with the international development of tea culture, the holding of international Tea Expo, World Tea Culture Festival and other activities has further promoted the exchange of tea culture and the development of tea industry.

（三）茶文化在现代社会的影响（The Influence of Tea Culture in Modern Society）

1. 健康生活方式的推广（Promotion of Healthy Lifestyle）

随着人们健康意识的提高，茶叶因其优异的抗氧化功能和其他有益健康的成分而被广泛推崇。研究表明，定期饮用茶叶可以帮助降低患心脏病和某些癌症的风险，同时对减肥和提高新陈代谢也有促进作用。因此，茶成了健康生活方式的一个重要组成部分，各种健康茶饮品牌和产品应运而生。

As people's health awareness increases, tea is widely admired for its excellent antioxidant function and other health-beneficial ingredients. Studies have shown that drinking tea regularly can help reduce the risk of heart disease and certain cancers, while also having a promoting effect on weight loss and improving metabolism. As a result, tea has become an important part of a healthy lifestyle, a variety of healthy tea brands and products have emerged.

2. 茶文化与旅游业的结合（The Combination of Tea Culture and Tourism）

茶文化的魅力吸引了众多旅游者前往各大产茶区和著名茶园进行体验式旅游。例如，中国的西湖龙井村、印度的大吉岭茶园等，都成了热门的文化旅游目的地。游客们不仅可以亲自体验采茶、制茶的乐趣，还能通过参观茶博物馆、参加茶艺表演等方式深入了解当地茶文化。

The charm of tea culture has attracted many tourists to the major tea producing areas and famous tea gardens for experiential tourism. For example, West Lake Longjing Village in China, Darjeeling tea gardens in India have all become popular cultural tourism destinations. Visitors can not only experience the fun of picking and making tea themselves, but also gain an in-depth understanding of local tea culture by visiting tea museums and participating in tea art performances.

3. 当代茶文化的国际交流（International Exchange of Contemporary Tea Culture）

随着经济全球化的发展，茶文化也在国际间进行广泛的交流与融合。世界各地的茶叶博览会、研讨会和文化节等活动促进了不同国家和地区之间的相互学习和交流。例如，国际茶叶委员会（ITC）定期举办的世界茶叶会议就是一个促进全球茶叶产业合作与发展的平台。此外，茶叶也成了一种重要的国际贸易商品，影响着全球经济格局。

With the development of economic globalization, tea culture is also widely exchanged and integrated in the international community. Events such as tea expos, seminars and cultural festivals around the world promote mutual learning and communication between different countries and regions. For example, the World Tea Conference organized regularly by the International Tea Council (ITC) is a platform to promote cooperation and development of the global tea industry. In addition, tea has become an important international trade commodity, affecting the global economic pattern.

总之，通过对茶文化在现代社会中影响的探讨，可以认识到茶不仅是一种饮品，更是一种文化现象，它在促进健康生活、推动旅游经济和加强国际交流方面发挥着重要作用。这些内容的学习有助于我们从更广阔的视角理解和评价茶文化的现代价值。

In short, through the discussion of the influence of tea culture in modern society, we can realize that tea is not only a drink, but also a cultural phenomenon, which plays an important role in promoting healthy life, promoting tourism economy and strengthening international exchanges. The study of these contents will help us to understand and evaluate the modern value of tea culture from a broader perspective.

赛证直通 Competitions and Certificates

一、基础知识部分（Basic Knowledge Part）

二维码 2-4-2：知识拓展

（一）专业名词解释（Explanation of Professional Terms）

1. 茶道（Tea Ceremony）：
2. 茶艺（Tea Art）：

（二）思考题（Thinking Question）

中国茶文化在世界传播的积极意义？

What are the etiquette requirements of tea art personnel?

二、技能操作部分（Skill Operation Part）

学生在实训室进行茶礼活动，让其分享感受和体验。

Students held tea ceremony activities in the training room, and share their feelings and experiences.

【单元四　反思与评价】Unit 4　Reflection and Evaluation

学会了：_____

Learned to: _____

成功实践了：_____

Successful practice: _____

最大的收获：_____

Difficulties encountered: _____

对教师的建议：_____

Suggestions for teachers：_____

模块三　咖啡及可可

单元一　咖　啡

只要世界上有咖啡，事情会有多糟？

——卡桑德拉·克莱尔（美国）

单元导入 Unit Introduction

咖啡在人类的历史发展及社会生活中，扮演着重要的角色，有着巨大的影响，也独具非凡魅力。那么，它究竟魅力何在？本单元将走进让人着迷的咖啡文化。

Coffee in the history of human development and social life, plays an important role, has a huge impact, but also unique charm. So what is its charm? In this unit, we will go into the fascinating coffee culture.

学习目标 Learning Objectives

➤ 知识目标（Knowledge Objectives）

（1）了解咖啡的起源。

To understand the origins of coffee.

（2）了解咖啡树和咖啡豆的结构。

To understand the structure of coffee plants and beans.

（3）掌握咖啡豆的种类。

To master the type of coffee bean.

（4）了解咖啡的主要成分。

To know the main ingredients of coffee.

（5）掌握正确的咖啡礼仪文化。

Master proper coffee etiquette and culture.

➤ 能力目标（Ability Objective）

能根据所学知识进行咖啡豆鉴别，能够掌握手冲咖啡的正确手法。

Can identify coffee beans according to the knowledge learned, can master the correct method of making pour-over coffee.

➢ **素质目标（Quality Objective）**

培养学生具备良好的职业素养、文化素养、工匠精神，以及学习能力。

Cultivate students with good professional quality, cultural quality, craftsman spirit, and learning ability.

专题一
大"咖"之谜：咖啡概述

【微课】
大"咖"之
谜：咖啡概述

一、咖啡的概念（The concept of coffee）

咖啡源于咖啡树，这是一种属茜草科的多年生常绿灌木或小乔木。咖啡树结的果实称为咖啡樱桃或咖啡浆果，包括红色果皮、薄层果肉、内果皮及果仁。这果仁就是咖啡豆。咖啡，正是用烘焙成熟的咖啡豆磨粉后制作而成的饮料。它不仅具有提神醒脑的功效，味道还醇香浓郁，深受全球人们的喜爱，与可可、茶并列为世界三大饮料

Coffee is derived from the coffee tree, which is a perennial evergreen shrub or small tree in the rubiaceae family. The fruit of the coffee tree, called the coffee cherry or coffee berry, consists of a red skin, a thin layer of flesh, an inner peel, and a kernel. The kernel is the coffee bean. Coffee is a beverage made from roasted and matured coffee beans. It not only has a refreshing effect, the taste is also mellow and rich, loved by people around the world, and cocoa, tea and listed as the world's three major drinks.

二、咖啡的起源（The Origin of Coffee）

经考证，目前已公认咖啡的诞生地为埃塞俄比亚的卡法（Kaffa）省。后来咖啡流传到世界各地，就采用其来源地"KAFFA"命名，直到 18 世纪才正式以"coffee"命名。虽然咖啡诞生于埃塞俄比亚，但有关咖啡起源的说法却不统一，具有可信度的说法分为两大类：一是广为流传的"牧羊人的传说"，另一个是伊斯兰教徒盛行的"奥马酋长的传说"。

After research, it has been recognized that the birthplace of coffee is Kaffa province in Ethiopia. Later, when coffee was spread around the world, it was named after its origin "KAFFA" until the 18 th century, when it was officially named after what we now call "coffee". Although we know that coffee was born in Ethiopia, but the origin of coffee has been described in different ways, there are mainly two credible legends: the "legend of the shepherd", which is widely spread, and the "legend of the Sheikh Omar", which is prevalent among the Muslims.

"牧羊人的传说"起源于黎巴嫩语言学家法司特·奈洛尼。在其 1671 年所著的《不知睡

眠的修道院》中记载：公元 6 世纪，在非洲埃塞俄比亚高原的卡法地区，有个名叫卡迪的牧羊人，到新草地去放牧，突然发现羊群蹦蹦跳跳，兴奋异常，即使入夜亦无法睡觉，于是跑到阿比西尼亚修道院求救。经修道院院长及修士调查发现山羊是吃了矮树丛上的红色果实，才显得特别兴奋，于是将果实采摘回去并煮成汤汁饮用，果然一夜无法入眠。于是院长把这种汤分派给做晚礼拜打瞌睡的僧侣饮用，效果极佳，这种提神药由此流传开来。

"The legend of the shepherd" originated from the Lebanese philologist Faust Neroni, who wrote in his 1 671 book "The Monastery of the Sleepless" that in the 6 th century AD, in the Kaffa region of the Ethiopian Plateau in Africa, a shepherd named Kadi went to the new grassland to graze, and suddenly found that the goats were jumping excitedly, and could not sleep even at night, so he ran to the Abyssinian monastery for help. The abbot and the monks investigated and found that the goats were excited because they had eaten the red fruit from the bushes; so they picked the fruit and boiled it into soup and drank it, and they were unable to sleep all night. So the abbot assigned this soup to the monks who dozed off in the evening church service to drink, the effect was excellent, and this refreshing medicine spread.

"奥马酋长的传说"是伊斯兰教徒阿布达尔·卡迪在其 1 587 年编撰之《咖啡由来书》中所记载的故事。1258 年，因犯罪而被族人驱逐出境的奥马酋长，流浪到离故乡摩卡很远的位于阿拉伯的瓦萨巴时，已经饥饿疲倦难行，当他坐在树根上休息时发现一只小鸟停在枝头，以一种从未听过极为悦耳的声音啼叫。他发现那只小鸟是在啄食枝头上的果实后，才叫出这美妙的啼声，于是便将那一带的果实采下，放入锅中加水熬煮，不久之后竟开始散发出浓郁的香味，饮用之后不但感觉醇厚可口，亦可解除身心疲惫，于是他采下许多这种神奇的果实，遇到病人便给他们熬汤饮之。由于其四处行善，国王与族人便赦免了他的罪行，请他回到故乡，并推崇为"圣者"。

The legend of Sheikh Omar is a story recorded by the Muslim Abdal Qadi in his 1 587 book "The Origin of Coffee". In 1258, Sheikh Omar, who was expelled by his clansman for crimes, wandered to Wazaba in Arabia, far from his hometown of Mocha, hungry and tired. When he sat on the roots of a tree to rest, he found a small bird resting on the branch singing with an extremely pleasant voice that he had never heard before. He found that bird was pecking at fruits on the branch before making this beautiful crowing, so he picked the fruits from that area, put them into a pot and boiled with water. Soon after, it began to emit a strong fragrance. It not only tasted good, but also relieved physical and mental fatigue after drinking it. So he picked many of these miraculous fruits and gave them to the sick when he met them to make soup and drink. Because of his good deeds, the king and his old friends forgave him for his sins, welcomed him back to his hometown, and revered him as a "holy man".

三、咖啡的传播历史（The Spread History of Coffee）

虽然非洲是咖啡的故乡，但 15 到 17 世纪，黑奴贸易盛行，大量非洲黑奴被贩卖至也门和阿拉伯半岛。途中，他们携带咖啡果充饥，使咖啡得以传入阿拉伯并被广泛种植，阿拉伯人非常珍视咖啡，在实现人工种植后，严禁将未剥壳的咖啡豆带出，以防其传播。然而，很快被殖民侵略者荷兰人经过精心策划，突破了这道障碍，从阿拉伯偷出咖啡种子和树苗，带到荷兰并成功实现大面积人工种植。

Although Africa was the home of coffee, the slave trade flourished between the 15th and 17th centuries, with large numbers of African slaves being sold to Yemen and the Arabian Peninsula. On the way, they carried coffee fruit to feed their hunger, so that coffee was introduced to Arabia and was widely cultivated, the Arabs cherish coffee very much, after the realization of artificial cultivation, it is strictly prohibited to take out the unpeeled coffee beans to prevent its spread. However, soon the colonial invaders, the Dutch, after careful planning, broke through this barrier, stole coffee seeds and saplings from Arabia, brought them to the Netherlands and successfully realized large-scale artificial cultivation.

1616 年，一株咖啡树经摩卡港转运到荷兰，使荷兰人在咖啡种植的竞争中取得上风。1658 年，荷兰人开始在锡兰（现斯里兰卡）培植咖啡。1699 年，荷兰人在爪哇岛建立了第一批欧式种植园。1715 年，法国人将咖啡树种带到了波旁岛（现留尼汪岛）。1718 年，荷兰人把咖啡带到了南美洲的苏里南，拉开了世界咖啡中心地区（南美洲）种植业飞速发展的序幕。1723 年，一个法国人将咖啡树带到马提尼克岛（北美洲）。1727 年，南美洲的第一个种植园在巴西帕拉建立，随后咖啡树在里约热内卢附近栽培。

In 1616, a coffee tree was transported to the Netherlands via the port of Mocha, giving the Dutch the upper hand in the competition for coffee cultivation. In 1658, the Dutch began cultivating coffee in Ceylon (now called sri Lanka). In 1699, the Dutch established the first European-style plantations in Java island in Indonesia. In 1715, the French brought coffee seeds to Bourbon Island (now called Reunion Island in France). In 1718, the Dutch brought coffee to Suriname in South America, initiating the rapid development of cultivation in the world's coffee center (South America). In 1723, a Frenchman, brought coffee trees to Martinique (North America). In 1727, the first plantation in South America was established in Para, Brazil, then coffee trees cultivated near Rio de Janeiro.

1730 年，英国人把咖啡引入牙买加，这之后富有传奇色彩的牙买加蓝山咖啡开始在蓝山地区生长。1750 年至 1760 年，危地马拉出现咖啡种植。1779 年，咖啡从古巴传入了哥斯达黎加。1790 年，咖啡第一次在墨西哥种植。1825 年，来自巴西里约热内卢的咖啡种子被带到了夏威夷岛屿，成为之后享有盛名的夏威夷可娜咖啡。1878 年，英国人在肯尼亚建立咖啡种植园区。1887 年，法国人带着咖啡树苗在越南建立了种植园。

In 1730, the British introduced coffee to Jamaica, after which the legendary Jamaican Blue Mountain coffee began to grow in the Blue Mountain region. From 1750 to 1760, coffee cultivation appeared in Guatemala. In 1779, coffee was introduced from Cuba to Costa Rica. In 1790, coffee was first planted in Mexico. In 1825, coffee seeds from Rio de Janeiro in Brazil were brought to the Hawaiian islands and became the later famous Hawaiian Kona coffee. In 1878, the British establishing coffee plantations in Kenya. In 1887, the French brought coffee seedlings and established plantations in Vietnam.

1896 年，咖啡开始登陆澳大利亚的昆士兰地区。从此以后咖啡种植的秘密一传十，十传百，成为公开的秘密。据史料记载，1884 年咖啡在中国台湾首次种植成功，从而揭开了咖啡在中国发展的序幕。中国内地最早的咖啡种植则始于云南省，20 世纪初一个法国传教士将第一批咖啡树苗带到云南的宾川县。在以后的近百年时间里，中国也只有部分地区种植咖啡。目前是中国的云南省、海南省种植咖啡较多。

In 1896, coffee began to land in the Queensland region of Australia. Since then the secret of coffee planting has spread from one to another and has become an open secret. According to historical records, in 1884 coffee was successfully grown for the first time in Taiwan in China, thus starting the development of coffee in China. The earliest coffee cultivation in Chinese mainland, began in Yunnan province, in the early twentieth century, when a French missionary brought the first coffee seedlings to Binchuan County, Yunnan Province. For nearly a hundred years, coffee was only grown in some parts of China. Currently, it is Yunnan and Hainan provinces in China that grow more coffee.

四、咖啡的产地（The Origin of Coffee）

从世界范围来看，咖啡喜欢温暖的环境，耐热不耐寒，光照充足，利于其生长，最适宜的种植温度在 16～21 ℃，由此以赤道为中心，从北纬 25° 到南纬 25° 形成了"世界咖啡黄金种植带"；从区域来看，目前主要的咖啡产区集中在亚、非、美三大洲。非洲主要产地：有阿拉比卡咖啡豆的诞生地埃塞俄比亚、肯尼亚、坦桑尼亚；亚洲主要产地：罗布斯塔产量最大的越南、盛产曼特宁、猫屎咖啡（图 3-1-1）的印度尼西亚、中国云南及海南；美洲主要产地：咖啡产量排名世界第一的巴西、盛产蓝山咖啡的牙买加，以及哥伦比亚、古巴、危地马拉。

图 3-1-1　猫屎咖啡（civet coffee）

From a worldwide perspective, coffee likes a warm environment, heat and cold resistance, sufficient light, conducive to its growth, the most suitable planting temperature in 16-21 degrees Celsius, which is centered on the equator, from 25° north latitude to 25° South latitude to form the "world coffee gold planting belt"; From the regional point of view, the main coffee producing areas are concentrated in Asia, Africa and the United States. Main production areas in Africa: Ethiopia, Kenya, Tanzania, the birthplace of Arabica coffee beans; Main producing areas in Asia: Vietnam, where Robusta produces the most, Indonesia, which is rich in Mantening and cat poop coffee (Figure 3-1-1), Yunnan and Hainan of China; Main producing areas in the Americas: Brazil, which ranks first in coffee production in the world, Jamaica, which is rich in Blue Mountain coffee, and Colombia, Cuba, Guatemala.

五、咖啡豆的种类（Types of Coffee Beans）

广义而言，咖啡豆主要分为阿拉比卡豆和罗布斯塔豆两大类。阿拉比卡咖啡豆是最常见和最受欢迎的类型，占全球咖啡市场的 70%，通常生长在较高海拔地区，呈椭圆形，糖分含量高，咖啡因含量较低，酸度较高，风味独特且有层次感。而罗布斯塔咖啡豆占全球市场的 30%，生长在较低海拔地区，形状圆润，糖分含量少，口感苦涩浓烈，咖啡因含量高，风味相对单一，常用于制作浓缩咖啡和咖啡因含量较高的饮品。

Broadly speaking, coffee beans are mainly divided into two categories: Arabica beans and Robusta beans. Arabica coffee beans are the most common and popular type, accounting for 70%

of the global coffee market, and are typically grown at higher elevations, oval in shape, high in sugar, low in caffeine, high in acidity, and unique and layered in flavor. Robusta coffee, which accounts for 30 percent of the global market, is grown at lower altitudes and has a rounded shape, low sugar content, a bitter taste, a high caffeine content and a relatively simple flavor, which is often used to make espresso and drinks with a higher caffeine content.

六、咖啡豆的成分（Ingredients of Coffee Beans）

咖啡豆的主要成分包括咖啡因、丹宁酸、脂肪、蛋白质、糖分、矿物质、纤维素和其他微量成分。

咖啡豆中的主要成分及其作用如下。

The main components of coffee beans include caffeine, tannin, fat, protein, sugar, minerals, fiber and other trace components.

The main components in coffee beans and their functions are as follows.

（一）咖啡因（Caffeine）

咖啡因是咖啡所有成分中最为人注目的。它属于植物黄质的一种，性质和可可内含的可可碱、绿茶内含的茶碱相同，烘焙后减少的百分比极微小。咖啡因的作用极为广泛，会影响人体脑部、心脏、血管、胃肠、肌肉及肾脏等各部位，适量的咖啡因会刺激大脑皮层，促进感觉判断、记忆、感情活动，让心肌机能变得较活泼，血管扩张血液循环增强，并提高新陈代谢机能，因也可减轻肌肉疲劳，促进消化（消化食品）液分泌。此外，由于它也会促进肾脏机能帮助体内将多余的钠离子排出体外外，所以在利尿作用提高下，咖啡因不会像其他麻醉性、兴奋性物积在体内，约两个小时便会被排泄掉。咖啡风味中的最大特质——苦味，也与咖啡因有关。

Caffeine is the most notable of all the ingredients in coffee. It belongs to the plant xanthine, the nature and cocoa contain theobromine, theophylline containing green tea is the same, to reduce the percentage of microscopic after baking. The effect of caffeine is extremely broad, will affect the human brain, heart, muscle, blood vessels, stomach, kidney and so on each place, moderate doses of caffeine can stimulate the cerebral cortex, promote sensory judgment, memory, emotional activities, make cardiac muscle function become more active, blood vessels dilated blood circulation enhanced, and improve metabolic function, also reduce muscle fatigue, promote digestion (digestion of food) fluid secretion. But because it will also promote kidney function help excess sodium in the body metabolism of the chemical composition of the water molecules out of the body, so under the diuretic effect improving, caffeine does not like other narcotic, excitatory in the body, in about two hours, will be drained off. The most intense flavor of coffee-bitterness, Is also related to caffeine.

（二）有机酸（Organic acid）

咖啡中包含多种有机酸，如枸橼酸、苹果酸、单宁酸等。适量的酸度有助于咖啡的口感平衡，但是，经提炼后单宁酸会变成淡黄色的粉末，易溶入水，经煮沸它会分解而产生焦梧酸，使咖啡味道变差。如果冲泡好又放上好几个小时，咖啡颜色会变得比刚泡好时浓，而且味道也较差，所以才会有"冲泡好最好尽快喝完"的说法。

Coffee contains many organic acids, such as citric acid, malic acid, tannic acid and so on. The right amount of acidity helps to balance the taste of coffee, but after being refined, tannic acid will turn into a pale yellow powder, which is easy to dissolve into water, and it will break down when boiled to produce pyrowulic acid, which makes the coffee taste worse. If it is brewed well and put for a few hours, the color of the coffee will become stronger than when it is just soaked, and the taste is also poor, so there is a saying that " brewed well and it is best to drink as soon as possible".

(三) 脂肪（Fat）

咖啡内含的脂肪，在风味上占极为重要的角色。经分析发现咖啡内含的脂肪有多种，而其中最主要的是脂肪酸和挥发性脂肪。脂肪酸是指脂肪中含有酸，其强弱会因咖啡种类不同而异。挥发性脂肪是咖啡香气的主要来源。烘焙过的咖啡豆内所含的脂肪一旦接触到空气，会发生化学变化，味道香味都会变差。

The fat of coffee contains accounting for a very important role in the flavor Analysis has found there are many type of fat in coffee, but one of the most main is fatty acid and volatile fatty acid fat. Fatty acid is contained in its strength in different types of coffee varies. Volatile fatty acid is a major source of coffee aroma. Once the fat contained in roasted coffee beans is exposed to air, it will undergo chemical changes and taste, taste will deteriorate.

(四) 蛋白质（Protein）

咖啡豆中含有丰富的蛋白质，主要是氨基酸，影响咖啡的风味和口感。卡路里的主要来源之一就是蛋白质，但像滴落式冲泡出来的咖啡，蛋白质多半不会析出来，所以咖啡的营养是有限的，这也就是咖啡会成为减肥食品的缘故。

Coffee beans are rich in protein, mainly amino acids, which affect the flavor and taste of coffee. One of the main sources of calories is protein, but like drip-brewed coffee, protein is mostly not out, so the nutrition of coffee is limited, which is why coffee will become a weight loss food.

(五) 糖分（Sugar）

咖啡豆中含有多种糖分，主要是葡萄糖、果糖和麦芽糖等。这些糖分在烘焙过程中会发生化学反应，形成咖啡的香气和风味，同时，大部分糖分会转为焦糖，为咖啡带来独特的褐色。

Coffee beans contain a variety of sugars, mainly glucose, fructose and maltose. These sugars undergo chemical reactions during roasting, forming the aroma and flavor of the coffee, while most of the sugars turn to caramelize, giving the coffee its distinctive brown color.

(六) 矿物质（Minerals）

咖啡所含的矿物质有石灰、铁（铁食品）质、硫黄、碳酸钠、磷、氯、硅等，因所占的比例对咖啡的风味影响并不大，只带来稍许涩味。

Coffee contains minerals such as lime, iron (iron food), sulfur, sodium carbonate, phosphorus, chlorine, silicon, etc., because the proportion of coffee is rarely affected by the flavor, only bring a little astringency.

（七）粗纤维（Fibre）

咖啡生豆含有纤维素，烘焙后会炭化，这种碳质和糖分的焦糖化互相结合，形成咖啡的色调。

Green coffee beans contain cellulose, which is charred when roasted, and this caramelization of carbon and sugar combines to give coffee its hue.

赛证直通 Competitions and Certificates

一、基础知识部分（Basic Knowledge Part）

（一）专业名词解释（Explanation of Professional Terms）

1. 咖啡因（Caffeine）：
2. 单宁酸（Tannin）：

（二）思考题（Thinking Question）

请简述咖啡豆的成分。

Please describe briefly the composition of coffee beans.

二维码 3-1-1：
知识拓展

二、技能操作部分（Skill Operation Part）

请以观察实物的方式，全方位对比分析阿拉比卡咖啡豆和罗布斯塔咖啡豆。

Compare and analyze Arabica coffee beans and Robusta coffee beans in all aspects by looking at the real thing.

专题二
浓情蜜"意"：意式咖啡

一、意式咖啡的概念（The Concept of Espresso）

【微课】
浓情蜜"意"：
意式咖啡

意式咖啡是由蒸汽压力咖啡机做出来的浓缩咖啡，被称为 Espresso，以意式浓缩咖啡为基底，配合牛奶、奶沫等，各种演变组合出来的咖啡饮品都统称意式咖啡。

Espresso is an espresso made by the steam pressure coffee machine, known as Espresso, espresso as the base, with milk, milk foam, etc., various evolution of the combination of coffee drinks are collectively referred to as Italian coffee.

二、经典意式咖啡的种类（The Type of Classic Espresso）

（一）意式浓缩咖啡（Espresso）

意式浓缩咖啡的基础、特点是高浓度、味道强烈，制作迅速。它是众多咖啡爱好者的首选。

Espresso is the basis of Italian coffee, is characterized by high concentration, strong taste, quick preparation . It is the first choice of many coffee lovers.

（二）摩卡（Mocha）

摩卡咖啡是意式花式咖啡的一种，其基底是浓缩咖啡。这种咖啡的特点在于其与巧克力糖浆、鲜奶和奶泡以特定比例混合，通常为 1：0.5：1.5：1。除了巧克力糖浆，摩卡咖啡中也可以加入白巧克力。其最显著的风味是可可的微苦焦香。为了增加视觉和口感上的享受，一些摩卡咖啡还会装饰奶油、可可粉和棉花糖。这些装饰通常放在奶泡上，以增强咖啡的特色和香气。

Mocha coffee is a kind of Italian fancy coffee, its base is espresso. This coffee is characterized by its specific mix of chocolate syrup, fresh milk and foam, usually 1：0.5：1.5：1. In addition to chocolate syrup, mocha coffee can also add white chocolate. Its most notable flavor is the slightly bitter char of cocoa. To add visual and taste pleasure, some mocha coffees are also decorated with cream, cocoa powder and marshmallows. These decorations are usually placed on the milk foam to enhance the character and aroma of the coffee.

（三）拿铁（Latte）

拿铁是由意式浓缩咖啡加牛奶制成的一种经典咖啡，浓郁香醇，口感顺滑。拿铁（Latte）在意大利语里原意为"牛奶"，所以，在意大利点拿铁咖啡不能单说 Latte，要说 Caffe Latte。否则送上来的就是一杯纯牛奶。

It is a classic coffee made from espresso and milk. It is rich and mellow and smooth in taste. Latte means "milk" in Italian, so ordering a Latte in Italy doesn't just say "Latte", it says "Caffe Latte". Otherwise, it's a glass of pure milk.

（四）玛奇朵（Macchiato）

玛奇朵是一种意式咖啡，在浓缩咖啡上只加一层奶泡，没有再加牛奶，因此奶香只停留在唇边，而浓缩咖啡的味道不会被牛奶稀释。它看起来像是缩小版的卡布奇诺，分量仅为卡布奇诺的三分之一。此外，玛奇朵还有另一个名字"天使之吻"。很多商家会在咖啡奶沫上添加一些网格状的焦糖，因此它也被称为焦糖玛奇朵。

Macchiato is an Italian type of coffee, characterized by a single layer of foam on the espresso, no milk added, so the milk flavor only stays on the lips, and the espresso flavor is not diluted by the milk. It looks like a scaled-down version of Cappuccino, with only a third of the weight. Macchiato is also known as "angel kiss". Many companies add a grid of caramel to the coffee foam, so it is also called caramel Macchiato.

三、意式咖啡工具（Espresso Tools）

制作意式咖啡的工具有：

Tools for making espresso are:

（1）磨豆机：用于将咖啡豆研磨成适当粗细的咖啡粉。手摇磨豆机省粉且不需要电源，适合家用，但效率较低；电动磨豆机则方便省时，适合家用和商用。

bean grinder: used to grind coffee beans into the appropriate thickness of coffee powder. The hand grinder saves powder and does not require power supply, which is suitable for home use, but the efficiency is low; The electric bean grinder is convenient and time-saving, suitable for home and commercial use.

（2）电子秤：帮助精确称量咖啡粉和水的比例，确保粉水比的精准控制。

electronic scale: help to accurately weigh the ratio of coffee powder and water, ensure the accurate control of the ratio of powder and water.

（3）布粉器：使咖啡粉表面达到平整，轻敲粉碗后使用布粉器转动，使咖啡粉在粉碗中更贴合。

powder distributor: make the surface of the coffee powder level, gently tap the powder bowl and then use the powder distributor to rotate, so that the coffee powder is more fit in the powder bowl.

（4）压粉器：将咖啡粉饼压实，确保咖啡粉的密度均匀，防止水流过快通过粉碗导致萃取不足。

powder press: compact the coffee powder to ensure the uniform density of the coffee powder and prevent the water from flowing too fast through the powder bowl and resulting in insufficient extraction.

（5）打奶缸／拉花缸：用于打奶和拉花，不同设计的缸适用于不同的操作需求。

milk cylinder/latte jar: Used for milk and latte. Different designs of the jar are suitable for different operation requirements.

（6）咖啡机：制作咖啡的核心设备，根据需求可选择商用机或家用机。

coffee machine: the core equipment for making coffee, according to demand can choose commercial machine or home machine.

（7）摩卡壶：摩卡壶作为一种过滤式萃取方式的家用咖啡器具，也深受意式浓缩咖啡爱好者的喜爱。

Moka pot: The moka pot, as a filter extraction method of home coffee utensils, is also loved by espresso lovers.

四、经典意式咖啡的制作流程（The Process of Making Classic Espresso）

经典意式咖啡的具体制作流程如下：

The specific production process of classic Italian coffee is as follows:

（1）准备咖啡豆和工具：首先，需要使用磨豆机将咖啡豆研磨成粉，确保咖啡粉的细度适中，以利于咖啡的萃取。同时，准备一个意式咖啡机，这是制作意式咖啡的关键工具。

Prepare coffee beans and tools: first, need to use a bean grinder to grind the coffee beans into powder to ensure that the coffee powder is moderate in fineness to facilitate the extraction of coffee. At the same time, get an espresso machine, which is a key tool for making espresso.

（2）填充咖啡粉：将研磨好的咖啡粉倒入咖啡机的过滤器中，使用捣棒将咖啡粉压平，确保咖啡粉分布均匀且紧密，但不要过度压实，以免影响咖啡的萃取。

Fill the coffee powder: pour the ground coffee into the filter of the coffee machine, and use a rammer to flat the coffee powder, ensuring that the coffee powder is evenly distributed and compact, but do not overcompaction, so as not to affect the extraction of coffee.

（3）启动咖啡机：装好过滤器后，启动咖啡机进行萃取。可根据咖啡机的不同，调整萃取时间，以确保萃取出适量的浓缩咖啡液。

Start the coffee machine: After installing the filter, start the coffee machine for extraction. Depending on the coffee machine, the extraction time can be adjusted to ensure that the right amount of espresso liquid is extracted.

（4）牛奶处理：对于需要加入牛奶的意式咖啡（如拿铁、卡布奇诺等），将牛奶倒入另一个容器中，使用蒸汽棒或蒸汽奶泡机将牛奶打成绵密的奶泡。这一步是为了在咖啡上形成奶泡，增加咖啡的口感和视觉效果。

Milk treatment: for espresso coffee that requires milk (such as latte, cappuccino, etc.), pour the milk into another container and use a steam stick or steam foam machine to beat the milk into a dense foam. This step is to form foam on the coffee, adding taste and visual effect to the coffee.

（5）融合和装饰：根据具体的意式咖啡类型，将牛奶和奶泡倒入装有浓缩咖啡的杯子中，然后进行融合或形成特定的图案。例如，卡布奇诺需要在杯中形成奶泡的"金圈"，拿铁则需要将奶泡均匀地铺在咖啡上。

Blending and decorating: depending on the specific type of espresso, pour milk and foam into a cup containing espresso and then blend or form a specific pattern. For example, a cappuccino needs to form a "golden circle" of milk foam in the cup, while a latte needs to spread the milk foam evenly over the coffee.

（6）完成和享用：最后，可以根据个人口味添加糖、奶油或其他配料，然后就可以享用一杯美味的意式咖啡了。

Finished and enjoyed: finally, you can add sugar, cream or other toppings according to your taste, and enjoy a delicious cup of espresso.

通过上述步骤，可以制作出各种口感的意式咖啡，从经典的浓缩咖啡到各种花式咖啡，如拿铁、卡布奇诺等。每一步都至关重要，从选择合适的咖啡豆到正确地萃取和牛奶处理，都是影响最终口感的关键因素。

Through the above steps, you can make a variety of tastes of Italian coffee, from classic espresso to various fancy coffee, such as latte, cappuccino and so on. Every step is crucial, from choosing the right coffee beans to the correct extraction and milk handling, and is a key factor in the final taste.

赛证直通 Competitions and Certificates

一、基础知识部分（Basic Knowledge Part）

（一）专业名词解释（Explanation of Professional Terms）

1. 意式咖啡（Espresso）：
2. 摩卡咖啡（Mocha coffee）：

（二）思考题（Thinking Question）

请简述经典意式咖啡的种类。

Please briefly describe the types of classic Italian coffee.

二、技能操作部分（Skill Operation Part）

请尝试冲泡玛奇朵。

Please try brewing macchiato.

专题三

巧"手"生花：手冲咖啡

一、手冲咖啡的概念（The Conception of Hand-brewed Coffee）

手冲咖啡，又称手工滴漏咖啡或手工冲泡咖啡，与利用机器制作的咖啡截然不同。它采用手工操作，利用热水与咖啡粉的接触、浸泡、滤过等过程精心制作而成。

Hand brewed coffee, also known as hand drip coffee or hand brewed coffee, is very different from machines made coffee. It uses manual operation, the use of hot water and coffee powder contact, soaking, filtration and other processes carefully made.

二、手冲咖啡的原理（The Principle of Hand-brewed Coffee）

手冲咖啡的原理是溶解和扩散。

咖啡豆在经过烘焙之后会发生化学反应，生成散发出咖啡香气和味道的咖啡物质成分，为了萃取这些咖啡物质成分，就需要把咖啡磨成粉，然后慢慢注水，溶化出咖啡里面的物质成分，称为"溶解"。

在日常的生活中，我们知道，要想更好的溶解，就需要把粉磨得更细一些，但在手冲咖啡方面如果把咖啡粉磨得很细就会堵住滤纸，所以就得把咖啡粉磨成像细白砂糖一样大小（中细研磨），但这样的话在溶解过程中就会溶解不均匀（咖啡物质有容易溶解的小分子和不容易溶解的大分子），就需要借助在萃取中的"闷蒸""搅拌""晃动盛放咖啡液的分享壶"（在注水溶解过程中会有浓度之差）来溶解，称为"扩散"。

The principle of pour-over coffee is "dissolution" and "diffusion".

After roasting, coffee beans will undergo a chemical reaction to produce coffee components that emit coffee aroma and taste. In order to extract these coffee components, it is necessary to grind the coffee into powder, and then slowly inject water to dissolve the components of the substances in the coffee, which is called "dissolving".

In daily life, we know that for better dissolution, we need to grind the powder finer, but when it comes to brewing coffee, if the coffee powder is very fine, it will block the filter paper. Therefore, it is necessary to grind the coffee powder to the size of fine white sugar (medium fine grinding), but this will dissolve unevenly during the dissolution process (coffee substances have small molecules that are easily dissolved and large molecules that are not easily dissolved), and it is necessary to use the "smothered" "stirred" "shaking the coffee liquid sharing pot" in the extraction process (there will be a difference in concentration during the water dissolution process). to dissolve called "diffusion".

三、手冲咖啡的特点（The Characteristics of Hand-brewed Coffee）

手冲咖啡在冲泡过程中，由于热水与咖啡粉的接触时间较短，能够最大限度地保留咖啡的原生态风味和香气，风味清新怡人；同时，由于热水与咖啡粉的接触面积大，能够充分提取咖啡的油脂和固体成分，使得咖啡味道更加醇厚；此外，手冲咖啡的冲泡过程具有高度的可调性，通过调整热水的温度、流速、浸泡时间等参数，可以控制咖啡的味道和口感，可以使咖啡呈现出层次分明的味道，具有前中后味之分；再者，通过选择不同品种、烘焙度、研磨度的咖啡豆，还能调整咖啡的酸甜度，使咖啡味道酸甜适中，并充分展现咖啡豆中的果香和花香成分，使得每一杯手冲咖啡都独具个性，具有独一无二的味觉享受。

In the brewing process, due to the short contact time between hot water and coffee powder, the original ecological flavor and aroma of coffee can be preserved to the maximum extent, and the flavor is fresh and pleasant. At the same time, due to the large contact area between hot water and coffee powder, the oil and solid components of coffee can be fully extracted, making the coffee taste more mellow; In addition, the brewing process of hand-brewed coffee has a high degree of adjustability, by adjusting the temperature of hot water, flow rate, soaking time and other parameters, you can control the taste and taste of coffee, can make the coffee present a distinct flavor, with the front, middle and after taste; Moreover, by choosing different varieties, roasting degree and grinding degree of coffee beans, the sour and sweet degree of coffee can be adjusted, so that the coffee taste is moderate, and the fruit and floral components in the coffee beans are fully displayed, making each cup of hand-brewed coffee has a unique personality and unique taste enjoyment.

四、手冲咖啡的制作过程及标准（Preparation Process and Standard of Pour-over Coffee）

（一）准备沸水（Prepare Boiling Water）

需预估使用水量并将其煮沸，其中包括冲洗器具所需的用量。然后，准备好滤纸，再将热水均匀地冲在滤纸上，使其全部湿润并紧紧贴附在滤杯上，随后倒掉分享壶内的热水。这一步骤是为了更好地还原咖啡的纯粹味道。

Estimate the amount of water to be used and boil it, including the amount needed to flush the appliance. Then, prepare the filter paper, and then evenly wash the hot water on the filter paper, so that it is all wet and tightly attached to the filter cup, and then pour out the hot water in the sharing pot. This step is to better restore the pure taste of coffee.

（二）准备咖啡粉（Prepare the Coffee Ground）

将研磨精细、看起来很像砂糖的质地的咖啡粉倒入滤杯中，轻轻拍平，然后准备冲泡，建议每2勺新鲜研磨的咖啡粉配合6盎司热水。

Pour the finely ground, sugar-like texture into a filter glass, pat it flat, and prepare to brew. recommends 6 ounces of hot water for every 2 tablespoons of freshly ground coffee.

（三）冲泡与萃取（Brewing and Extraction）

这是关键步骤，要更好释放咖啡的精髓，必须耐心细致且精准。

Which is a key step, to better release the essence of coffee, must be patient and meticulous and precise.

（1）首先要进行闷蒸，轻轻注入少量热水确保所有咖啡粉都被润湿。时间为30秒左右，以排除二氧化碳，让后续水流可以更好地萃取咖啡。此时水温对于咖啡的萃取非常重要，一般来说，控制在85～95℃之间比较合适。

The first thing to do is to steam-steam, lightly pour in a little hot water to make sure all the ground coffee is moistened. The time is about 30 seconds, to remove carbon dioxide, so that the subsequent water can better extract the coffee. At this time, the water temperature is very important for the extraction of coffee, in general, it is more appropriate to control between 85-95 degrees Celsius.

（2）接着是注水冲煮：整个冲煮过程可以分为三段。第一段注水后等待水量下降到一半再进行第二段注水，最后注入第三段水量。总萃取时间大约为3分钟。注水时可以采用绕圈式注水法，即先从小水流开始绕圈注入，然后逐渐加大水流。在冲泡过程中，保持稳定的水流和适当的注水高度是关键。

Followed by water boiling: The whole boiling process can be divided into three stages. After the first phase of water injection, wait for the water volume to drop to half, then carry out the second phase of water injection, and finally inject the third phase of water. The total extraction time is approximately 3 minutes. The circular water injection method can be used when flooding, that is, the small water flow starts in a circle, and then gradually increases the water flow. In the brewing process, maintaining a stable water flow and appropriate water injection height is the key.

（3）冲泡完成后尽快享用。因为咖啡含有单宁酸，当温度下降，单宁酸释放出来，咖啡酸味加重，口感会变差。而且在温度较低的时候，人的味觉对酸味更加敏感。

Enjoy as soon as possible after brewing. Because coffee contains tannins, when the temperature drops, tannins are released, coffee acidity is increased, and the taste will deteriorate. And at lower temperatures, the human palate is more sensitive to acidity.

赛证直通 Competitions and Certificates

一、基础知识部分（Basic Knowledge Part）

（一）专业名词解释（Explanation of Professional Terms）

1. 手冲咖啡（Hand-brewed coffee）：
2. 溶解（dissolve）：

（二）思考题（Thinking Question）

简述手冲咖啡的原理。

Describe the principle of making hand-brewed coffee.

二维码 3-1-3：
知识拓展

二、技能操作部分（Skill Operation Part）

请尝试练习手冲咖啡。

Please try making hand-brewed coffee.

微课视频：
"趣"创咖啡：
创意咖啡制作

专题四
"趣"创咖啡：创意咖啡制作

一、创意咖啡的概念（The Concept of Creative Coffee Concept）

创意咖啡是一种结合了创新和创意的咖啡饮品，它突破了传统咖啡的制作方式和口味，通过融入各种新颖的元素和理念，为消费者带来全新的味觉体验。

Creative coffee is a kind of coffee drink that combines innovation and creativity, it breaks through the traditional way of making coffee and taste, by integrating a variety of novel elements and ideas, brings consumers a new taste experience.

创意咖啡是一种创新的咖啡概念，它不仅仅是一杯普通的咖啡，更是一种艺术和文化的体现。它在制作方法、图案设计和口感创新等方面都有独特之处。

Creative coffee is an innovative coffee concept, which is not only a cup of ordinary coffee, but also a reflection of art and culture. It has unique features in production method, pattern design and taste.

二、创意咖啡的起源（The Origin of Creative Coffee）

创意咖啡的起源可以追溯到人们对健康饮品的追求以及对传统咖啡的改良需求。在追求健康生活的背景下，无因咖啡（即不含咖啡因的咖啡）应运而生，旨在满足那些希望享受咖啡风味而不愿摄入咖啡因的人群。无因咖啡的发明与整体性的健康生活理念相契合，通过化学方法去除咖啡豆中的咖啡因，同时保留其香气，为人们提供了一种健康的咖啡替代饮品。随着时间的推移，人们对咖啡的制作和享用方式不断创新，尤其是在咖啡的制作工艺和食材选择上。这种趋势推动了咖啡行业的持续发展与创新，促使人们将咖啡与更多创意元素相结合，制作出各种独具特色的新品咖啡，创意咖啡因此得以蓬勃发展，也具有自身鲜明的特点。

The origin of creative coffee can be traced back to people's pursuit of healthy drinks and the need to improve traditional coffee. In the context of the pursuit of healthy living, causeless coffee (that is, coffee without caffeine) came into being, aiming to satisfy those who want to enjoy the flavor of coffee without consuming caffeine. The invention of coffee without a cause aligns with the holistic concept of healthy living by chemically removing caffeine from coffee beans while retaining their aroma, providing people with a healthy alternative to coffee. Over time, people have continued to innovate in the way coffee is made and enjoyed, especially in the process of making coffee and the choice of ingredients. This trend has promoted the continuous development and innovation of the coffee industry, prompting people to combine coffee with more creative elements to produce a variety of unique new coffee, creative coffee has been able to flourish, but also has its own distinctive characteristics.

三、创意咖啡的特点（Features of Creative Coffee）

（一）风味多样性（Flavor Diversity）

创意特调咖啡注重口味的创新，每一款都具有自身独特的风味。

Creative special coffee pays attention to the innovation of taste, each has its own unique flavor.

（二）口感丰富性（Rich Taste）

创意咖啡通过充满惊喜的多元组合，让咖啡的味道更加丰富多彩。

Creative coffee through the combination of surprises, makes the coffee taste more colorful.

（三）视觉艺术性（Visual Artistry）

创意特调咖啡不仅在口味上追求创新，更在视觉效果上下功夫，让每一杯咖啡都成为独有的艺术品，不仅提升了咖啡的观赏性，也增加了饮用的乐趣。

Creative special coffee not only pursues innovation in taste, also works hard on visual effect, makes every cup of coffee become a unique work of art, not only improves the appreciation of coffee, also increases the pleasure of drinking.

（四）健康营养性（Health and Nutrition）

创意咖啡在追求口感的同时，也注重健康与营养的平衡。通过选择低糖、低脂的食材和调料，以及采用健康的制作工艺，使得特调咖啡在满足味蕾的同时，也能为身体带来益处。

Creative coffee in the pursuit of taste, but also pay attention to the balance of health and nutrition. Through the selection of low-sugar, and low-fat ingredients and seasonings, and the use of healthy techniques, makes it possible to satisfy the taste buds as well as .

（五）无限可能性（Infinite possibilities）

随着人们对咖啡品质和口感的追求不断提高，创意咖啡也在不断研发和创新以能够持续满足消费者的新需求，这使得创意咖啡充满了无限可能。

With the continuous improvement of people's pursuit of coffee quality and taste, creative coffee is also in constant research and development and innovation to continue to meet the new needs of consumers, makes creative coffee full of infinite possibilities.

四、创意咖啡的制作原则（Principles Of Creative Coffee Making）

（一）了解咖啡差异（Understanding Coffee Differences）

即了解不同咖啡的差异及不同做法给咖啡口感带来的影响（简单来说，就是明白什么是咖啡、掌握咖啡制作方法，懂得品鉴咖啡）。

That is, understanding the differences between different coffees and the impact of different practices on the taste of coffee (In short, understanding what coffee is, mastering coffee making methods, and knowing how to taste coffee).

（二）掌握食材特性及食用方法（To Master the Characteristics of Food Materials and eating methods）

创意咖啡在创新上有无限可能，但如果在咖啡中加入不对的食材则可能适得其反，破坏咖啡的口感和品质。因此，正确选择食材，是提升咖啡风味的关键。

Creative coffee has unlimited possibilities in innovation, but if the wrong ingredients are added to the coffee, it may backfire and destroy the taste and quality of the coffee. Therefore, the right choice of ingredients is the key to enhance the flavor of coffee.

（三）创意与生活实际结合（The Combination of Creativity and Practical Life）

创意咖啡的创新需与人们的饮食习惯、兴趣爱好及文化属性紧密结合。贴近生活、深入了解顾客需求，同时尊重并融入多元文化，这既是创意咖啡创新的源头，也是其稳固的基石。

The innovation of creative coffee needs to be closely combined with people's eating habits, interests and cultural attributes. Close to life, in-depth understanding of customer needs, while respecting and integrating into the multicultural, this is both the source of creative coffee innovation, but also its solid cornerstone.

（四）故事性和文化性相结合（The Combination of Story and Culture）

创意咖啡不仅是一款饮品，它也需要具有故事性和文化内涵，这样才能更有温度和吸引力，从而更容易被市场接受。

Creative coffee is not only a drink, it also needs to have a story and cultural connotation, so that it can be more warm and attractive, and thus easier to be accepted by the market.

（五）制作工艺与材料选择独特性（Unique Production Process and Material Selection）

创意咖啡的制作工艺和材料选择需要独具特色，这是创作独特咖啡风味的诀窍。一般来说，可以通过独特的萃取技术和巧妙的材料搭配，打造出令人耳目一新的创意咖啡，从而让每一杯咖啡都充满惊喜和个性。

The production process and material selection of creative coffee need to be unique, which is the secret of creating a unique coffee flavor. In general, unique extraction techniques and clever material combinations can be used to create a refreshing creative coffee, so that every cup of coffee is full of surprises and personality.

（六）口感与健康相平衡（Balance Between Taste and Health）

创意咖啡在追求口感的同时，也要注重健康与营养的平衡，千万不能盲目追求口感而忽视了健康，要做到选材健康、制作健康，保证咖啡在满足味蕾的同时，也不影响身体健康。

Creative coffee in the pursuit of taste, but also pay attention to the balance of health and nutrition, must not blindly pursue taste and production health, to achieve healthy selection of materials, healthy process, to ensure that coffee to meet the taste buds at the same time, does not affect the health of the body.

赛证直通 Competitions and Certificates

一、基础知识部分（Basic Knowledge Part）

（一）专业名词解释（Explanation of Professional Terms）

创意咖啡（Creative Coffee）：

（二）思考题（Thinking Question）

简述创意咖啡的原则。

Briefly describe the principles of creative coffee.

二维码 3-1-4：
知识拓展

二、技能操作部分（Skill Operation Part）

请结合所学知识，自创一款创意咖啡。

Please combine your knowledge to create a creative coffee.

专题五
匠心甄选：咖啡品鉴

【微课】
匠心甄选：
咖啡品鉴

一、咖啡品鉴基础知识（Coffee Tasting Basics）

品鉴咖啡，得出一杯咖啡的最终评价，需要从以下维度进行考量。

Tasting coffee and getting a final evaluation of a cup of coffee need to be considered from the following dimensions.

（一）外观（Appearance）

一杯优质咖啡的油脂应该是厚实平滑、富有光泽，充满了咖啡油脂的香气。如果颜色过浅或过深，可能表明萃取不足或过萃。

A cup of good coffee oil should be thick and smooth, glossy, full of coffee oil aroma. If the color is too light or dark, may indicate insufficient or over-extracted extraction.

（二）香气（Aroma）

优质的咖啡应该能闻到新鲜水果的酸甜、轻微的焦苦香，以及迷人的花香和特别的香辛料香。

Good quality coffee should be able to smell the sweet and sour of fresh fruits, slightly scorched aroma, as well as a charming floral scent and a special spicy aroma.

（三）酸度（Acidity）

酸度是指咖啡在入口时口水的分泌程度，清新明亮的自然果酸为佳，如果酸味不自然或过于尖锐，则表明咖啡品质不佳。

Acidity refers to the degree of saliva secretion of coffee in the mouth, fresh and bright natural fruit acid is better, if the sour taste is not natural or too sharp, indicates that the coffee quality is not good.

（四）醇度（Alcohol）

醇度是咖啡在口中的触感，包括重量、厚度和饱满度。醇度并不是通过味觉直接品尝出的某种特定味道，而是一种综合的感受，涉及咖啡的黏稠度、多汁感及在口中的"重量"。与舌头的感受相关，优质的咖啡在舌头上留下醇厚和微微的重量感，而劣质的咖啡则可能感觉"浮游"在舌尖。

Alcohol how the coffee feels in the mouth, includes weight, thickness and fullness. alcohol is not a specific taste directly detected by the palate. is a combination of sensations involving the consistency of the coffee, the juiciness of the coffee, and its "weight" in the mouth. is related to how the tongue feels. a good cup of coffee will leave a mellow and slightly weighted sensation on the tongue, whereas a bad cup of coffee may feel "floating" on the tip of the tongue.

（五）口感（Taste）

类似于葡萄酒中的"酒体"，是咖啡在口中可以感受到的"厚度"，优质的咖啡会让人感觉浓厚，劣质的则让人感觉清淡。

Is similar to the "wine body" in wine, is the "thickness" of coffee can be felt in the mouth, quality coffee will make people feel strong, inferior coffee will make people feel light.

（六）风味（Flavor）

优质的咖啡可以让人联想到生活中的美好事物，如水果、花香、蜜糖等，不同的品种、产地、精制方式都可以造就咖啡不同的风味特征。

High quality coffee can make people think of the good things in life, such as fruits, flowers, honey and so on. different varieties, origin and refining methods can create different flavor characteristics of coffee.

（七）余味（Aftertaste）

好的咖啡滋味充足，也有着更绵长美妙的余味，但如果余味中有感受到霉变、烟尘的味道，那咖啡品质也大打折扣了。

Good coffee taste is sufficient, also has a longer and wonderful aftertaste, but if you feel the aftertaste of mildew, smoke, the quality of the coffee is also greatly reduced.

（八）整体表现（Overall Performance）

即从喝到高温第一口到低温最后一口，咖啡所体现出来的比如酸度，body 感，平衡性，余韵等的综合品质带给品鉴者的主观整体感受。

From the first sip of coffee at high temperature to the last sip at low temperature, the overall subjective feeling brought to the taster by the comprehensive qualities such as acidity, body feeling, balance and lingering.

二、品鉴方法（Tasting Method）

（一）闻香味（Smell The Scent）

1. 干香（Dry Aroma）

将新鲜研磨的咖啡粉凑近口鼻处，能感知到来自不同咖啡产地的不同咖啡的干香。如在拉丁美洲咖啡中将闻到类似于坚果、黑巧克力类型的味道；非洲产区的咖啡则多有花草、水果类型的风味。

Put freshly ground coffee close to your mouth and nose, and you will feel the dry aroma of different coffees from different coffee regions. For example, in Latin American coffee, you will smell similar to nuts, dark chocolate type flavor; however, coffee from African producing areas has more herbal and fruit-type flavors.

2. 湿香（Wet Aroma）

将制作好的咖啡倒入杯中，闭上眼睛，慢慢地闻它的湿香，也许你会闻到坚果香、巧克力香、果香或花香。

Pour the prepared coffee into the cup, close your eyes, and slowly smell its wet aroma, perhaps you will smell nutty, chocolate, fruity or floral.

3. 劣质咖啡香味（Inferior Coffee Aroma）

劣质香味：工业香精气味、腐朽味道、泥土味等。

Inferior corona: industrial flavor smell, rotten taste, earthy taste, etc

4. 优质咖啡香味（Excellent Coffee Aroma）

优质香味：咖啡的香气中包含了酒香、花香、果香等丰富的气味。

Excellent aroma: the aroma of coffee contains rich aromas such as wine, flowers, and fruit.

（二）观颜色（Observing the Color）

品鉴咖啡，颜色是重要的判断依据。劣质咖啡往往浑浊不清澈，这可能是由于咖啡豆品质不佳或冲泡方式不恰当所致。相反，优质咖啡则会呈现出深棕色的诱人色泽，且清澈明亮、透明度良好，这样的咖啡往往代表着上乘的品质和恰当的冲泡手法。

Tasting coffee, color is also an important basis for judgment. Poor quality coffee is often cloudy and unclear, which can be due to poor quality beans or improper brewing methods. On the other hand, a good quality coffee will show an attractive dark brown color, and clear, bright and transparent, such coffee often represents superior quality and proper brewing method.

（三）尝风味（Taste the Flavor）

当咖啡喝到口中的时候，能感到它的风味。品鉴咖啡的要点便是如何分辨咖啡的甜、咸、酸、苦，以及醇度，即通常说的黏稠度，指咖啡液表现出来的圆熟、芳醇、味道的浓厚。一杯水与一杯咖啡的区别就在于，水没有味道，不黏稠，但咖啡液则富含油脂，拥有很好的醇度。

When you drink coffee in your mouth, you can feel its flavor. The main point of coffee tasting is how to distinguish the sweet, salty, sour, bitter, and alcohol of coffee, that is, usually said the consistency, refers to the mellow, aromatic, and strong taste of coffee liquid. The difference between a cup of water and a cup of coffee is that the water has no taste and is not viscous, but the coffee liquid is rich in oil and has a good alcohol.

一杯拥有良好醇度的咖啡会让香气在口中停留得更久，尾韵更加绵长。

A cup of coffee with a good body will make the aroma stay in the mouth longer, and the end rhyme will be longer.

1. 甜味（Sweet Taste）

劣质：像苦瓜，过于苦涩。

Bad quality: like bitter melon, too bitter.

优质：甘甜，类似蔗糖、红糖、焦糖，满口生津的感觉。

Excellent quality: sweet, similar to sucrose, brown sugar, caramel, full mouth feeling.

品鉴：带有甜度的咖啡就像一种水果，在咖啡豆中蕴含着果糖。

Taste: coffee with its sweetness is like a fruit, and contains fructose in the beans.

2. 干净度（Cleanliness）

劣质：杂味多。

Inferior quality: mixed taste.

优质：味道纯净清新。

Quality: pure and fresh taste.

品鉴：味道较为杂乱的或有不良味道混入的则不够纯净。

Tasting: The taste is more messy or has bad taste mixed is not pure enough.

3. 苦味（Bitter Taste）

劣质：舌根发苦，粗糙的"苦涩"。

Bad quality: bitter at the base of the tongue, rough "bitterness".

优质：黑巧克力般苦。

Quality: Dark chocolate bitter.

品鉴：精品咖啡不存在单纯的"苦"，一般都会粗浅地用"苦"来形容坚果味。

Taste: specialty coffee does not exist simple "bitter", generally we will use "bitter" to describe the nutty taste.

4. 酸味（Acidity）

劣质：腐酸，尖锐的酸，酸度很高，较为刺激。

Inferior: humic acid, sharp acid, high acidity, more irritating.

优质：果酸，清新的酸，圆润的酸，柔和的酸。

Excellent quality: fruit acid, fresh acid, round acid, soft acid.

品鉴："甜很容易，好酸难得"。优质咖啡的酸味应该是包含在咖啡内，而不是明显且尖利的。

Taste: "sweet is easy, good sour is rare". The sour taste of good coffee should be contained in the coffee, not obvious and sharp.

5. 咸味（Saltiness）

品鉴：咸味分很多种，香气会让人想到咸味较重的食物，联想到类似虾条这一类带有咸味的零食，口感上就是盐的味道。

Tasting: salty taste can be divided into many kinds, the aroma will make people think of salty food, similar to shrimp strips with salty snacks, the taste is the taste of salt.

6. 回味（Aftertaste）

品尝的最后一点就是回甘，指咖啡在口腔中和咽喉处，带来的余韵。喝下咖啡后会有一个味道从喉咙处返回来。有的回味很持久清晰，有的则很短暂模糊。

The last point of tasting is the sweetness, which refers to the coffee in the mouth and throat, to bring an aftertaste. Coffee always has a taste coming back from the throat after drinking it. Some aftertastes are long and clear, while others are short and fuzzy.

此外，还可以从鼻前嗅觉和鼻后嗅觉两个角度品鉴咖啡。鼻前嗅觉是在闻咖啡粉时感受到的香气；鼻后嗅觉则是喝下咖啡后，在口腔内感受到的香气。通过这些方法，可以更深入地品鉴咖啡，从而判断其品质和风味。

In addition, coffee can also be tasted from the perspective of prenasal smell and postnasal smell. The prenasal smell is the aroma felt when smelling coffee powder, while the postnasal smell is the aroma felt in the mouth after drinking coffee. Through these methods, it is possible to taste the coffee more deeply, so as to judge its quality and flavor.

赛证直通 Competitions and Certificates

一、基础知识部分（Basic Knowledge Part）

（一）专业名词解释（Explanation of Professional Terms）

1. 干香（Dry Aroma）：
2. 湿香（Wet Aroma）：

（二）思考题（Thinking Question）

简述咖啡品鉴的判断方法。
Briefly describe the method of judging coffee tasting.

二维码 3-1-5：
知识拓展

二、技能操作部分（Skill Operation Part）

请任意品鉴一款咖啡，并写出结论。
Please taste any coffee and write the conclusion.

专题六

饮"咖"之道：咖啡文化及礼仪

【微课】
饮"咖"之
道：咖啡文化
及礼仪

一、咖啡杯的用法（The Use of Coffee Cups）

端杯时手指不从杯耳过是较为得体的方式（图 3-1-2、图 3-1-3）。通常，餐后饮用的咖啡杯多为小杯设计，其杯耳较小，手指一般无法穿过，然而，其他场合，比如在咖啡厅，如果使用的是较大的咖啡杯，则需特别注意礼仪细节：切忌用手指穿过杯耳来端杯。正确的做法应该是，优雅地使用拇指和食指捏住杯把，轻轻端起杯子，然后，即可享用咖啡。

It is more appropriate to hold the cup without passing the finger over the ear (Figure 3-1-2, Figure 3-1-3). Usually, the coffee cup after a meal is mostly a small cup design, the cup ears are small, the fingers generally can not pass through, however, other occasions, such as in the coffee shop, if the

use of a larger coffee cup, you need to pay special attention to the etiquette details: do not put your fingers through the cup ears to hold the cup. The correct approach should be to gently pick up the cup using the handle of the thumb and forefinger, and then enjoy the coffee.

图 3-1-2 正确拿咖啡杯 (hold the coffee cup correctly)

图 3-1-3 错误拿咖啡杯 (hold the coffee cup mistake)

二、咖啡加糖的方法 (The Way to Add Sugar of Coffee)

咖啡加糖有多种方式。如果是砂糖，可用咖啡匙舀取，直接加入杯内；如果是方糖（图 3-1-4），先用糖夹子把方糖夹在咖啡碟的近身一侧，再用咖啡匙把方糖加进杯子里；如果是糖包则可以贴着杯壁慢慢倒入咖啡中。

There are many kinds of coffee candies. If it is granulated sugar, it can be scooped with a coffee spoon and added directly into the cup. If it is a sugar cube (Figure 3-1-4), first use a sugar tongs to put the sugar cube on the side of the coffee saucer, and then use a coffee spoon to add the sugar cube into the cup. If it is a sugar packet, you can slowly pour it into the coffee.

图 3-1-4 咖啡加糖方法 (method of adding sugar to coffee)

三、咖啡勺的使用 (The Use of Coffee Spoon)

许多人喜欢用咖啡勺（图 3-1-5）喝咖啡或碾碎方糖，这是错误的。咖啡勺的主要功能是加糖或搅拌咖啡，但是搅拌咖啡的力度和幅度不要太大，以免过度搅拌破坏了咖啡的原有风味。而在饮用时，则应将咖啡勺取出，以免影响品饮体验及风度。

Many people like to drink coffee with a coffee spoon (Figure 3-1-5) or crush sugar cubes, which is wrong. The main function of the coffee spoon is to add sugar or stir coffee, but the intensity and amplitude of stirring coffee

图 3-1-5 咖啡勺 (coffee spoon)

should not be too large, so as not to over-stir and destroy the original flavor of coffee. When drinking, the coffee spoon should be taken out to avoid affecting the drinking experience and demeanor.

四、冷却咖啡的方法（A Way to Cool Coffee）

如果咖啡太烫，最好的方法是等咖啡自然冷却，再开始饮用，或者用咖啡勺慢慢搅拌咖啡，切记不要直接用嘴吹凉咖啡，这样很不文雅。如果等待时间太长，还可以在咖啡中加入一些冰块，以便快速降低咖啡的温度。

If the coffee is too hot, the best way is to wait for the coffee to cool naturally, then start drinking, or slowly stir the coffee with the coffee spoon, remember not to blow the coffee directly with the mouth, this is very inelegant. If the waiting time is too long, you can also add some ice cubes to the coffee in order to quickly reduce the temperature of the coffee.

五、咖啡杯碟的用法（The Use of Coffee Cup and Saucer）

盛放咖啡的杯碟应当放在饮用者的正面或右侧，杯耳应指向右方。喝咖啡时，一般而言只需端起杯子就好，但是，如果没有餐桌可以依托，就可以用左手端碟子，右手持咖啡杯耳，这样会比较安全且得体（图 3-1-6）。

The cup or saucer should be placed on the front or right side of the drinker, with the cup ear pointing to the right. When drinking coffee, generally speaking, you only need to pick up the cup, but if there is no table to rely on, you can use the left hand to end the dish, the right hand to hold the coffee cup ear, which will be safer and more appropriate (Figure 3-1-6).

图 3-1-6 咖啡杯碟的用法（the use of coffee cup and saucer）

六、品尝咖啡的方法（The Way to Taste Coffee）

喝咖啡前可先喝一口清水清洁口腔，以便更好地品尝咖啡的风味。同时，正式场合，不要大口喝咖啡，也不要一次喝太多杯，并且，喝咖啡时不要发出声响。

Before drinking coffee, you can drink a mouthful of water to clean your mouth, so as to better taste the flavor of coffee. At the same time, formal occasions, do not drink coffee, do not drink too many cups at a time, and do not make noise when drinking coffee.

七、适度续杯咖啡（Refill your coffee in moderation）

在品味咖啡时，建议连续饮用的量不要超过两杯，以保持适度的享受。当需要续杯时，也应注意细节礼仪，无需将咖啡杯从咖啡碟中拿起，直接添加即可，这样的举止更显文雅得体。

When tasting coffee, it is recommended to drink no more than two cups in a row to maintain moderate enjoyment. When you need to refill a cup, you should also pay attention to the details of etiquette, no need to pick up the coffee cup from the coffee saucer, directly add it, such behavior is more elegant and decent.

八、搭配点心 (Snack With)

通常在社交场合，饮用咖啡都会佐以点心，此时，千万不要一手端着咖啡杯，一手拿着点心，吃一口喝一口地交替进行。这样很不雅观，正确的方法是，饮咖啡时放下点心，吃点心时则放下咖啡，这样才得体大方。

Usually in social occasions, drinking coffee will be accompanied by snacks, at this time, do not hold a coffee cup in one hand, hold a snack in the other hand, eat and drink alternately. This is very unelegant, the correct way is to put down the dim sum when drinking the coffee, and put down the coffee when eating the dim sum, which is decent and generous.

九、咖啡待客礼仪 (Coffee Hospitality)

当邀请客人喝咖啡时，让客人自己添加奶或糖，是对客人口味的尊重，千万不要冒昧帮忙。如果客人非常懂咖啡，主人还可以为之准备一杯凉开水，让客人可以在咖啡和水交替中更好的品味咖啡。

When inviting guests to drink coffee, let the guests add their own milk or sugar, is to respect the taste of the guests, do not be presumptuous to help. If the guest knows coffee very well, the host can also prepare a cup of cold water for him, so that the guest can better taste the coffee in the alternations of coffee and water.

十、呈送咖啡 (Deliver Coffee)

常规社交礼仪都遵循女士优先，有两位以上女士在场时，则以先远后近为原则。切忌在呈送咖啡时，妨碍对方的交谈或阻挡客人交流的视线。如果需要麻烦客人，应先说"打扰一下"并说声"谢谢"。咖啡端到客人面前时，咖啡杯耳和勺把应指向客人的右手边。

The usual social etiquette is ladies first, but When there are more than two women present, the principle is far first than near. Do not obstruct the conversation or the view of the guest when presenting the coffee. If you need to trouble a guest, you should say " excuse me" and " thank you" first. When the coffee is brought to the guest, the coffee cup ear and spoon handle should point to the guest's right hand side.

赛证直通 Competitions and Certificates

一、基础知识部分 (Basic Knowledge Part)

思考题 (Thinking Question)

谈谈如何优雅地饮用一杯咖啡。

Talk about how to drink a cup of coffee gracefully.

二维码 3-1-6：
知识拓展

二、技能操作部分（Skill Operation Part）

请练习咖啡饮用礼仪。

Practice your coffee etiquette.

 【单元三　反思与评价】Unit 3　Reflection and Evaluation

学会了：＿＿＿＿＿＿＿＿＿＿＿＿＿＿＿＿＿＿＿＿＿＿＿

Learned to:＿＿＿＿＿＿＿＿＿＿＿＿＿＿＿＿＿＿＿＿＿

成功实践了：＿＿＿＿＿＿＿＿＿＿＿＿＿＿＿＿＿＿＿＿

Successful practice:＿＿＿＿＿＿＿＿＿＿＿＿＿＿＿＿＿

最大的收获：＿＿＿＿＿＿＿＿＿＿＿＿＿＿＿＿＿＿＿＿

The biggest gain：＿＿＿＿＿＿＿＿＿＿＿＿＿＿＿＿＿

遇到的困难：＿＿＿＿＿＿＿＿＿＿＿＿＿＿＿＿＿＿＿＿

Difficulties encountered:＿＿＿＿＿＿＿＿＿＿＿＿＿＿

对教师的建议：＿＿＿＿＿＿＿＿＿＿＿＿＿＿＿＿＿＿＿

Suggestions for teachers:＿＿＿＿＿＿＿＿＿＿＿＿＿＿

单元二　可　　可

"沿着赤道寻找，你一定会发现可可"。

赤道附近的地区通常能够提供可可树生长的条件，因此成了可可树种植的理想场所。

单元导入 Unit Introduction

可可与咖啡、茶并称为世界三大饮料作物，是世界上极受欢迎的食物——巧克力的原材料。那么，究竟是谁最早发现可可，并且食用可可的呢？可可又是如何征服人类味蕾，继而走向全球的呢？这个单元将带你走进可可的世界。

Cocoa, together with coffee and tea, is known as one of the world's three major beverage crops and is the raw material for the extremely popular food-chocolate. So, who was the first to discover cocoa and consume it? How did cocoa conquer human taste buds and then go global? This unit will take you into the world of cocoa.

学习目标 Learning Objectives

➤ 知识目标（Knowledge Objectives）

（1）了解可可的起源和传播。
Understand the origin and spread of cocoa.
（2）掌握可可的生产国和消费国。
Master the producing and consuming countries of cocoa.
（3）了解可可制品。
Understand cocoa products.
（4）熟悉鉴别可可粉品质的方法。
Be familiar with the methods of identifying the quality of cocoa powder.

➤ 能力目标（Ability Objectives）

（1）能介绍可可的主要制品。
Able to introduce the main products of cocoa.
（2）能介绍巧克力的发展史。
Able to introduce the history of chocolate development.
（3）能鉴别可可粉品质。
Able to identify the quality of cocoa powder.

➤ 素质目标（Quality Objective）

培养学生具备良好的职业精神、专业精神和工匠精神。
Cultivate students with a good professional spirit, professional spirit, and craftsman spirit.

专题一
非同小"可"：可可概述

【微课】
非同小"可"：
可可概述

你知道可可长什么样子吗？生活中接触过可可制品吗
（图3-2-1）？

_____

Do you know what cocoa looks like? Have you ever come
into contact with cocoa products in your life (Figure 3-2-1)?

_____

图3-2-1 可可（cocoa）

一、认识可可豆（Undertanding Cocoa Beans）

可可豆是一种热带植物，属于梧桐科可可属。可可果实呈椭圆形或卵圆形，果皮较厚，表面有明显的纵向沟纹，颜色从绿色到黄色、红色或紫色不等，成熟时变为橙黄色或深红色。每颗果实中含有 20～50 粒可可豆，种子呈扁圆形，长 2～3 厘米，初时为白色或淡紫色，经过发酵和干燥后变为深褐色。

Cocoa beans are a tropical plant belonging to the genus Theobroma in the family Malvaceae. The cocoa fruit is oval or egg-shaped, with a thick rind that has distinct longitudinal grooves. The color of the fruit ranges from green to yellow, red, or purple, and it turns orange-yellow or deep red when ripe. Each fruit contains 20-50 cocoa beans. The seeds are flat and round, about 2-3 centimeters in length. Initially, they are white or pale purple, but after fermentation and drying, they turn dark brown.

可可树喜欢温暖湿润的热带气候，适宜生长温度为 20～30 ℃，年降水量需要达到 1 500～2 000 毫米。在不同地域生长的可可豆会有不同的风味，有的带点果香，有的则有烟熏风味。当今可可豆主要的产地主要有中南美洲、西非及东南亚三地。

Cocoa trees thrive in warm and humid tropical climates, with an optimal growing temperature of 20-30 ℃ and an annual rainfall requirement of 1 500-2 000 millimeters. Cocoa beans grown in different regions can have varying flavors, with some exhibiting fruity notes and others having a smoky taste. Today, the main regions for cocoa bean production are Central and South America, West Africa, and Southeast Asia.

可可豆是一种神奇的天然食材，富含多种营养成分，对健康有着诸多益处。黄烷醇是可可豆中最重要的抗氧化物质之一，它能够中和自由基，保护细胞免受氧化损伤，从而延缓衰老、降低慢性疾病的风险。还能改善血管功能和大脑功能。可可豆中还含有可可碱和咖啡因，具有轻微的兴奋作用，能够刺激神经系统，带来愉悦感和放松感。此外，可可豆中的苯乙胺被称为"爱情分子"，是一种天然的神经递质。它能够刺激大脑释放内啡肽，带来幸福感和愉悦感，这也是为什么用可可豆制成的巧克力常被称为"爱情食品"的原因。

Cocoa beans are a miraculous natural ingredient, rich in various nutrients, offering numerous health benefits. Flavanols are one of the most important antioxidants in cocoa beans. They can neutralize free radicals, protect cells from oxidative damage, thereby delaying aging and reducing the risk of chronic diseases. They also improve vascular and brain functions. Cocoa beans also contain theobromine and caffeine, which have a mild stimulant effect, stimulating the nervous system to bring a sense of pleasure and relaxation. Additionally, phenylethylamine in cocoa beans, known as the "love molecule," is a natural neurotransmitter. It can stimulate the brain to release endorphins, bringing feelings of happiness and pleasure, which is why chocolate made from cocoa beans is often referred to as a "love food."

二、可可的传播（The Spread of Cocoa）

可可的起源可以追溯到古代美洲，尤其是中美洲和南美洲的热带雨林地区。玛雅人是最早将可可豆加工成饮品的文明之一。他们将可可豆磨碎，加入水、辣椒和其他香料，制成一种苦味浓郁的饮料。阿兹特克人继承了玛雅人的可可文化，并将其进一步发展。

The origin of cocoa can be traced back to ancient America, particularly the tropical rainforest regions of Central and South America. The Maya were one of the earliest civilizations to process cocoa beans into a beverage. They ground the cocoa beans, mixed them with water, chili peppers, and other spices to create a rich, bitter drink. The Aztecs inherited the cocoa culture from the Maya and further developed it.

1528 年，西班牙探险家赫尔南·科尔特斯将可可豆带回西班牙，这是可可豆首次正式进入欧洲。最初，可可饮料仍然保留了苦味风格，但西班牙人很快对其进行了改良，加入了糖、蜂蜜和香料，使其更符合欧洲人的口味。之后，科尔特斯在西班牙推广可可饮料，在美洲殖民地推广可可种植。他在墨西哥等地建立了可可种植园，鼓励当地居民种植可可豆，以满足欧洲市场的需求。这一举措为可可豆的规模化生产和贸易奠定了基础。

In 1528, Spanish explorer Hernán Cortés brought cocoa beans back to Spain, marking the first official introduction of cocoa to Europe. Initially, the cocoa beverage retained its bitter flavor, but the Spaniards quickly modified it by adding sugar, honey, and spices to better suit European tastes. Subsequently, Cortés promoted the cocoa beverage in Spain and encouraged the cultivation of cocoa in the American colonies. He established cocoa plantations in places like Mexico and encouraged local residents to grow cocoa beans to meet the demands of the European market. This initiative laid the foundation for the large-scale production and trade of cocoa beans.

三、可可的生产国和消费国（Cocoa Producing and Consuming Countries）

（一）生产国（Producing Countries）

全球大部分的可可产量来自西非国家，尤其是科特迪瓦和加纳，这两个国家通常占据全球可可豆产量的一半以上。全球超过 35% 的可可产自科特迪瓦，该国有超过 600 万人从事可可种植行业，可可收入占国内生产总值的 22%，出口量比重超过 50%。这也使得科特迪瓦成为非洲为数不多的贸易顺差国。除了非洲的几个国家，亚洲的印度尼西亚近年来可可的种植发展非常迅猛。

Most of the global cocoa production comes from West African countries, especially Côte d'Ivoire and Ghana. These two countries usually account for more than half of the global production of cocoa beans. More than 35% of the world's cocoa is produced in Côte d'Ivoire, which has more than 6 million people engaged in the cocoa cultivation industry. Cocoa revenue accounts for 22% of the country's GDP, and the export volume accounts for more than 50%. This also makes Côte d'Ivoire one of the few African countries with a trade surplus. In addition to several African countries, Indonesia in Asia has seen a rapid development of cocoa cultivation in recent years.

可可生产国大都将可可以原料形式出口，对可可的精细加工较少，导致附加值和利润很低。种植了一辈子可可的非洲农民，可能很少甚至根本就没有品尝过巧克力。不过近年来，随着可可产业的不断发展，可可的原产地加工率也在不断地提高，如在第一大可可生产国——科特迪瓦，在 2015 年建立起了第一家由法国投资的巧克力生产工厂。2019 年，中国企业承建科特迪瓦最大自有可可加工厂，助力科特迪瓦可可产业加速发展，这也是中科共建"一带一路"的重点合作项目。

Cocoa-producing countries mostly export cocoa in the form of raw materials and have less refined processing of cocoa, resulting in low added value and profits. African farmers who have been planting cocoa for a lifetime may seldom or even never have tasted chocolate. However, in recent years, with the continuous development of the cocoa industry, the rate of local processing of cocoa has been continuously increasing. For example, in Côte d'Ivoire, the first-largest cocoa-producing country, In 2015, the first chocolate production factory invested by France was established. In 2019, a Chinese enterprise undertook the construction of Côte d'Ivoire's largest self-owned cocoa processing plant, accelerating the development of Côte d'Ivoire's cocoa industry. This is also a key cooperation project under the "Belt and Road" initiative jointly built by China and Côte d'Ivoire.

（二）消费国（Consuming Countries）

可可的主要消费国大都位于欧美国家和地区。荷兰是第一大可可进口国，世界上最大的可可交易集散地位于荷兰的第一大城市——阿姆斯特丹，这里每年成交的可可豆大约占据世界产量的五分之一。荷兰进口了大量可可，将其加工成巧克力等可可制成品后再出口到国外，从而创造了大量外汇收入，其利润比众多的可可种植国高出许多。美国则是可可第二大进口国，每年进口大量的可可豆用于生产可可饼、可可酒、巧克力、糖果等加工产品供国内消费和出口。德国的饮料产业、糖果产业非常发达，可可的消费量自然也不小，位居全球第三位。

The main consuming countries and regions of cocoa are located in Europe and the United States. The Netherlands is the largest importer of cocoa, and the world's largest cocoa trading hub is located in Amsterdam, the largest city in the Netherlands. The cocoa beans traded here each year account for about one-fifth of the world's production. The Netherlands imports a large amount of cocoa, processes it into chocolate and other cocoa products, and then exports them to other countries, thus creating a large amount of foreign exchange income. Its profits are much higher than those of many cocoa-producing countries. The United States is the second-largest cocoa importer, importing a large amount of cocoa beans each year for the production of cocoa cakes, cocoa wine, chocolate, candy, and other processed products for domestic consumption and export. The beverage and confectionery industries in Germany are very developed, and the consumption of cocoa is naturally not small, ranking third in the world.

四、可可制品（Cocoa Products）

最初，可可是被人们磨成粉末用水熬煮冲泡饮用。在对可可食用方式的研究中，人们发现了让可可发挥独特风味的形式——巧克力。巧克力是欧洲人几百年来对可可的一种创造，这种创造，包括形式、口味、食用方式等多个方面。1815 年，荷兰人范·豪尔顿发明了碱化技术，使巧克力苦味减低，口感更加类似今天喝到的热巧克力。经过这种处理的巧克力从此被称为"荷兰巧克力"。随着时间的推进，巧克力中的香料添加越来越少，制作工艺也不断发生改进。1828 年，豪尔顿发明了一种螺旋压力机，使用这种压力机，他第一次将可可脂从可可豆中分离。随后有人发现，可可脂可以使可可豆磨成的糊状物更光滑，更容易跟糖混合在一起；而可可豆榨完脂后剩下的可可粉，则可以使可可饮料的风味更浓厚。这个发明使巧克力质量大大提高。从此以后固体巧克力终于告别了粗糙和易碎，开始精致起来了。1875 年，瑞士人丹尼尔·彼得把奶粉和炼乳加进了巧克力，发明了牛奶巧克力，从而改变了巧克力苦味的历史。

Initially, cocoa was ground into powder and boiled with water for drinking. In the study of edible forms of cocoa, people discovered a form that gives cocoa its unique flavor—chocolate. Chocolate is a creation of Europeans for hundreds of years, which includes aspects such as form, taste, and eating methods. In 1815, a Dutchman named Van Houten invented the process of alkalization, which reduced the bitterness of chocolate and made its taste more similar to the hot chocolate we drink today. Chocolate processed in this way has since been known as "Dutch chocolate". Over time, the addition of spices in chocolate has decreased, and the manufacturing process has been continuously improved. In 1828, Van Houten invented a screw press, which for the first time, allowed the separation of cocoa butter from cocoa beans. It was then discovered that cocoa butter can make the paste made from cocoa beans smoother and easier to mix with sugar, while the remaining cocoa powder after the cocoa beans have been pressed can make the cocoa drink more flavorful. This invention greatly improved the quality of chocolate. Since then, solid chocolate has finally moved away from being rough and fragile and has become more refined. In 1875, the Swiss Daniel Peter added milk powder and condensed milk to chocolate, inventing milk chocolate, thus changing the history of chocolate being bitter.

赛证直通 Competitions and Certificates

一、基础知识部分（Basic Knowledge Part）

二维码 3-2-1：
知识拓展

思考题（Thinking Question）

说说可可的主要生产国和消费国。
Talk about the main producing and consuming countries of cocoa.

二、技能操作部分（Skill Operation Part）

分析可可的营养成分，并介绍其益处。
Analyze the nutritional components of cocoa and introduce its benefits.

专题二
历历"可"辨：可可粉品鉴

你知道可可粉（图 3-2-2）可以用来制作哪些食品吗？

_____

Do you know what kind of foods can be made with cocoa powder (Figure 3-2-2)?

_____

图 3-2-2　可可粉（cocoa powder）

一、可可粉的类别（Categories of Cocoa Powder）

可可粉是一种由可可豆经过发酵、烘焙、研磨等工艺制成的粉末，广泛应用于食品加工和烹饪中。它不仅是制作巧克力、蛋糕、饼干等甜点的重要原料，还可以用来调制热可可饮料，深受人们喜爱。可可粉按其含脂量分为高、中、低脂可可粉；按加工方法不同分为非荷兰可可粉和荷兰可可粉。

Cocoa powder is a powder made from cocoa beans through processes such as fermentation, roasting, and grinding. It is widely used in food processing and cooking. Not only is it an essential ingredient for making desserts like chocolate, cakes, and cookies, but it can also be used to prepare hot cocoa beverages, making it highly popular among people. Cocoa powder is divided into high, medium, and low-fat cocoa powders according to their fat content; and into natural (non-Dutch) and Dutch-processed cocoa powders based on different processing methods.

（一）按含脂量分（By Fat Content）

1. 高脂可可粉（High-fat Cocoa Powder）

高脂可可粉保留了更多的可可脂，脂肪含量通常在 22% ～ 24%。这种可可粉通常在制作巧克力时使用，因其脂肪含量较高，能提供丰富的口感和浓郁的巧克力风味。

High-fat cocoa powder retains more cocoa butter, with a fat content typically ranging from 22% to 24%. This type of cocoa powder is commonly used in chocolate production, as its higher fat content provides a rich texture and intense chocolate flavor.

2. 中脂可可粉（Medium-fat Cocoa Powder）

中脂可可粉的可可脂含量介于高脂和低脂之间，通常在 10% ～ 22%。这种类型的可可粉适合用于制作一些需要适中口感和风味的甜点和烘焙食品。

Medium-fat cocoa powder has a cocoa butter content between high-fat and low-fat, usually between 10% and 22%. This type of cocoa powder is suitable for making desserts and baked goods that require a moderate flavor and texture.

3. 低脂可可粉（Low-fat Cocoa Powder）

低脂可可粉在加工过程中去除了大部分的可可脂，脂肪含量通常在 10% 以下。这种可可粉适合那些需要控制脂肪摄入量的人群，或者在制作清淡口感的甜点时使用。低脂可可粉在食谱中可以作为高脂可可粉的替代品，但可能影响最终产品的口感和风味。

Low-fat cocoa powder has most of its cocoa butter removed during the processing, with a fat content usually below 10%. This type of cocoa powder is suitable for people who need to control their fat intake or for making desserts that require a lighter taste. Low-fat cocoa powder can be used as a substitute for high-fat cocoa powder in recipes, but it may affect the final product's taste and texture.

（二）按加工方法分（By Processing Method）

非荷兰可可粉和荷兰可可粉的主要区别在于加工过程中是否使用了一种称为"荷兰化处理"的方法。荷兰化处理是一种通过碱处理来改变可可粉的 pH 值的工艺，这样做可以减少可可粉的酸度，使其颜色更加鲜艳，口感更加温和。

The main difference between natural (non-Dutch) and Dutch-processed cocoa powder lies in whether a process called "Dutch processing" is used during the processing. Dutching is a process that alters the pH of cocoa powder through alkali treatment, which reduces the acidity, makes the color more vibrant, and results in a milder flavor.

1. 非荷兰可可粉（Natural Cocoa Powder）

没有经过碱性处理，颜色较浅，保留了可可豆的自然酸味和果香，味道更酸、更苦，适合于巧克力烘焙和某些特定的甜点，如巧克力蛋糕、布朗尼等。

Not treated with alkali, it has a lighter color and retains the natural acidity and fruity aroma of the cocoa beans, with a more acidic and bitter taste. It is suitable for use in chocolate baking and specific desserts, such as chocolate cakes, brownies.

2. 荷兰可可粉（Dutch Cocoa Powder）

经过碱处理，酸度降低，口感更加温和，颜色更加红艳。这种可可粉适合那些不喜欢太强烈酸味的人群，常用于液体巧克力或烘焙食品。例如，需要颜色鲜艳的烘焙产品中，像红色天鹅绒蛋糕。

Treated with alkali, it has reduced acidity and a milder flavor, with a more reddish color. This type of cocoa powder is suitable for people who do not prefer a strong acidic taste and is often used in liquid chocolate or baked goods. For example, it is often used in baked products that require a vibrant color, such as red velvet cakes.

二、可可粉的鉴别（Identification of Cocoa Powder）

可可粉是食品原料，国标对它的生产给出了严格的规定。在没有检测仪器的情况下可以通过简单的方法判断其品质，主要从颜色、气味、细度和含脂量等几个方面着手。

Cocoa powder is a food ingredient, and national standards have strict regulations on its production. In the absence of testing equipment, a simple method can be used to judge its quality, mainly from aspects such as color, smell, fineness, and fat content.

（一）观颜色（Observe The Color）

天然可可粉，即非荷兰可可粉，其颜色应该是浅棕色，若呈现棕色甚至是深棕色的天然可可粉很可能是添加了可可皮或其他食用色素。

Natural cocoa powder, that is, non-Dutch cocoa powder, should be light brown. Natural cocoa powder that appears brown or even dark brown is likely to have added cocoa skin or other edible pigments.

碱化可可粉，即荷兰可可粉，其颜色应该是棕红色。若为深棕色或棕黑色，那定是碱化过重，灰粉含量过多所致。

Alkali cocoa powder, that is, Dutch cocoa powder, should be reddish-brown. If it is dark brown or brown-black, it is definitely due to excessive alkalization and too much ash content.

（二）闻气味（Smell The Aroma）

将一小撮可可粉放在手心，用鼻子深吸。优质的可可粉应具有浓郁的巧克力香气，可能还带有一些坚果、果香或其他复杂的香气。

Place a small pinch of cocoa powder in the palm of your hand and take a deep sniff. High-quality cocoa powder should have a rich chocolate aroma, which may also have some nutty, fruity, or other complex aromas.

天然可可粉的气味是天然的可可香味，是淡淡的清香。浓香或是焦味的可可粉则为品质较差的粉。碱化可可粉的气味应该是正常的可可香味，其香气比天然可可粉的要浓一些，但并没有焦味。若碱化可可粉的香气太浓或是有焦味，则为品质较差的粉。

The aroma of natural cocoa powder is the natural fragrance of cocoa, which is a light and refreshing scent. Cocoa powder with a strong fragrance or a burnt smell is of poorer quality. The aroma of alkali cocoa powder should be the normal aroma of cocoa, which is more intense than that of natural cocoa powder, but without a burnt smell. If the aroma of alkali cocoa powder is too strong or has a burnt smell, it is also of poorer quality.

（三）看质地（Check The Texture）

可可粉的细度直接影响烘焙和饮品口感。细度高的可可粉更容易溶解或混合，口感更细腻，而粗颗粒的可可粉可能会影响成品的质地。

首先，可目测观察，细度高的可可粉看起来细腻且颜色均匀，没有明显的颗粒感。粗颗粒的可可粉则会看到明显的颗粒或结块，颜色可能不均匀。

其次，通过触感测试，用手指轻轻捏取少量可可粉，感受其质地。细腻的可可粉触感柔滑，粗颗粒可可粉能感觉到颗粒感，甚至有轻微的摩擦感。

最后，还可通过溶解来测试。将可可粉加入温水中搅拌，细度高的可可粉能快速溶解，形成均匀的液体，无明显沉淀。粗颗粒的可可粉溶解较慢，可能会有沉淀或颗粒残留。

The fineness of cocoa powder directly affects the texture of baked goods and beverages. Cocoa powder with a high degree of fineness dissolves or mixes more easily, resulting in a smoother texture, whereas coarser cocoa powder may affect the final product's consistency.

Firstly, you can visually inspect the cocoa powder. High-fineness cocoa powder appears smooth and has a uniform color without noticeable graininess. Coarser cocoa powder, on the other hand, may show visible grains or clumps, and the color might appear uneven.

Secondly, you can perform a tactile test. Gently pinch a small amount of cocoa powder between your fingers to feel its texture. Fine cocoa powder will feel smooth and silky, while coarser cocoa powder will have a noticeable graininess and may even produce a slight gritty sensation.

Lastly, you can test the cocoa powder by dissolving it. Add the cocoa powder to warm water and stir. High-fineness cocoa powder will dissolve quickly, forming a uniform liquid with no noticeable sediment. Coarser cocoa powder will dissolve more slowly and may leave behind sediment or undissolved particles.

（四）辨别含脂量（Identify The Fat Content）

取少量的可可粉于手掌心，两手对搓，含脂量高的会有明显的油腻感，含脂量较低的则不明显。含脂量 10% 以上的可以用这个方法测试。含脂量在 8% 以下的基本上感觉不到油腻感。

Place a small amount of cocoa powder in the palm of your hand and rub your hands together. Cocoa powder with a high-fat content will have a noticeable greasy feel, while cocoa powder with a lower-fat content will not be as noticeable. This method works well for testing cocoa powder with a fat content of more than 10%. Essentially, there will be no noticeable greasy feel for cocoa powder with a fat content of less than 8%.

赛证直通 Competitions and Certificates

一、基础知识部分（Basic Knowledge Part）

二维码 3-2-2：
知识拓展

专业名词解释（Explanation of Professional Terms）

1. 可可粉（Cocoa Powder）：
2. 荷兰化处理（Dutch Process）：

二、技能操作部分（Skill Operation Part）

试鉴别实物可可粉的优劣。

Please identify the quality of the actual cocoa powder.

3

附录一
"术"说酒类：酒类英文术语

一、酒类品种 (Types of Alcoholic Beverages)

白酒	Baijiu(Chinese spirits)	白葡萄酒	White Wine	威士忌	Whisky(or Whiskey)
啤酒	Beer	伏特加	Vodka	龙舌兰酒	Tequila
红酒	Red Wine	朗姆酒	Rum	力娇酒	Liqueur
香槟	Champagne	金酒	Gin	果酒	Fruit Wine
清酒	Sake	梅酒	Plum Wine	鸡尾酒	Cocktail
波特酒	Port Wine	雪利酒	Sherry	马提尼	Martini
苦艾酒	Vermouth	白兰地	Brandy	葡萄酒	Wine

二、酿造工艺 (Brewing process)

发酵	Fermentation	蒸馏	Distillation	陈化	Maturation	陈年	Aging
桶陈	Barrel Aging	酿造	Brewing	酿酒	Mead Maker	糖化	Mashing
煮沸	Boiling	加酒花	Hopping	去梗	Stems	压榨	Crating
澄清	Clarification	酒曲	Fermenting agent	蒸馏器	Distillery	酒精度	Alcohol by volume
酒窖	Cellar	基酒	Base wine	混酿	Blending	过滤	Filtration
发酵	Fermentation	蒸馏	Distillation	陈酿	Aging	老熟	maturation

三、品鉴方法 (Approach to Tasting)

酒体	Body	单宁	Tannins	矿物质感	Minerality	风味	Flavor
酸度	Acidity	果香	Fruitiness	复杂性	Complexity	回味	Aftertaste
香气	Aroma	浓郁	Richness	橡木桶香	Oak barrel fragrame	浑浊	Haze
圆润	Roundness	酒精感	Alcoholic	酒石酸	Tartaric Acid	酒腿	Legs
优雅	Elegance	结构	Structure	瓶差	Bottle Variation	醒酒	Decanting

四、酒水服务场景中英文对话 (Beverage Service Scene English Dialogue)

(一) 场景一：客户询问葡萄酒推荐

客户：你好，我想问一下有什么葡萄酒推荐吗？

Customer: Hello, could you recommend a wine for me?

服务员：您好，我们这里有一款来自波尔多的红葡萄酒，口感丰富，非常适合搭配红肉类菜肴。

Waiter: Hello, we have a red wine from Bordeaux that has a rich palate and pairs well with red meat dishes.

(二) 场景二：客户询问威士忌品饮方式

客户：我想尝试一些威士忌，有什么建议吗？

Customer: I'd like to try some whisky. Do you have any suggestions?

服务员：当然，您可以先闻香，然后小口细品，感受威士忌的多层次风味。

Waiter: Of course, you can start by smelling the aroma, then take a small sip and savor the multi layered flavors of the whisky.

(三) 场景三：广泛适用于酒吧或餐厅酒水点单服务

服务员：晚上好，先生／女士。欢迎来到我们餐厅。请问今晚您需要点些什么饮品？

Waiter: Good evening, sir/madam. Welcome to our restaurant. May I take your order for drinks this evening?

客户：是的，我不太确定要点什么。你能推荐一款与我们的餐点搭配的葡萄酒吗？

Customer: Yes, I'm not quite sure what to have. Could you please recommend a wine that pairs well with our meals?

服务员：当然可以。如果您点的是牛排，我会推荐我们的赤霞珠。它酒体饱满，单宁丰富，非常适合牛排的风味。

Waiter: Certainly. If you're having the steak, I would recommend our Cabernet Sauvignon. It has a rich body and full tannins that complement the flavors of the steak very well.

客户：听起来不错。我还需要给我妻子点一瓶白葡萄酒。她更喜欢轻盈且果味浓郁的酒。

Customer: That sounds great. I'll also need a white wine for my wife. She prefers something light and fruity.

服务员：这样的话，我们的霞多丽会是个很好的选择。它是一款酒体轻盈、酸度清爽且果味丰富的白葡萄酒。

Waiter: In that case, our Chardonnay would be a good choice. It's a white wine with a light body, crisp acidity, and plenty of fruit flavors.

客户：太好了。请给我们各来一瓶。

Customer:Excellent. We'll have a bottle of each, please.

服务员：绝佳的选择。我马上为您拿来。您需要先点一些开胃菜吗？

Waiter: Wonderful choice. I'll bring them right over. Would you like to start with any appetizers?

客户：我们先点一份奶酪拼盘，然后再继续点主菜。

Customer:We'll start with the cheese platter and then proceed to the main course.

服务员：非常好。我会为您准备这些，并很快为您提供您的饮品。

Waiter:Very well. I'll prepare that for you and have your drinks served shortly.

客户：谢谢！

Customer:Thank you!

附录二
"叙"说茶类：茶类英文术语

一、茶叶分类专业词汇（Specialized Vocabulary for Tea Classification）

茶	Tea	绿茶	green tea	红茶	black tea
白茶	white tea	花茶	jasmine tea	黑茶	dark tea
黄茶	yellow tea	不发酵茶	non-fermented	后发酵茶	post-fermented
半发酵茶	partially fermented	全发酵茶	complete fermentation		

绿茶分类术语	蒸青绿茶	steamed green tea	粉末绿茶	powered green tea
	银针绿茶	silver needle green tea	原形绿茶	lightly rubbed green tea
	松卷绿茶	curled green tea	剑片绿茶	sword shaped green tea
	条形绿茶	twisted green tea	圆珠绿茶	pearled green tea
普洱茶分类术语	陈放普洱	age-puer	渥堆普洱	pile-fermented puer
乌龙茶分类术语	白茶乌龙	white oolong	条形乌龙	twisted oolong
	球形乌龙	Spherial oolong	熟火乌龙	roasted oolong
	白毫乌龙	white tipped oolong		
红茶分类术语	工夫红茶	Congou black tea	碎形红茶	shredded black tea
熏花茶分类木语	熏花绿茶	scented green tea	熏花普洱	scented puer tea
	熏花乌龙	scented oolong tea	熏花红茶	scented black tea
	熏花茉莉	jasmine scented green tea		

二、制茶工艺词汇（Tea Processing Terminology）

茶树丛	tca bush	茶园	tea garden	采青	tea harvesting
茶青	tea leaves	萎凋	withering	日光萎凋	sun withering
室内萎凋	indoor withering	静置	setting	搅拌（浪青）	tossing
发酵	fermentation	氧化	oxidation	杀青	fixation
蒸青	steaming	炒青	stir fixation	烘青	baking
晒青	sunning	揉捻	rolling	轻揉	light rolling

续表

重揉	heavy rolling	布揉	cloth rolling	干燥	drying
炒干	stir-dry	烘干	baking	晒干	sunning
渥堆	piling	精制	refining	筛分	screening
剪切	cutting	整形	shaping	风选	winnowing
拼配	blending	紧压	compressing	覆火	re-drying
陈放	aging	加工	added process	焙火	roasting
熏花	scenting	调味	spicing	茶饮料	tea beverage

三、茶叶名称词汇（Tea Names Vocabulary）

铁观音	iron mercy goddess	桂花乌龙	osmanthus oolong
人参乌龙茶	ginseng oolong	茉莉花茶	jasmine tea
玫瑰绣球	rose bulb	工夫红茶	gongfu black
烟熏红茶	smoke black	熟火乌龙	roast oolong
清茶	light oolong	安吉白茶	Anji white leaf
六安瓜片	Lu'an leaf	凤凰单丛	Fenghuang unique bush
茶粉	tea powder	抹茶	fine powder tea
白毫乌龙	white tipped oolong	武夷岩茶	Wuyi rock
黄山毛峰	Huangshan Mountain fuzz tip	龙井	dragon well
珠茶	peart tea	青沱	age cake puer
君山银针	Junshan Mountain silver needle	白毫银针	white tip silver needle
玉露	long brow jade dew	大红袍	robe tea
水仙	narcissus	肉桂	Cassia tea
白牡丹	white peony	碧螺春	green spiral

四、茶具名称词汇（Tea Set Terminology）

有流茶碗	spout bowl	茶席	Tea Ceremony	茶壶	tea pot
壶垫	tea pad	茶船	tea plate	茶盅	tea pitcher
盖置	lid saucer	奉茶盘	tea serving tray	茶杯	tea cup
杯托	cup saucer	杯盖	cup cover	茶巾	tea towel
茶巾盘	tea towel tray	茶荷	tea holder	茶拂	tea brush
定时器	timer	煮水器	water heater	水壶	water kettle
茶车	tea cart	坐垫	seat cushion	茶具袋	tea ware bag
地衣	ground pad	煮水器底座	heating base	茶托	tea ware tray
个人茶道组	personal tea set	冲泡盅	brewing vessel	水盂	tea basin
盖碗	covered bowl	茶匙	tea spoon	茶器	tea ware
热水瓶	thermos	茶叶罐	tea caddy	茶具	tea set
茶桌	tea table	侧柜	side table	茶碗	tea bowl

五、泡茶术语 (The term for brewing tea)

备具	prepare tea ware	备水	prepare water	温壶	warm pot
备茶	prepare tea	识茶	recognize tea	赏茶	appreciate tea
温盅	warm pitcher	置茶	put in tea	闻香	smell fragrance
第一道茶	first infusion	计时	set timer	烫杯	warm cups
斟茶	pour tea	备茶杯	prepare cups	分茶	divide tea
端杯奉茶	serve tea by cups	第二道茶	second infusion	持盅奉茶	serve tea by pitcher
收杯	collect cups	去渣	Take out brewed leaves	赏叶底	appreciate leaves
涮壶	rinse pot	归位	return to seat	清盅	rinse pitcher

六、茶叶品质鉴别术语 (Terminology for Tea Quality Assessment)

乌润	black bloom	乌而油润。此术语也适用于红茶和乌龙茶干茶色泽
半筒黄	semi-yellow	色杂，叶尖黑色，柄端黄黑色
黑褐	black auburn	褐中带黑。此术语也适用于压制茶汤色、叶底色泽，及乌龙茶和红茶干茶色泽
棕褐	brownish auburn	褐中带棕。此术语也适用于压制茶汤色、叶底和红茶干茶色泽
青黄	blueish yellow	黄中泛青，原料后发酵不足所致
褐红	Brownish red	红中带褐
橙红	orange red	红中泛橙色。此术语也适用于乌龙茶汤色
暗红	red dull	此术语也适用于红茶汤色
棕	brownish red	红中泛棕，似咖啡色。此术语也适用于红茶干茶色泽及红碎茶茶汤加奶后的汤色
棕黄	brownish yellow	黄中泛棕。此术语也适用于红碎茶干茶色泽
红黄	reddish yellow	黄中带红
金黄	golden yellow	以黄为主，带有橙色，有深浅之分
清黄	clear yellow	茶汤黄而清澈
软亮	soft and bright	叶质柔软，叶色透明发亮
清醇	clean and mellow	茶汤味新鲜，入口爽适
甘鲜	sweet and fresh	鲜洁有甜感
粗浓	coarse and heavy	味粗而浓

七、茶馆服务常用英语 (Common English Phrases for Tea House Service)

(一) 茶礼

Good morning!/Good afternoon!/Good evening! Welcome to the ta house.

您好！欢迎光临！

Can I help you?

请问需要什么帮助？

This way, please.

请往这边走。

Do you want a separate room?

您需要包厢吗？

How about seats near the window? You can see beautiful scenes outsides.

靠窗的座位行吗？从这里可以看到室外的景色。

This is the tea menu, please make your choices.

这是我们茶馆的茶单，请随意挑选。

Excuse me, would you please tell me which kind of tea you prefer?

打扰了，我现在可以知道您喜欢喝哪种茶吗？

I'd like to recommend you the famous Yixing-Mingding tea.

我向您推荐著名的宜兴茗鼎茶。

Sorry for having kept you waiting for so long. Here is Mingding tea you ordered. I hope you will like it.

对不起，让你们久等，这是你们点的茗鼎茶，请品尝。

China is the hometown of tea and cradle of tea culture.

中国是茶的故乡，茶文化的发祥地。

It's virtue of Chinese people to serve tea to guests.

客来敬茶是中国人的美德。

Hello, everybody. Now, I am preparing Fujian oolong tea for you.

你们好，现在我为大家冲泡福建乌龙茶。

The first infusion is ready, I hope you will like it.

第一壶茶冲泡好了，请各位慢用。

Here are the refreshments you want. If you want something else, please feel free to let me know.

这是您要的茶点，还有什么需要的，请尽管吩咐。

Excuse me, it is better to add some water.

打扰了，我给壶里加点水。

Here are the tea refreshments you have ordered. Enjoy yourself.

你们要的几种茶点已经都上齐了，请慢慢品尝。

Excuse me, Sir, you are wanted to the phone. Would you please go to the counter?

打扰了，先生，有您的电话，请到柜台那边去接。

Well,150 Yuan in total. 200 Yuan, thank you. Here is your change, 50 Yuan.

结账吗？好，一共是150元整，收您200整，找您50元。

Thanks for coming. Hope to see you again!

谢谢光临，欢迎下次再来！

（二）茶艺

To prepare a good cup of tea, you need fine tea, good water, beautiful cup, nice people and proper environment.

泡一杯好茶，要做到茶美、水美、器美、人美、环境美。

There are three stages when water is boiling. At the first stage, the bubbles look like crab eyes; at the second, the bubbles look like fish eyes; finally, they look like surging waves.

烧水时，一沸为"蟹眼"，二沸为"鱼眼"，三沸称作"沸波鼓浪"。

The water boiling between the crab-eye stage and the fish-eye stage is the best for preparing tea

泡茶用的开水，一般以"蟹眼已过鱼眼生"时最好。

We should use big fire to make water boil quickly 烧水要做到活火快煎。

The water that has been boiling for a long time is not good. 水老（即已沸波鼓浪多时）不理想。

A cup of good tea requires skills in preparing. 好茶还需巧冲泡。

Natural mountain spring water is best for tea 泡茶用的水，以天然的山泉水为上。

Today, we prepare tea with water from Taihua Mountain. 今天我们选用的是太华山的泉水。

（三）预约

Albert: Hi, May I make a reservation please?

阿尔伯特：你好，我可以预订吗？

Waitress: Sure. What time would you like?

服务员：当然可以。您想订几点？

Albert: Is Friday 6 pm available?

阿尔伯特：周五晚上 6 点可以吗？

Waitress: Yes, how many of you?

服务员：可以，您几位？

Albert:4 persons.

阿尔伯特：4 位。

Waitress: Sure. Do you have any special request?

服务员：好的。您有什么特别要求吗？

Albert: En, I would prefer seats near the window in non-smoking area.

阿尔伯特：嗯，我想要非吸烟区的靠窗座位。

Waitress: Let me check. Yes, there is a table meets your need. May I have your name and telephone number please?

服务员：让我查一下。是的，有一张桌子符合您的需求。能告诉我您的姓名和电话号码吗？

Albert: Yes. My name is Albert Wu, and my telephone number is 136×××××××.

阿尔伯特：好的。我叫阿尔伯特·吴，我的电话号码是 136×××××××。

Waitress: Thank you, Mr. Wu. Now let me confirm the details. You would like to have a table for four on Friday evening at 6 pm, and prefer window seats in non-smoking area. Your telephone number is 136×××××××. Is this information correct?

服务员：谢谢，吴先生。现在让我确认一下细节。您想要预订周五晚上 6 点的一张四人桌，并希望是非吸烟区的靠窗座位。您的电话号码是 136×××××××。这些信息对吗？

Albert: Yes, it is.

阿尔伯特：是的，没错。

Waitress: Your reservation has been made and we will keep it for you until 6:30 pm. Please contact us if there is any change. Thank you for choose our tea house, Wish you have a nice day!

服务员：您的预订已经完成，我们将为您保留到晚上 6:30。如果有任何变动，请与我们联系。感谢您选择我们的茶馆，祝您有个愉快的一天！

Albert: Thank you and wish you have a nice day too. Bye.

阿尔伯特：谢谢，也祝您有个愉快的一天。再见。

附录三
"专"说咖啡：咖啡类英文术语

一、常见的咖啡豆（Common Coffee Beans）

单品	Single Estate	肯亚特极	Kenya AA
综合	Blend	曼特宁	Sumatra Mandheling
耶加雪菲	Ethiopia Yirgacheffe	巴西咖啡	Brazilian Coffee
肯尼亚	Kenya	哥伦比亚咖啡	Colombian Coffee
蓝山咖啡	Blue Mountain Coffee	猫屎咖啡	Kopi Luwak

二、最常见的咖啡分类（Most Common Coffee Classifications）

低因咖啡	Decaffeinated Coffee	半脱因咖啡	Half-decaf Coffee
现磨咖啡	Fresh Ground Coffee	速溶咖啡	Instant Coffee
手冲咖啡	Pour Over Coffee	冷萃咖啡	Cold Brew Coffee
黑咖啡	Black Coffee	冰滴咖啡	Cold Drip Coffee
法式滴滤咖啡	French Coffee	挂耳包	Drip Bag Coffee

三、最常见的咖啡器具（Most Common Coffee Brewing Equipment）

摩卡壶	Moka Maker	滴滤杯	Dripper
法压壶	French Press	美式滤泡壶	Chemex
爱乐压	AeroPress	虹吸式咖啡壶	Siphon/syphon
蛋糕杯	Kalita Wave	土耳其咖啡壶	Cezve
越南壶	Vietnamese Pot	美式滴漏咖啡壶	Electric Coffee Maker

四、咖啡馆点单常见饮品类型（Common Drink Types Ordered at Coffee Shops）

拿铁咖啡	Coffee Latte	焦糖玛奇朵	Caramel Macchiato
卡布奇诺	Cappuccino	馥芮白 / 平白咖啡	Flat White
美式咖啡	Americano	欧蕾咖啡	Coffee Au Lait
摩卡咖啡	Coffee Mocha / Mocha	半拿铁 / 布雷卫咖啡	Breve
玛奇朵	Macchiato	阿芙佳朵	Affogato

五、咖啡的制作和种类（Coffee Preparation and Varieties）

榛果	Hazelnut	肉桂覆盆子	Cinnamon Raspberry
香草	Vanilla	薄荷	Peppermint
焦糖	Caramel	樱桃	Cherry
太妃坚果	Toffee nut	巧克力	Chocolate
橙子	Orange	芒果	Mango

六、咖啡的文化和术语（Coffee Culture and Terminology）

蒸汽棒	Steam Wand	研磨度	Grind Size
蒸煮牛奶	Steamed Milk	压粉	Tamping
特色饮品	Featured Drink	每日特供	Daily Special

附件一
创意鸡尾酒设计表格 Creative cocktail design table

鸡尾酒名称 Cocktail name		日期 Date	
分量 Amount	配方 Ingredients		
装饰物 Garnish			
杯子 Glasses			
创意鸡尾酒描述 Description of the cocktail			

参考文献

［1］ 缪佳作，倪晓波.酒水知识［M］.北京：清华大学出版社，2016.
［2］ 边昊，朱海燕.酒水知识与调酒技术［M］.2版.北京：中国轻工业出版社，2016.
［3］ 贺正柏，祝红文.酒水知识与酒吧管理［M］.3版.北京：旅游教育出版社，2014.
［4］ 林小文.调酒知识与酒水出品实训教程［M］.北京：科学出版社，2014.
［5］ 杨经洲，童忠东.红酒生产工艺与技术［M］.北京：化学工业出版社，2014.
［6］ 何立萍，卢正茂.酒吧服务与管理［M］.北京：中国人民大学出版社，2012.
［7］ 徐明.酒水知识与酒吧管理［M］.北京：中国经济出版社，2012.
［8］ 王钰.酒水知识与服务技巧［M］.北京：中国铁道出版社，2012.
［9］ 王森.就想开间小小咖啡馆［M］.北京：中信出版社，2012.
［10］ 马特.酒水知识与文化［M］.北京：清华大学出版社，2019.
［11］ 李海英.葡萄酒的世界与侍酒服务［M］.武汉：华中科技大学出版社，2021.
［12］ 黄梅.酒水与咖啡的品鉴和调制［M］.大连：大连海事大学出版社，2019.
［13］ 约翰·贝克.葡萄酒品鉴与收藏［M］.北京：中国轻工业出版社，2021.
［14］ 托马斯·克莱顿.咖啡的化学与文化［M］.北京：机械工业出版社，2021.
［15］ 张国辉.白酒文化与酿造工艺［M］.成都：四川科学技术出版社，2021.
［16］ 王明.茶的盛宴：从文化到养生［M］.北京：北京大学出版社，2022.
［17］ 陈明亮.中国名酒发展史［M］.武汉：湖北人民出版社，2022.
［18］ 张丽娜.巧克力文化：历史与全球影响［M］.成都：四川人民出版社，2023.
［19］ 李玉华.黄酒品鉴与健康养生［M］.上海：上海古籍出版社，2023.
［20］ 张凌云.中华茶文化［M］.2版.北京：中国轻工业出版社，2024.